Anxiety, Modern Society, and the Critical Method

Studies in Critical Social Sciences

Series Editor

David Fasenfest

(*Wayne State University*)

VOLUME 182

New Scholarship in Political Economy

Series Editors

David Fasenfest

(*Wayne State University*)

Alfredo Saad-Filho

(*King's College London*)

Editorial Board

Kevin B. Anderson (*University of California, Santa Barbara*)
Tom Brass (*formerly of SPS, University of Cambridge*)
Raju Das (*York University*)
Ben Fine ((*emeritus*) *SOAS University of London*)
Jayati Ghosh (*Jawaharlal Nehru University*)
Elizabeth Hill (*University of Sydney*)
Dan Krier (*Iowa State University*)
Lauren Langman (*Loyola University Chicago*)
Valentine Moghadam (*Northeastern University*)
David N. Smith (*University of Kansas*)
Susanne Soederberg (*Queen's University*)
Aylin Topal (*Middle East Technical University*)
Fiona Tregenna (*University of Johannesburg*)
Matt Vidal (*Loughborough University London*)
Michelle Williams (*University of the Witwatersrand*)

VOLUME 4

The titles published in this series are listed at *brill.com/nspe*

Anxiety, Modern Society, and the Critical Method

Toward a Theory and Practice of Critical Socioanalysis

By

Joel Michael Crombez

BRILL

LEIDEN | BOSTON

Cover illustration: Bust of Karl Marx, 1939, by S.D. Merkurov, at the Fallen Monument Park (Muzeon Park of Arts) in Moscow, Russia. Photo courtesy of Alfredo Saad-Filho.

Library of Congress Cataloging-in-Publication Data

Names: Crombez, Joel Michael, author.
Title: Anxiety, modern society, and the critical method : toward a theory and
 practice of critical socioanalysis / Joel Michael Crombez.
Description: Leiden ; Boston : Brill, 2021. | Series: Studies in critical social
 sciences, 2666-2205 ; vol. 182 | Includes bibliographical references and index.
Identifiers: LCCN 2020057236 (print) | LCCN 2020057237 (ebook) |
 ISBN 9789004445574 (hardback) | ISBN 9789004445581 (ebook)
Subjects: LCSH: Critical theory–Social aspects. | Anxiety–Social aspects. |
 Technology–Social aspects. | Civilization, Modern–1950-
Classification: LCC HM480 .C76 2021 (print) | LCC HM480 (ebook) |
 DDC 152.4/6–dc23
LC record available at https://lccn.loc.gov/2020057236
LC ebook record available at https://lccn.loc.gov/2020057237

Typeface for the Latin, Greek, and Cyrillic scripts: "Brill". See and download: brill.com/brill-typeface.

ISSN 2666-2205
ISBN 978-90-04-44557-4 (hardback)
ISBN 978-90-04-44558-1 (e-book)

Copyright 2021 by Koninklijke Brill NV, Leiden, The Netherlands.
Koninklijke Brill NV incorporates the imprints Brill, Brill Hes & De Graaf, Brill Nijhoff, Brill Rodopi, Brill Sense, Hotei Publishing, mentis Verlag, Verlag Ferdinand Schöningh and Wilhelm Fink Verlag.
All rights reserved. No part of this publication may be reproduced, translated, stored in a retrieval system, or transmitted in any form or by any means, electronic, mechanical, photocopying, recording or otherwise, without prior written permission from the publisher. Requests for re-use and/or translations must be addressed to Koninklijke Brill NV via brill.com or copyright.com.

This book is printed on acid-free paper and produced in a sustainable manner.

Contents

Preface   VII
Acknowledgements   XI
List of Illustrations   XII

Introduction   1

PART 1
From Traditional to Modern Society: The Critical Method of Early Modern Social Thought

Introduction to Part 1   19

1 **Scenic Landscape 1**   25
　1　Traditional Life and the Totalizing Logic of Christianity   25
　2　A Revolution in Space and of the Mind   30
　3　Enlightenment and the Birth of Modernity   39

2 **Critical Methods 1**   48
　1　From Philosophy to Social Theory: (Hegel, Feuerbach and) Marx's Critical Method and the Totalizing Logic of Capital   48
　2　The Individual, the Social, and the Knot: Toward the Durkheimian Symptom of Anomie   65
　3　Totalizing the Mind and Spirit: Weber and the Unholy Union of Religion and Capital   78
　4　From Psychoanalysis to Socioanalysis: Anxiety, Repression, and Talk Therapy in Freud   95

Conclusion to Part 1   119

PART 2
Technology, Modern Wars, and the Rise of Consumer Culture: The Frankfurt School Revisits the Critical Method

Introduction to Part 2   127

3 Scenic Landscape 2   139
    1   The Totalizing Logic of Capital Comes of Age: The Path to Technological Embeddedness and Mass Society   139
    2   The Darker Side of Modernity: World War   159

4 Critical Methods 2   197
    1   The Psychosocial Origins of the Frankfurt School   197
    2   From the Critical Method to Critical Theory and Negative Dialectics   208
    3   A Lesson for Socioanalysis: Anxiety and the Social Vicissitudes of Technology and War in Mass Society   232

    Conclusion to Part 2   254

PART 3
Anxiety-Dreams of Posthuman Futures: Sorting through the Discourses of 21st Century Life

    Introduction to Part 3   261

5 Scenic Landscape 3   269
    1   The Postmodern Rupture in Modern Society   269
    2   We Are All Cyborgs Now: Life in the Mass   285
    3   The Coming Tide: Automation, Artificial Intelligence, and Space Colonization   302

6 Critical Methods 3   318
    1   Critical Socioanalysis: Setting Up   318
    2   Discourses of the Psyche and the Self: A Lacanian Framework   321
    3   The Other Side of Socioanalysis: A Guide for Talk Therapy   335

    Conclusion to Part 3   357

    Conclusion   359

    Bibliography   367
    Index   406

Preface

Several years ago, Harry F. Dahms and a group of his graduate students, myself included, met to discuss the potential for reinvigorating a critical practice within sociology. In this meeting he referred us to a short piece on "Alienation" (2008a), in which he had written:

> Individuals cannot actively overcome alienation, because it is an inherently social condition that is at the very core of modern society. Yet we may be able to take steps toward *recognizing* the power of alienation over our lives and existence ... Compounding layers of alienation undermine our ability to recognize the intrinsic relationship between the growing potential for destruction that comes with the pursuit of prosperity. In analogy to psychoanalysis, sociology must embrace the possibility of and need for *socioanalysis* as one of its greatest yet unopened treasure troves. Socioanalysis in this sense involves therapeutically enabling the individual to recognize how, in addition to psychological limitations and barriers, there are societal limitations and barriers that both are built into and constitute our very selves as social beings ... Whether sociologists in the future will make a truly constructive contribution to the lives of human beings and their efforts to overcome social problems indeed may depend on our ability and willingness to meet the challenge of circumscribing the thrust and purpose of socioanalysis, above and beyond the confines of what Freud erroneously ascribed to psychoanalysis, neglecting that many mental problems are expressions of the contradictions of the modern age. (pp. 41–42)

The topic intrigued me, but at the time I was focused on researching how changes in spatial awareness trigger revolutions in the cultural conceptions we have of our species and how placing earth in its cosmic setting and making a turn toward the creation of virtual spaces has amplified the process of our becoming posthuman; a process that began in the commodification and machinic disciplining of labor under capital.

When I completed that research project, I noticed a stark contrast in how well it resonated with people, and it had less to do with whether they were a sociologist or not and more to do with their age. Those born post-1980 generally accepted the posthuman thesis as an apt description of the material reality that produces the anxiety they experience, while those born prior to 1980 often expressed anxiety over the implications of such a thesis, denying it less on a material basis than on an ideological one. The implications are that in a post-human reality there are so few, if any, avenues to develop one's individuality that it limits the ability of becoming an active agent of social change, and it also means that if this transformation effects all of us, regardless of our personal opposition to it, then there can be no social basis for resisting the coming posthuman world as it is a mass phenomenon. What had changed around 1980 that made this dividing line emerge was the invention and mass cultural adoption of the personal computer, which has had the effect of changing the way that people see and interact with reality by overwhelmingly restructuring their everyday life activities. For those born after 1980, in the advanced[1] modern societies of the West, they have always known this technologically mediated life, they were born into it. Just as with the older generations, they have anxiety, the difference between the two is that for the younger generations anxiety is linked to the uncertainty of what is coming, whereas for the older generations there is anxiety about the uncertainty of what has been lost and how this will effect the future as they cling to avenues of hope that largely foreclosed well before 1980. This divide even persists among many social scientists, even though the tradition of critical social thought has for nearly 200 years argued that capital and informationalization are the historical causes of the diffusion of the individual and the social, and the posthuman thesis is only the latest iteration of tracking how our identities and societies have been structured by those logics in modern society. While many social scientists borrow from the classical thinkers, their failure to work through their anxiety and confront the reality of the material conditions today is visible in their reification of the concepts and desire to statically apply them without taking into account the historically dynamic nature of their object.

Anxiety began to appear to me as not only a prevailing and rising problem in modern/postmodern societies, but as a serious obstacle in the social sciences to theorizing the concrete gravity of modern society and its effects on the future. To gain a better understanding of the causes and effects of anxiety,

---

1   I do not use the word "advanced" as a value judgement to imply that western modern societies are superior or to imply the counter that eastern societies are primitive. Rather, I use this term in a pathological sense, such as when a medical professional informs a patient that their disease is in an "advanced" form. Western societies have an advanced form of modernity in that the symptoms emanating from the modern condition, as diagnosed by Marx, Durkheim, Weber, and Freud, are in a more acute form in those societies. However, it must also be recognized that with developments such as climate change, societies in the global South are experiencing advanced symptoms of a different kind that result from the modern configuration at an alarmingly higher rate than those in the West.

I began to study psychoanalysis, first through Freud, then through the Frankfurt School, and finally through the Neo-Freudian, Jacques Lacan. It became clear from the efforts made by the Frankfurt School that publishing works to expose these logics is not sufficient today, especially since we now live in an age when so few, in and out of academia, are willing to work through difficult and complex ideas. Resistance to reading complicated texts is only underscored by the reception one gets at mentioning the name Jacques Lacan in social science circles, and yet, there is something in all these thinkers that rings true once one works through their texts. If one wants to break out of the thought patterns that society conditions in us, then it requires working through those logics historically, and this work cannot be done by someone else for you. It must be done by every person who wishes to gain a better understanding of why their lives are limited and why they experience anxiety when they face this fact. Since mainstream psychoanalysis has likewise shunned Lacan's work in favor of thought that aligns with modern society and continues to ignore most of the implications of how society structures the psyche, it became clear to me that it was time to work on developing socioanalysis as a necessary practice that would create an artificial setting in which one could confront the contradictions of how they conceive of their self and its relation to society beyond the institutional and methodological frameworks available today.

In the 1990s a practice that goes by the name socio-analysis was developed by Alastair Bain and others who formed the Australian Institute of Socio-Analysis as a consultancy for working on organizational group-dynamics. Their work is based on that of the British psychoanalyst W. R. Bion ([1961] 2004), who established what they have called socio-analysis "in the Northfield Experiments during the Second World War ... [and] Bion's explorations of group behavior at the Tavistock Clinic in the 1940s" (Bain, 1999, p. 1). Bain explains:

> Northfield was a military hospital, situated in the Midlands, with the task of treating soldiers who had developed psychiatric problems, with the aim of getting them back into the war. Bion was responsible for the "Military Training and Rehabilitation Wing" ... [T]he focus of Bion's attention was on the properties of the group as a whole. The group had its own dynamics and was not simply an aggregate of individuals [and he treated them accordingly, by questioning the roles within a hospital where the staff and doctors are considered "well" and the patients are, according to their social role, "ill"] ... Bion in *Experiences in Groups* ([1961] 2004) ... makes explicit the significance of the unconscious in group behavior, the stance he was working from during the Northfield Experiment was

> to make hypotheses about unconscious functioning at the level of the group [and get them to internalize a group super-ego that would regulate the group dynamic]. (pp. 4–6)

From these psychoanalytic roots, the "socio-analysts" have added organizational and institutional theories, as well as group relations and social systems thinking. The goal of this work is not to confront the contradictions of society when they emerge as contradictions of the self, but to find ways to create cohesion in group dynamics. For Bion this involved getting soldiers psychologically ready to return to war. For the new "socio-analysts" this means getting teams in corporate and other institutional environments to function more like a well-oiled machine by aligning the group's thought processes. These goals are fully in line with the mainstream practices of the social sciences because they affirm the status quo and work to smooth out the wrinkles in the totalizing logics of modern society. Therefore, the practice suggested by Dahms is decidedly different than this version of socio-analysis.

What I develop in this book aligns with Dahms's suggestion and is more aptly called critical socioanalysis as it is an extension of the critical method. The goal is not to make people more well adapted to modern society, but to help them have a better grasp of what modern society is doing to them as a result of the contradictions that exist between their concept of the self and that of society. Critical socioanalysis reveals this relationship as it examines the horizon of the possible through the work accomplished by the analysand (i.e. a term borrowed from psychoanalysis to describe the person who enters analysis), thereby illuminating the source of their anxiety and what it is signaling to them. As such, the method that I outline here for critical socioanalysis is as different in form and intention from that which goes under the name socio-analysis, as critical sociology is from mainstream sociology.

Acknowledgements

Too often theorizing is an individual practice conducted in solitude. It is my belief, however, that the best social theory is itself a social project. Although written alone, for much of the good to be found in these pages I owe an enormous debt of gratitude to the conversations I have had with fellow theorists, although any mistakes remain my own.

First and foremost, the intellectual mentorship, friendship, and inspiration of Harry F. Dahms impacted and influenced this research project in all the best ways. Likewise, I owe special thanks to Daniel Krier for his support and guidance in making this book a reality and to Steven Panageotou for his friendship and our many intellectual exchanges.

I am also incredibly grateful for the discussions I have had over the years—in seminars, at conferences, over beers, and in the many virtual spaces we all now occupy—on this project and related topics with Michelle Brown, Jim Block, François Debrix, Allen Dunn, Amy Elias, Jeffrey Halley, David Harvey, Emily Landry, Lauren Langman, Rhiannon Leebrick, Andrew Long, Jean-Luc Nancy, Alexander Stoner, Michael J. Thompson, Charles Walton, Mark P. Worrell, and Cindy Zeiher. And although they passed away before seeing this book come to fruition, I am forever grateful for the intellectual example and unparalleled encouragement I received for this project from R. Scott Frey and Thomas Calhoun.

All the years researching and writing this project were made better by the constant companionship and love of my wife, Sarah Crombez. She deserves more credit than words can express.

Illustrations

1   Key to mathemes (psychoanalysis)   326
2   The four discourses (psychoanalysis)   326
3   Schema for translating the discourses   326
4   The "fifth" discourse   335
5   Key to mathemes (socioanalysis)   335
6   The four discourses (critical socioanalysis)   335

Introduction

Anxiety is one of the fastest growing mental illnesses in western societies and it is empirically linked to many pressing social issues, including economic distress and a rise in drug use and abuse. Historically, anxiety has primarily been conceived of as belonging to the domain of psychology. But it holds that if anxiety is social in nature, then treatments which stem from a perspective rooted in biological psychology—the dominant perspective of the day—can only treat the manifest symptoms. Rather than address the failure of the biomedical model to treat the root causes of the problem, mainstream psychology resigns itself to mere management of these symptoms. This represents a conflict of interest in a discipline which receives funding and social prestige by supporting the prevailing model in the interests of the power structure over and above the interests of those who suffer from these affects. As outlined below, this is not a war over disciplinary boundaries, as many practitioners and researchers working in psychology recognize and attempt to address the critical significance of this problem in their own science. Sociology does not have the answers or the cure to this problem either; after all, sociology suffers from many of the same problems and its practice is likewise deserving of a critical gaze. However, I do suggest that if we employ this critical gaze, then we will be better equipped to recognize the necessity of developing a new practice which combines elements from sociology and psychology to confront the current material situation in which we find ourselves. My research suggests a way forward, one that is built on these interdisciplinary foundations to get at both the societal forces that cause anxiety and the ways it manifests itself in individuals across modern societies.

The broad thesis of this book is that as political economy and technology shape social and identity structures they play a significant role in the rise and spread of anxiety in modern societies and, as such, we must account for how they shape our thoughts and our thoughts about our thoughts. By exploring the social roots of anxiety this text will demonstrate how the concept, manifestation, and impact of anxiety has transformed throughout the history of modern (and postmodern) societies. Because the predominant biomedical model of treating anxiety in psychology fails to sufficiently account for these root causes of anxiety, I propose a theory and method for improving the diagnosis and treatment of anxiety that mobilizes sociological knowledge as the basis of a new clinical practice. I call this practice critical socioanalysis. It shares common elements with psychoanalysis, including a foundation in talk therapy which places the onus for defining the ailment on those who suffer from it, while creating a space and time for guided conversations with the self to work through anxiety, locate its object, and understand how and why it arises. Following Fromm ([1941] 1969; [1955] 1990), where it differs from traditional Freudian psychoanalysis is in the theoretical structure that guides the conversation. Rather than focusing on the psychic structure of sexual repression rooted in childhood development, critical socioanalysis holds that it is the social structures that shape our thoughts and actions throughout the life course as direct consequences of the logic of capital and the technologization of our reality.

What follows is a comparative historical analysis of critical, social, and psychoanalytic theories, that track the effects of political economic and technological development on modern and postmodern societies and the subjects who live in them. I argue that through their common methodological approaches we can combine elements of their critical methods to track the symptoms arising from the contradictions of life in modern/postmodern societies and, in so doing, confront the myriad ways that modern society alters, shapes, and controls individual and collective thought patterns. Following the trajectory of negative thought through the classics of social theory, the first generation of the Frankfurt School, and some strands of mid-to-late 20th century French theory, I employ Alfred Lorenzer's depth hermeneutics methodology—which combines elements of sociological and psychoanalytic methods in the critical tradition—to create a scenic understanding of how their critiques represent systematic attempts to counter totalizing systems of thought which congeal anxiety within the individual and the collective. Attention is paid to the sociohistorical evolution of their methods and the dynamic relationship between identity structure and social structure that is central to their substantive works from the dawn of the modern age until the late-20th century. Building on this foundation I diagnose the impacts on the self and society brought about by the current nexus of political economy and technology in posthuman developments such as artificial intelligence, automation, and space exploration in the early 21st century. Finally, I turn to the method of talk therapy for working through the anxiety that these transformations generate, to understand how they respond to and trigger anxiety and what the passage through anxiety entails. At that stage, I propose a theory for sorting through the discourses that come to structure our speech patterns in modern/postmodern societies and how to reveal them in analysis. These discourses serve as a model for critical socioanalysis to uncover how these structural changes in society reproduce a constant state of anxiety, so that researchers, those in positions of power in modern and postmodern societies, and people suffering from anxiety, can come to understand how their anxiety is constituted by the co-dissolution of the individual and the social, and the implications this has for how the subjects of these societies choose to channel their anxieties.

The identification of anxiety, to describe a variety of unpleasant affects, is an increasingly common phenomenon in modern and postmodern societies. Although it is a term that has been used throughout the history of modern society, in the last few decades it entered everyday use. The word can be traced to Latin roots, but it has assumed a specific etymological meaning which distinguishes it from other mental states (such as angst and fear) only in the context of modernity. Despite this narrowing of meaning, the concept continues to have a broad range of application, as experts and laypeople speak of things like generalized anxiety, social anxiety, and performance anxiety to explain a variety of psychic states that effect peoples' abilities to think and act. It is of paramount importance that we have a clear understanding of anxiety, especially as the term is relied upon heavily to explain world events, such as, the 2016 election of Donald Trump, with concepts such as status anxiety (Mutz, 2018; Gidron & Hall, 2017), racial anxiety (Morgan & Lee, 2018), economic/class anxiety (Prins, Bates, Keyes, & Muntaner, 2015; Fuchs, 2017),[1] or some combination of these in a general catch-all: cultural anxiety (Wuthnow, 2018).

Thought to originate within the subject, anxiety is an affect that causes mental and physical distress. According to the Anxiety and Depression Association of America (2018), anxiety disorders are the most common diagnosis that confronts contemporary psychology. "In DSM-5," the American Psychiatric Association's most recent authoritative guide for diagnosing and classifying psychiatric disorders, "*anxiety (French: anxiété; German: Angst)* is defined as the anticipation of future threat; it is distinguished from *fear (peur; Furcht)*, the emotional response to real or perceived imminent threat" (Crocq, 2015, p. 319). Prior to its codification in the DSM the concept was derided by some psychologists as being a reified metaphor with "no unequivocal definition" (Sarbin, 1964, p. 630). This was evidenced in the first two editions of the DSM, which treated anxiety under the broad categories of psychoneurotic disorders (DSM-I, 1952) and neuroses (DSM-II, 1968).[2] With the publication of DSM-III (1980), the guide signaled a major paradigm shift in psychology from a predominantly psychodynamic approach to one firmly rooted in biological psychology (Paris & Phillips, 2013); along with that shift came a specific chapter on anxiety disorders. This trend toward the biomedical model of psychology was further cemented in DSM-IV (1994) and DSM-5 (2013), as was the treatment of anxiety

---

[1] Fuch's piece stands out in that he presents a theory of anxiety and critically evaluates it in his argument.
[2] For a detailed history see Crocq, 2015.

as a specific (i.e. discrete) mental disorder. This development turned anxiety into a legitimate diagnosis for mental health care professionals that made it subject to psychopharmacological treatments and, therefore, eligible for insurance payouts. Although its formal codification in DSM appears to place anxiety on solid scientific foundations and legitimate it as a medical diagnosis, the increased alliance between psychology and neuroscience has unintended social consequences, including diagnostic drift, because of this approach.

Many psychologists, especially those from the psychodynamic perspective, debated and challenged the paradigmatic shift in DSM-III, and after more than 30 years of evidence to support some of their challenges, they saw an opportunity to address them once again with the writing of the DSM-5 (Greenberg, 2014). Despite their attempts, the concerns were still largely ignored. The problem was not simply the ambiguous nature of some definitions and their broad applicability in developing medical diagnoses for modern ailments, it was that in following the biomedical model the solution to these diagnoses were maintained to be predominantly biochemical in origin and, therefore, psychopharmacological in treatment. The effect of this ambiguously defined concept was that it could, and did, lead to vast increases in the number of medicated people. Rather than see a decrease in these mental disorders we have only witnessed an increase, evidenced by the 22% rise in the number of Americans taking anti-depressant and anti-anxiety medications from 2001 to 2010, with 1 in 4 women on these medications and a 43% increase in men ages 20–44 taking them over the same time period (Medco Health Solutions, Inc, 2011). Conflicts of interest were raised, and subsequently dismissed, when those who developed these definitions received funding from the very pharmaceutical companies who profit from the increased number of those on these prescriptions (Welsh, Klassen, Borisova, & Clothier, 2013). If the strategy is to lower rates of mental illness in our societies, then this data shows just how catastrophic a failure it is. On the other hand, if the purpose is to increase the profitability of pharmaceutical companies and their shareholders, equate disciplinary prestige with the increased alignment of science and the interests of business and industry, and to medicate the populace, then it has been a smashing success (Greenberg, 2011; Whooley, 2017).

Although the definition of anxiety provided by the DSM appears straightforward and to the point, these definitions are socially constructed, influenced by powerful social structures, and have significant ramifications for patients seeking treatment (Wakefield, 2013). In pursuing a shift away from psychodynamic approaches and those informed by social psychology, the advancement of understanding anxiety as a complex problem of our times has suffered as have the people who attempt to cope with these affects in modern societies.

While biological psychology and neuroscience have much to contribute to the debate—and certainly in cases of neurochemical imbalance they provide a necessary treatment protocol—they offer only a partial view on this affect, and experience what the historian Russell Jacoby (1975) refers to as "social amnesia: the repression of critical thought" in the name of a distorted understanding of progress (p. 150).

By exclusively focusing on the manifestation of anxiety, psychology has failed to adequately address its root causes. It has sacrificed the study of latent forces for those whose manifest content is overwhelmingly visible on the surface level. The key word in the DSM-5 definition is "anticipation," which is a state of mind within the subject. By focusing on the subject's experience to the exclusion of external factors, this ignores the social dynamics that influence our mental states as both ourselves and our societies are co-constructed. While it is perhaps not surprising that psychology would ignore or at the least downplay the foundational theories of sociology, it is more surprising that this definition ignores the work of prominent social psychologists, like Erving Goffman (1959; 1966; 1981; 1982), who demonstrated in detail the social construction of the self and its psychological impact on how we think, talk, and respond in social situations. The problem, therefore, is less with what mainstream psychology says than with what it fails to say. Complicating any easy fix by adopting an interdisciplinary approach, however, is that mainstream sociology suffers from the same problem.

In 1960, the psychologist Vincent M. Murphy presented a paper at the annual meeting of the American Catholic Sociological Society where he suggested that anxiety was indeed the concept through which psychology and sociology overlapped. However, he spoke as a psychologist, first and foremost, advocating not for a realignment of the disciplines or the need for a novel approach to anxiety that worked with the methodological and theoretical insights of both. Rather he proposed a clear division of labor in which sociology should serve a supporting role to psychology and gather, largely a-theoretical, quantitative data at the social level that psychologists could then use at the institutional level to justify their treatments and effect public policy. However, if the root of anxiety is social, then we must ask how psychological treatments could ever do anything more than mask the true nature of anxiety and why psychology would be better suited to take the lead in treating the problem of anxiety over sociology? In the 60 years since Murphy's proposal the sciences largely remain, in problematic fashion, in the same situation as he saw them because the scientific contributions continue to be affected by the biomedical model that dominates psychological science and the quantitative models that dominate sociological science. However, even though the sciences have largely remained as he wished, his proposal has at best been accomplished in a lackadaisical manner with no clear research agenda forming between the disciplines to specifically address this widespread problem of anxiety. The result is that there have been no major policy initiatives to address the causes of anxiety and it continues to be predominantly seen as a wholly negative phenomenon by mainstream psychologists and sociologists, who treat it as the cause of dissatisfaction in modern societies rather than the effect of living in modern societies. By holding onto this perspective, it allows them to continue to advocate for anxiety residing in the domain of medical psychology where it is treated with regular and continuous doses of psychotropic drugs, the use and costs of which are tracked by sociology.

The data overwhelmingly suggests, however, that anxiety is a problem of modern society, and therefore should be studied more closely by sociology. The economist Seth Stephens-Davidowitz (2016), writing for the New York Times, did an analysis of Google search terms and found that searches for anxiety climbed 150% in the United States from 2004–2016. While some of this can simply be attributed to the increase in internet usage over this time due to the spread of smart phones, he found two important correlations in his data. First, searches on anxiety rose the most in states hit hardest by the economic downturn following the Great Recession in 2008; and they also correlated to states that had the highest rate of opioid prescription abuse. Dealing with the persistence of anomie in modern societies, the populace has been conditioned to medicate the symptoms away rather than work to change the conditions that produce this anomie. While sociology has tracked economic downturn and drug use, by not following a critical model, it has done relatively little to explain the causal link between these social issues and the effects they have on the identity structure, therefore, it has not sufficiently provided the necessary tools for the public to think through the causes of their mental traumas.

The sheer number of mentions of anxiety in media, on social media, in self-help guides and commercials advertising new pharmaceutical treatments for it, suggests that modern societies, not just modern subjects, have a serious anxiety problem. Several prominent sociological theorists have recognized a link between the processes of modernization and anxiety (Giddens, 1991; Beck, 1992; Pahl, 1995) and the intensification of anxiety under conditions of postmodernity (Kroker and Cook, 1988; Bauman, 1993; 1995; 1997). While many excellent contributions in the history of science and the sociology of mental illness have discussed the problems outlined above at length (Harrington, 2019; Horwitz, 2003; 2020; Horwitz and Wakefield, 2012a; Karp, 2002; 2007; 2016; Karp and Sisson, 2009; Scull, 2016; 2019), rarely has there been a strict focus on anxiety (Horwitz, 2013; Horwitz and Wakefield, 2012b). The situation remains as Iain Wilkinson (1999) wrote, when he hit the nail on the head over twenty years ago: "many sociologists presume to analyse the putative causes of anxiety; but few by comparison have ventured to study the condition of anxiety itself to explain the link between its public causes and personal effects" (p. 446). Identifying this link is precisely the task of the sociological imagination, according to Mills' definition ([1959] 2000), and reflects the kind of critical sociology laid out in the classical works of Marx, Weber and Durkheim, and the critical psychology of Freud.

If the classics recognized that the task of sociology requires the linkage of social structures—those objective but intangible forces institutionalized by modern societies and internalized by its subjects—to identity structures—those forces of socialization which vary in space and time to shape personality and behavior—then why has so much contemporary sociology ignored or failed to successfully engage and explain this link?

One reason is that mainstream sociology, like psychology, has largely abandoned the debate between positive and critical approaches to the science. Rather than resolve or prolong the debate, mainstream sociology has eschewed the label of positivism while retaining many of its defining characteristics. In modeling itself after the natural sciences in the pursuit of professionalization among the sciences, mainstream sociology does not question the philosophical implications of the science in terms of how it embodies and distorts the values and norms that it claims to explain. Just as Marx critiqued political economy for assuming as natural the very foundations which had yet to be explained, mainstream sociology commits the same error with the objects of its attention. Implicit in mainstream sociology, and the social sciences which have followed in this trend, is an ignorance of the socio-historical context in which the research is conducted and as a result, rather than addressing "the issue of whether and how prevailing norms and values are reconcilable with societal transformations currently occurring, approaches in each discipline follow a trajectory of "progress" according to priorities that mostly tend to be the function of agendas and designs carried over from the discipline's very own past" (Dahms, 2008b, pp. 11–12). In sociology, this has meant the dominance of empirical studies which are content with the task of converting surface level phenomena into mere information on them. The result is a sociology that "dissects its subjects according to the sectors of society to which they simultaneously belong ... It does not progress beyond classificatory enumeration (taxonomy), the interdependence of these areas ([i.e. social structures and identity structures]) is not comprehended" (The Frankfurt Institute of Social Research, [1956] 1972, p. 5). In other words, the prevailing model of mainstream sociology is incapable of performing the task needed to illuminate how anxiety is produced by the co-construction of self and society, because it does not pay attention to the conditions under which it performs its tasks.

The standard introduction to sociology has for some time now consisted in introducing students to the work of C. Wright Mills' *The Sociological Imagination* ([1959] 2000). This was how I was first introduced to the subject and it is how the various introductory textbooks I have taught in my own classes introduce the subject. The focus is not on the larger critique of the Parsonian method that makes up the book's core, but rather on the principles so eloquently put forth in the chapter titled "The Promise." Like many other students when I was first exposed to the ideas he put forward in that chapter my curiosity was piqued, and my professor spoke of all of the ways that sociologists have used their sociological imagination to study seemingly everything under the sun. However, I dug deeper into the text in the years that followed, once I too had been enamored by the promise, and I discovered that a vast majority of sociology has in fact not delivered on Mills' promise. While their work may express something that we could call a "sociological imagination" many fail in the basic task laid out by Mills when he plainly stated that: "The sociological imagination enables us to grasp history and biography and the relations between the two within society" (p. 6). What this promises us is that sociology can provide us with the tools to see ourselves in the world and see the world in ourselves, to understand how even when we feel so detached from public issues, ones that may not even relate to us in a straight forward manner by virtue of our class, race, ethnicity, religious preference, nationality, gender, sexual identity, personality type, age, or any other ascribed status that we are born with, that something out there unites us: that public problems have an impact on our private lives. This is the ambitious promise of sociology, and yet, as one who chose to spend their life in this vocation, as I read sociological journals, attend professional conferences, and speak to my colleagues and mentors, it is all too clear that sociology is as fragmented as the society it studies and that it has so far failed in delivering on this promise (Lemert, [1995] 2004). This state of affairs is all the more troubling because it is so clear that modern societies are in desperate need of a sociological intervention.

What is remarkable about Mills' idea for what constitutes a sociological practice is that it manages to capture the essence of the classics of sociology in a simple prose that is accessible to those who are not yet versed in the technical language of sociological theory. Although sociological theories that illuminate modern society cannot stay at this simplistic level of language, the goal should be to excite the desire of those who are exposed to these ideas so that they can in turn learn the more complicated language of critical social theory and train their thoughts as a result to think sociologically about their lives. Mills' summation of the discipline captures how Marx, Durkheim, and Weber engaged with the content of political economy to uncover the structure of modern society while at the same time explaining exactly how these structures impacted the lives of modern subjects. One might object, however, and point to just how radically the times have changed, not only since the classics developed sociology as a science in the late 19th and early 20th century but even since Mills was writing in the mid-20th century. The complexity of modern society with a globalized system of capital, a networked populace, and a pace of change that is ceaselessly accelerating, are all factors that play a role in the difficulty that sociology has in being practiced in the same manner as it was before. However, what then is the purpose of sociology if it is only to churn out empirical reports that capture small artificially segmented portions of our reality and reduce them to static snapshots? And who would such a sociology primarily benefit if it does not and cannot manage to link the most pressing issues in our societies to the most intimate and personal concerns of individuals living in those societies? The answers that I would propose to these questions are: first, that kind of sociological practice would primarily serve the interests of capital by providing easily digestible sound bites from which politicians, business leaders, and others in positions of power can sort through and pick those that best support their ideological perspectives (Bauman, 2014) and, second, this brand of sociology would affirm the system of informationalization (Garfinkel, [2008] 2016; Lash, 2002) whose object and goals are technical in nature and seek to represent all of reality as a digital code that may allow computers to run more sophisticated simulations on which to model reality (Baudrillard, [1992] 1994; Crombez and Dahms, 2015).

The development of a critical socioanalysis seeks a return to the impetus for the discipline of sociology that was captured by Mills and developed by the classics. These pressing methodological and theoretical questions must form a basis for any sociological investigation that wishes to avoid these traps and maintain a commitment to a project of social justice. Therefore, the first goal of understanding the socio-historical causes of anxiety is to provide meta-methodological and meta-theoretical answers as to how we can conduct this kind of research today. However, these questions cannot be fully answered a priori; rather they must be answered alongside the research and as a result of the research. Given the complexity of advanced modern (or, postmodern) societies, how must we study society to ensure that we capture the ways in which the structure of society impacts the structure of identity? And, how can we make sure that our work is not merely an affirmation of the status quo but rather a penetrating critique that illuminates the sources of domination? I suggest that the only way to accomplish this is to engage in a process of auto-critique that subjects both method and theory to the same critical lens that is being applied to the object of study. The end result then is not only a report on empirical data but a contribution to the methodological and theoretical literature that has been sharpened against the material reality in which we live.

The fact is, however, that our material reality is so dynamic that we cannot know with certainty what reality will look like as our lives on planet earth become ever more embedded in and dependent on technology. However, we can be certain the changes will be profound, and they will happen at a pace that continues to eclipse our ability to understand their full impact before path dependency sets in. Since the Cold War years when the mass injection of public and private capital into the development of novel technologies became a central principle of modern societies (Bridgestock, et al., 1998), technical advances have outpaced the human subject's ability to process the myriad changes, adjust to them, cope with the loss of old forms, contemplate the consequences of the new forms, and determine the best direction to steer our history (for two vastly different perspectives on this see Fukuyama, 2002 and Kroker, 2014). Working through the history of modern society reveals how it became such fertile ground for the production of anxiety as it underwent a shift from supporting the development of the individual and the social to their perversion in mass society. Life in this mass society is experienced and perceived in an atomistic sense but all are nonetheless fatefully tied to the directionality of the mass even as the illusion of individualism gains in strength. Profound as these transformations are, their effects are largely left unexamined as our "selves" and societies are caught in the gravitational pull of this mass life, built upon our own material interventions.

Yet we plow full steam ahead, welcoming the technologies of mass society into our work places, our homes, our bodies, and our minds, with the promise that they will care for more of the burdensome and tedious aspects of our daily lives and solve our global crises (Featherstone and Burrows, 1995). On the surface they promise us more time to be ourselves and solutions to our planetary problems that do not require us to drastically change our unsustainable lives. This is the ecstasy of technology. In most instances and for many people, however, technology has done anything but this. Rather it has amplified artificial divides, intensified alienation, accelerated environmental degradation, and broken down the skills that facilitate our interactions with each other and our surroundings to the point that what it means to be human and whether or not we still are human is up for debate, as is the very politics of life itself (Rose, 2007). This is the anxiety. In the darkest version of events, these advances have created a number of existential risks that not only threaten our persons but our planet (Bostrom 2002; 2013; 2014). And while those outcomes effect the entirety of the planet, at the level of the subjects of modern society these changes are being experienced as the loss of employment, the loss of conversation skills, the loss of purpose, the loss of pleasure, and the loss of the ability to survive and live in a meaningful manner. In short, we are experiencing both the ecstasy and the anxiety of technical vertigo, as technology seduces and enchants us by opening up new worlds and imagined realities, while destabilizing our hold on material reality and undermining what historically we have come to think of as the characteristics that once made us human. My hypothesis is that as these processes accelerate there will be a corresponding rise in anxiety that is not treatable by the typical psychological cures because it originates at the level of the social structures that weave the fabric of modernity. Baring significant changes to the structure of modern society and the abolishment of capitalism—or likewise if those changes were to occur and humanity would have to reconfigure its consciousness to cope with a completely new mode of social organization—this anxiety must be treated in a novel fashion that combines the elements of psychoanalysis and critical sociological theories to rise up to the Millsian challenge of connecting the history of modernity with personal biographies.

For this reason, it is necessary to reconstruct the history of the critical approach to the discipline so that it can explain the link between the social causes of anxiety and the ways that it manifests within individuals. To do so is to follow the birth of sociology from its roots in modern society through its various transformations. Given that the current system of power is still bound to the logic of capital and uses technological means to secure that power, my focus is on the tradition of critical theory that has substantively engaged with a critique of those logics. Specifically, I trace what I identify as the three totalizing logics (religion, capital, and information) that have shaped our collective life under their influence. Combining sociological and psychological insights, my research begins from the premise that *anxiety is a future-oriented affect that intensifies between the moment a person starts struggling with the ability to locate their "self" in their social milieu and the moment when the fiction of the self, experienced as their individual reality, becomes irreconcilable with the given social reality.* As such it recognizes that anxiety is experienced differently by different people as they come closer to recognizing the truth of the reality of our condition, and this is significantly impacted by whether the person has the social and material capital to influence social forces or if they are left to react to those forces. In this sense my research is representative of the absurdist sociology advocated for by the sociologist Stanford M. Lyman (1997), when he wrote that figuring out the links between the self and society, is in the final analysis, "a problem for the actor to solve" (p. 42). This doesn't mean that the social scientist plays no role in this process, simply that when we are dealing with a concept like anxiety we must turn to its manifestations at the level of the subject and imbue them with the sociological imagination needed to make the links between their self and their society manifest in their own minds. This is the task of critical socioanalysis.

The methods that I use to establish the foundations of critical socioanalysis are linked to the substantive areas I will examine. Critical theory is a method that was developed by studying the links between the self and society as they relate to the structures in modern society, especially as they are shaped by political economy and technology. As Dahms (2017b) has convincingly argued, critical theory represents a mode of radical comparative-historical research. He suggests that what is required to conduct critical theory is an "intimate familiarity with at least two instances of modern society ... to avoid conflating features that are characteristic of the version of modern society with which they are most familiar, with features that apply to all modern societies" (p. 179). This characteristic of having lived for extended periods of time and having familiarized oneself to the extent that both societies could be said to have had an impact on the theorist's development has been a rather remarkable, although often unremarked upon aspect of the intellectual development of many critical theorists. Beginning with Marx whose political radicalism led to his exile from a number of countries he once called home, to Weber whose visit to America rose him from an intellectual slumber caused by personal loss and mental distress, to the members of the Frankfurt School who were exiled in America to avoid the Nazis, to the French "post-modernists" Jean-François Lyotard and Jean Baudrillard who spent several years teaching in California. Each of their theories contains an element that is comparative historical whether it is explicitly acknowledged or not and whether it is that of comparing spaces, times, or both.

From the standpoint of my own development as a critical theorist, although I was born in the suburbs of Detroit, Michigan, I grew up in Ecuador and as an adult I lived for extended periods of time in several countries across Latin America, which allowed me to experience what Gloria Anzaldua ([1987] 1999) referred to as a life in the borderlands. What this means is that one has a foot in two cultures and as a result is always in some ways alienated from both. The advantage is that the person who lives in the borderlands is able to see things about each culture that would otherwise remain invisible to those for whom they have always constituted the norm of the society. In light of this project, which in many ways is a comparative historical analysis of how anxiety has changed from a time when the species was decidedly human to a time when the species is decidedly posthuman, my background provides a useful referent in this critical tradition in that it allows me to see features of American society that are particularly American and delineate those from features that are common to modern societies in general. Furthermore, my background working in the technology industry in both American and Latin American cultures and having it as an object of study in academia allows me to approach the subject of technology from a multifaceted perspective.

The way that I conduct this comparative historical research and develop a theory of anxiety for critical socioanalysis is two-fold. On the one hand, as the concern is textual I trace out the ways that previous generations of critical theorists diagnosed modern society and the symptoms they uncovered that produce anxiety, by turning to their texts and reading them within the social context in which they were written. On the other hand, I turn to the available historical data provided by scientific, governmental, public and private sources, to trace out the scenic landscape in which they were writing.

My methodological approach for this project was developed by the German sociologist and psychoanalyst, Alfred Lorenzer, which he called depth-hermeneutics. Using this method, I examine the general scenic landscape of the material conditions in which these critical thinkers developed their psychosocial selves and then analyze the texts in that context. Lorenzer developed this method as a combination of psychoanalysis, which plunges beneath the conscious register of language to the hidden unconscious layers, and a critical sociological perspective that recognizes that society has a hidden undercurrent that is often unrecognizable or inaccessible to the non-sociologist who does not see the code of the structure (i.e. one who has an under- or undeveloped sociological imagination). "A text interpretation is … not a psychoanalysis of the author. The point is far more the working out of typical interaction forms, the "inner panorama of middle class life experiences" … The staged understanding of a text interpretation is the understanding of typical scenes and not, as in therapy, the individual scenes found in personal history" (Leithäuser, 2013, p. 67). Summarizing his position, Lorenzer

> points to the hermeneutic (interpretational) nature of the form of understanding which is used in psychoanalytical therapeutic practice, generalizes it in what he calls the "scenic understanding," and comments the nature of this transformation. The particular value of these thoughts mainly becomes evident in reflecting a research practice by means of researchers subjectively engaging in the interpretation, tracing aspects of social relations which are not immediately visible and may be not even conscious for social actors.
>
> OLESEN and WEBER, 2013, p. 33

This method is a forerunner to socioanalysis, however, rather than applying it in a setting of talk therapy as I propose, it is applied to texts in order to flesh out the impact that the structure of the author's society and culture had on the way that they chose to present the contents of their work in writing. The goal of this method is to construct what Lorenzer (2016) referred to as a scenic understanding which he developed as part of a method of cultural analysis that aims to get the full scope of the interconnections that exist between the structures of individuals and the societies in which they develop. In other words, this is a method for understanding how society is reflected in these texts and helps illuminate the psychological impact that living in that society had on their creation; that is, it illuminates the scenic understanding of the material reality in which the writing of the text was made possible. Since my task is to trace out the spirit of anxiety that grows in modern societies, this method will allow me to trace this out through these bodies of theory by grouping them together under the same scenic umbrella of negative critical thought, the purpose of which is to illuminate the dark side of modernity and diagnose the modern/postmodern condition. It is this shared scenic level that unites the theorists considered in this text across the spaces and times in which they wrote. The process of reading texts is therefore one in which "the life plans found in the literature are examined in connection with collectively valid norms and values … [and the] analytical interest lies in the conflict between unconscious wishes and the values valid within society" (Leithäuser, 2013, p. 67). In other words, Lorenzer's depth-hermeneutics provides a method that has the same goals as critical theory.

One function of this depth-hermeneutics is to perform a task that the German philosopher Hans-Georg Gadamer suggested rather provocatively in his book *Truth and Method* ([1975] 2006), namely that each generation must return to the classics and re-appropriate them for their own uses. He explains that this is needed because our

> understanding it ([the text]) will always involve more than merely historically reconstructing the past "world" to which the work belongs. Our understanding will always retain the consciousness that we too belong to that world, and correlatively, that the work too belongs to our world. (p. 290)

What this means is that applying a hermeneutic approach to the classics is also a form of comparative historical research if we are explicit about how each of these perspectives effects the reading. On the one hand then this hermeneutic approach must perform an immanent critique, which means that it "attacks social reality from its own standpoint, but at the same time criticizes the standpoint from the perspective of its historical context" (Antonio, 1981, p. 338). And on the other hand, it must read into the texts the threads that resonate with the reality in which the critic is embedded. This allows each generation to find something new in the context of these texts that would not have been possible had history followed an alternative course.

My sample is composed of the history of critical social theory moving from the early modern classics (Marx, Durkheim, Weber, and Freud), to the inter-, intra-, and post-war writings of the first generation of the Frankfurt School (Horkheimer, Adorno, and Marcuse), to the post-war French theorists (primarily Baudrillard, Virilio, Lyotard, and Lacan), which are supplemented with a variety of government, academic, and public reports, analyses, studies, and documentation on the history of modern and postmodern societies. Drawing on these I alternate between chapters that trace out the scenic landscape of modern society and those that critically evaluate the theories developed within those landscapes. Simultaneously I follow how anxiety was theorized at each stage to arrive at a model for critical socioanalysis.

Chapter 1 sets the scenic landscape of the early modern world by examining the ways that it was shaped by the totalizing logic of religion and how capital and a scientific mindset in the enlightenment eroded its foundation. Chapter 2 tracks the origin of the critical method in the theories of Karl Marx, Emile Durkheim, Max Weber, and Sigmund Freud and outlines the diagnostic criteria for the symptoms of alienation, anomie, the Protestant ethic, and repression. Chapter 3 turns to the scenic landscape created by the totalizing logic of capital in the World War years as technology became embedded in modern society through the effects it had on war and capital accumulation. Chapter 4 follows the evolution of the critical method in the theories of the first generation of the Frankfurt School to see how they interpreted these material changes in the form of increased levels of instrumental and technical rationality and the diffusion of the social. Chapter 5 focuses on the spread of technology into everyday life with the rise of the computer age, how this impacted the thought of the French theorists, and the effects this had on the critical method once societies began to feel the effects of the totalizing logic of information, including the emergence of cyborg subjects and the massification of society. Finally, chapter 6 outlines the requirements for critical socioanalysis— starting with the method for tracking psychological discourses in analysis as provided by Lacan—and provides a model for tracking the discourses of mass society in the socioanalytic setting.

PART 1

From Traditional to Modern Society: The Critical Method of Early Modern Social Thought

∴

Introduction to Part 1

Claiming to be modern is to claim differentiation, distinction, and rupture with the past and tradition. To be modern is to be oriented to the future, but to a future that is limited by the scope of our ability to make decisions, to act, and to engage our agency. Because we are subjects of history, born in a society, our identity structure must be understood as one that is co-constructed alongside the social structure of our given society. In other words, we are born into a world whose features were determined before our arrival and whose impact begins to shape our personal lives while we are still in the womb.

If we are to learn what choices are available to us and how we are to respond to them, it is paramount that we learn how and why modern structural forces shape social reality in such a way as to limit the choices available to us above and beyond natural limitations, such as our biological makeup. We must also recognize how our minds, as the product of a historical psychosocial development, are structured to respond to the choices available to us. In other words, we must come to understand how our thoughts are themselves the product of our psychosocial conditioning. Doing so is a precondition to our being able to think for ourselves about modern society, our psychosocial complexity, and how we can take actions that harmonize the two with the values we most cherish. To understand ourselves, then, we must understand what made the world the way it is and the impact of the "circumstances directly found, given and transmitted from the past" (Marx, [1852] 2017, p. 37). Understanding traditional, or, premodern life, is the precondition for understanding modern life because contained within the concept of the modern is a tacit understanding that it only gains meaning through comparison. Sociology, which was developed as the study of modern societies, is therefore a comparative discipline that begins with an analysis of precisely how life in modern society differed from the traditional modes of life that it replaced.

Learning of our own boundaries and limitations, particularly as we come to realize how they change with sociohistorical circumstances and how they originate in a contradictory social reality that presents itself at one moment as fixed and another as fluid, can itself be both the cause and the effect of anxiety. On the one hand, asking these questions may cause us anxiety as we learn just how dictated and controlled our lives are by social forces over which we exercise little or no control. We may romanticize past modes of social organization or dream of futures not yet constructed, but it is in the here-and-now that we are forced to live our lives. As we come to learn how we are dominated by certain forces in society a tension arises between the source of domination and the knowledge that being subjected to domination requires at a minimum a level of our compliance and *obedience* (Weber, 1978), even when we can see no alternative but to live under the weight of its power. Which is to say that at a certain level of analysis we must learn to live with the knowledge that we continuously acquiesce to the visible and invisible forces which compel us to live under conditions of domination. Understanding this knowledge and gaining the full weight of its meaning is bound to unravel some of our last threads of innocence as we lay ourselves bare to the sober reality of our condition. On the other hand, this process may be an effect of anxiety, as those who suffer from this affect may be more likely to plumb the depths of their character and ask questions concerning their psychosocial makeup and that of the world in which they live than those who are unafflicted or unaware of how they are afflicted by the social causes of anxiety. However, being precise about the meaning of anxiety is a challenge given that it is experienced subjectively, and so, to speak of it as an objective social reality is to understand it only as the result of historical processes.

If there is any object to which we can apply the term anxiety from a psychosocial perspective with any congruency of meaning, "it can only be," as Weber ([1904-05] 2011) says of social concepts, "a specific historical case ... a complex of relationships in historical reality. We join them together from the vantage point of their cultural significance, into a conceptual unity" (p. 76). The goal is not to trap anxiety in a general abstraction that proposes a static definition that can be haphazardly applied to dynamic social processes as we study the self and society in conditions of modernity. Uncovering how social forces penetrate individuals' identity structures to create the seeds of anxiety is the task of the present investigation and is a question of history, but gaining a full understanding of the ways anxiety manifests within individuals is the task, as I will argue in the following chapters, of the socioanalytic session which requires a setting where the subject of analysis, or 'analysand' to borrow the term from psychoanalysis, engages in an intimate and sustained conversation with their 'self' to explore those connections with the socioanalyst serving as a catalyst for that conversation. One of the objects of the socioanalytic session, therefore, is to define anxiety anew with each analysand. The goal of the present text is to develop a theory of anxiety that is of a historically unique character so that it may be used to structure the diagnostic component of socioanalysis; which is to say, it requires the development of a theory that recognizes the dynamic nature of anxiety by revealing its historical development alongside the development of the method for its diagnosis.

Provisionally, however, the term *anxiety* has a certain acceptance in our society and general usage, which at this stage in the investigation will partially illuminate the object of study: *anxiety is a future-oriented affect—sharing this temporal orientation with modernity—that intensifies between the moment a person starts struggling with the ability to locate their "self" in their social milieu and the moment when the fiction of the self, experienced as their individual reality, becomes irreconcilable with the given social reality.* Like a character in a Don DeLillo or Thomas Pynchon novel, this anxiety of social origins can in some ways be thought of as an 'anxiety of obsolescence' (Fitzpatrick, 2006). It does not arise when one reflects on the past and develops a sense of nostalgic yearning, which would be more akin to depression—which is an affect oriented to the past—rather anxiety is developed when one looks toward the future, at what might be, and fails or becomes overwhelmed at the thought of locating themselves within a fiercely dynamic and negentropic future. If then, the study of modern society is the study of how modern society differs from past forms of life, sociology is at one and the same time the study of anxiety, as the affect is bound to arise when we are compelled to change ourselves to match the changes in society. This provisional way of conceiving anxiety recognizes that it is experienced differently by different people and is significantly impacted by whether the person has the social and material capital to influence social forces and partially tailor them to their will or if they are left to react to those forces. Anxiety warns us when action is required, but if we have lost the ability to channel our anxiety then it coagulates within us and we fail to heed its warnings as we are internally consumed by the affect rather than the external realities that triggered it. Although we will only come to understand anxiety if we understand the historical processes that agitate this condition, we know that anxiety manifests itself at the crossroads of the 'psychosocial divide' (Cavalletto, 2016), and as the co-constructed structures of society and identity are transformed in modernity so too are the manifest and latent sources of anxiety.

The sources of anxiety retain many similarities in modernity as they did in traditional settings, such as the everyday anxieties pertaining to immediate needs and the reproduction of life. Then there are those that we might call—to borrow a term from the French Annales School (Braudel, 1958)—anxieties of the 'longue durée,' such as those periodically arising from environmental catastrophe, famine, plagues, and war. Premodern anxiety was generally limited to situations when the actor was confronted with a range of possibilities and once the field of possibilities began to narrow into concrete avenues, they would have to take immediate action in their environment to preserve their life. For example, in the general unease of moving through an area where dangerous animals were known to be present, or in a time when the food stores were running low and there was no sign of rain in sight, or when conflicts between groups suggested that war may result, or in the moment of childbirth when the question of assuming parental identity hung in traumatic suspense of the newborn and mother's immanent health. In other words, the causes were manifestly obvious and were often intimately tied to the human in and as nature. What distinguished this affect from fear is that the object was not necessarily going to present itself, rather the anxiety arose at the moment of uncertainty over the range of possibilities that might arise from the given conditions. If those scenarios actualized, then anxiety would give way to fear as it would assume a definite and immediate object upon which the affect could latch. In modern societies, however, these kinds of scenarios, while still arising in everyday life, are less prevalent. Humanity has split itself from nature and occupies a world of artifice having mostly tamed the threat of wild animals in modern societies, unevenly deployed industrial technologies to prevent the conditions that give rise to famine, and developed modern medicine that ensures a much higher rate of survival in childbirth. Now the causes of anxiety that are endemic to our populations are less often immediately given and obvious. In this text, therefore, when I refer to anxiety, I am referring to anxiety at this higher level, that is, a form of anxiety that is less geared toward those aspects of nature and chance and is more oriented toward the artificial conditions of the social fabric in which we live our lives.

One distinction from traditional lives that bears a resemblance to the kind of modern anxiety that I mean here was the source of domination over their orientation toward the future. Future oriented worldviews in premodern life were dominated by religious narratives that displaced material anxieties onto a spiritual reality. These religious narratives, particularly those supplied by Christianity, represented a totalizing logic which in the last instance demanded that faith triumph over anxiety, thereby encouraging material inaction in favor of spiritual action. This logic allowed for a material social order dominated by the power of the Catholic Church, and it kept internal threats to that order to a minimum by displacing anxiety onto a supposedly spiritual parallel reality. Questioning the Church's power and taking action that ran counter to its narrative was to risk social and/or physical death. The inversion of religious salvation as the salve of anxiety was damnation as unfettered and eternal anxiety. Anxiety thereby came to have two domains, the first material and the second spiritual. The claim of spiritual reality is that it transcends the mundane and profane earthly existence, but since the spiritual reality is only superimposed upon material reality, what was supposed to alleviate the material anxieties merely replaced them with anxieties of a higher order. Rather than alleviating anxiety, Christianity, by displacing it from the given and immediate material circumstances, served to amplify it as a mode of social control through threats against an infinite existence in the spiritual plane. Thereby, Christianity came to control anxiety in both directions and exert its influence on the whole of material relations in the Western world. For non-believers there was anxiety over whether they could live authentic lives in the material realm without the threat of social death that came with refusing the Church its power, and for believers there was anxiety that the challenges of modernity would make a lifestyle centered on their values wither away. With a decline in the power of Christianity's totalizing logic at the birth of modernity, anxiety seemingly entered a brief period in which it flowed once again; that is, until a new totalizing logic in the form of capital came to realize the power of controlling anxiety as a means to further its own ends.

While many early thinkers of modernity thought that the crack in the totalizing logic of the Church meant that anxiety could be alleviated through pure unbridled progress and a secularization of religious principles (see for example, Saint-Simon, 1825, and Comte, 1858), those who critically evaluated the role of religion recognized that just because the institution was weakened, this did not mean that another institution could not wield the same power to equally, if not more, devastating results. The critical method that the early critically inclined theorists of modern society developed began with a critique of religion as the totalizing logic that dominated and controlled the thoughts and actions of the human species. Developing the critical method was only possible with a keen eye toward history, the structures of power and how they are wielded, and an unflinching critique of the self and the society in which one develops their mental abilities and thought patterns.

The goal of the first chapter is to obtain a scenic understanding of the psychosocial milieu that led to the circumstances which necessitated the early classics of modern social theory—namely, Karl Marx, Emile Durkheim, Max Weber, and Sigmund Freud—to develop a critical method of psychosocial analysis, and how that milieu came to be historically. The goal of the next chapter is to show how that history impacted their analyses of the links between identity and social structure, and how they deployed their methods to diagnose in their specificity the modern anxieties that emerged from this rupture with the traditional world. Rather than engage with and reconstruct all of the ideas leading up to this in detail, which would fill several volumes, for the purpose of this text I will develop a scenic understanding or a sketch of the broad psychosocial landscape that contributed to these ideas as represented by certain figures and events that shaped the material world and the construction of meaning made of that world in the West. The figures that I have chosen to highlight are by no means the only ones who contributed to and shaped the conversation, but through a review of their contributions and the history surrounding them the scenic landscape will develop in our minds as a panoramic image of space and time in which they were embedded. As the critical method is itself a historical method, the scenic understanding of those who developed this mode of thought is found in the material and intellectual histories that built their world and the circumstances that they found themselves in. By fully immersing ourselves in this landscape we will have a better understanding of how and why the critical method came to be developed and how it became an indispensable tool for diagnosing and understanding the anxieties that proliferate in modern societies.

CHAPTER 1

Scenic Landscape 1

1 Traditional Life and the Totalizing Logic of Christianity

Life may be called traditional when, over the course of many generations, the activities, experiences, thoughts, and attitudes of those who live that life persist with little change. Traditional life was static in a broad sense, insofar as those living hundreds of years apart could upon examination be said to have led comparable lives. For much of human history, traditional life consisted in a struggle to meet daily needs: food, shelter, warmth. According to the French sociologist Jacques Ellul ([1954] 1964), in premodern times "work was a punishment, not a virtue ... the rule was to work only as much as absolutely necessary in order to survive ... with the leisure time devoted to sleep, conversation, games, or, best of all to meditation" (pp. 65-66). This points to a general cultural ideal that prevailed in the premodern context, insofar as hard labor was not glamorized as an end unto itself. Rather work was primarily thought of in terms of necessity and need, not surplus and want. Materially, however, once traditional life advanced beyond the nomadic lifestyles and entered a more sedentary phase, this ideal only describes the life of those who held positions of privilege and had the power and resources to force slaves and employ servants to conduct labor on their behalf. Life for the underprivileged, the urban and rural poor, was often dominated by the constant daily struggle against nature and the hard labor needed to reproduce the bare existence of their life and those who dominated their existence.

The seeds of modern society gestated in these conditions of inequality and domination, in what is referred to as the Middle, Medieval, or Dark Ages, running from the fall of the Western Roman Empire circa 500 CE, to the dawn of the Renaissance in the 1400s. In that millennium, life in Western Europe changed, but at a glacial pace. Techniques and technologies for agricultural development improved since the sedentary agricultural lifestyle developed in antiquity but "a very high portion of the population—probably around 90 percent—was needed to raise food" (Singman, 1999, p. 65). Humanity's attachment and reliance on nature dictated their lives to a high degree. Change was primarily concentrated in newly developing urban centers, whereas the mass of the population had to live in rural areas on land dedicated to the cultivation of crops. When change and anxiety touched the lives of the rural poor, it tended to be caused by mass crises that took the shape of wars, famines, and plagues sweeping the land.

In the fourth century, Christianity began its meteoric rise in the West and unwittingly created the preconditions for the emergence of modern life and the critical method. Historically religion has always contained a foundation for the emergence of a totalizing logic to structure the minds of humanity because its narrative structure claims to answer questions of origin, end, and everything in between. *A totalizing logic arises when a structural force works continuously to extend its dominance across space and reproduce itself over time through processes that imprint the structure as a code on the collective consciousness.* It does this through a sustained program of discipline and education, coercion and cooptation, that is built on the claim that this logic provides the authoritative epistemological and ontological answers for interpreting reality and constructing authentic and socially acceptable worldviews. Educational efforts are reinforced with promises of rewards for following the logic and threats of punishment if it is ignored. Totalizing logics craft cultural narratives that are reinforced through campaigns of collective coercion. Agreement with the logic often reaps some immediate benefits from the community in the here-and-now, but because the logic and compliance with it are taken as the natural state of affairs the benefits are often not felt or interpreted as such until the system of logic is challenged, and those benefits are removed. This system of policing frequently compels those who reject the logic and those who are skeptical of it to acquiesce to its authority. For, in the moment they voice their skepticism or rejection of the logic, the social pressure to conform, which is generally hidden from view, suddenly emerges in visible ways. Those who challenge the dominant totalizing force at any given time in history are met with overt threats and actions ranging from social exclusion or social death, as in the case of religious ex-communication, or physical death, when militaries and zealots align themselves with these logics to craft laws which make adherence to the totalizing logic mandatory under pain of corporeal punishment.

The institution, in this case the Catholic Church, does not need to do all the policing itself. In a system of totalizing logic, those whose minds are totalized according to its stated norms and values act on behalf of the institution and police it without always needing to be ordered to do so; in this sense, totalizing systems become autopoietic and self-reinforcing. When the totalizing logic of religion confronts something it cannot answer, it has a built-in system to dissuade the pursuit of alternative answers that do not fit within its framework, often blaming the questions as being deceptive and against the will of God, the natural order, or a sense of decency. Christianity was the first religion to successfully become a totalizing force in the Western world because they treated this process as a divine commandment that was tied to the structure of their eschatological belief (see for example, Matthew 24:14 and Mark 16:15). Those who are absorbed into a totalizing logic must adapt their way of thinking to survive, while those born into a world that is under the spell of a totalizing logic fail to account for its artifice and assume that it is the natural and right way of thinking.

Although there is some debate about whether the Western Roman emperor Constantine completed a spiritual conversion to Christianity based on a genuine belief in the system, there is no doubt that he was the major influence on it becoming the totalizing logic that would dominate the Western world and its minds for the next 1000 plus years. His contribution to the Edict of Milan in 313 CE was a major step toward putting an end to the state sanctioned persecution of Christians (Drake, 1995). Then in 325, he organized the first Council of Nicaea, which brought together the various Christian factions to reach some consensus on common beliefs (Elliot, 1992). With these two actions the path was opened for the Church to expand its influence and power and set the political and cultural agenda for the Middle Ages. Until the culture war was won against Roman paganism, Christians were still subject to persecution, but as an institution the Church gained power and influence through the purchase and inheritance of property and the scales were tipped in its favor in the culture war because it gained legitimacy through the alliance with the Roman empire. This alliance was mutually beneficial, as Constantine had an empire to run and "[r]ealizing the bankruptcy of the policies of his predecessors and the growing strength of the Church, Constantine believed that Christians could do more to further his goals than could the adherents of the traditional religion" (Wilken, 2012, p. 83). However, the benefits were much shorter lived for the empire than for Christianity, as by 476, when Romulus Augustus was deposed, Rome had been overrun by Germanic tribes in the West and the Huns in the East (Wood, 1987; Schnee, 2016). Unable to police the lands and sufficiently control them through military force and cultural appeals, the decline of the Western Roman empire was reaching its completion. Christianity, however, had in that brief period gained in strength and influence, and since it was only a marriage of convenience that the Church shared with Rome, it did not share the same fateful decline of supremacy.

Rather than being fatal to Christianity, the collapse of the Roman empire helped cement the status of the Church as a central world power and a cultural force to be reckoned with. Karatani (2014) argues that "Christianity was indispensable ... as the sole ideology preserving the identity of the empire in the world that emerged following the collapse of the actual empire" (p. 147). What transpired over the next millennium, as territorial borders in Europe ebbed and flowed with shifts of political power and fortune, was the complete development of Christianity into the first totalizing logic of the West. Although there were occasional reports of Christian cults that strayed from Church orthodoxy and borrowed and infused pagan religious rituals in their worship (for example, see Sloterdijk, [1983] 1987, pp. 257–260), the Church ultimately brought stability of cultural narrative and provided an epistemological and an ontological worldview which instilled a grand narrative in the mind of the people and oriented their thoughts and actions in the material world to their religious identity. Following in the footsteps of Judaism, which was the first religion that when "their state had fallen, they did not abandon [their] god" (Karatani, 2014, p. 139), Christianity consolidated power in a way that transcended the state by decoupling the narrative that linked God to the head of any particular state. Rather than turning it into an ethnic identity, like Judaism which expanded its numbers through matrilineal inheritance, Christianity relied on conversion of the masses and offered inclusivity by conformance to its standards and practices, and, occasionally, by coopting local practices that aligned with the Christian tradition (Viola & Barna, 2002; Dreyer, 2012). By claiming to have their own spiritual kingdom led by the pope, it allowed for a more fluid and continuous reproduction of their logic across political and cultural divides. To maintain this dual power structure with the state, they opted for a model that reinforced the idea of private property—preempting the totalizing logic of capital that would replace their unrivaled dominance in modernity—by becoming massive land holders in Europe and by constructing a mythology around the sacred nature of those land holdings (Thurston, 1911). This tactic allowed the Church to retain ownership and cultural hegemony despite the transferences of political power. The partial separation of church and state was originally of great benefit to the Church and it became enshrined in a state sanctioned legal code ratified in the Magna Carta in 1215 (Daniell, 2013). After nearly a millennium of success with this strategy the Church saw its power as absolute.

By the 1300's a series of devastating crises ravaged the populace in Western Europe, throwing many traditional lives into a state of chaos. Three stand out for the massive disruption they caused. First, was the Great European Famine of 1315–17, when violent flooding and freezing temperatures caused crops to fail. Scarcity of grain led to massive increases in the cost of food, and without the cooperation of nature or state intervention in the market to prevent tragedy, up to 15% of the population starved to death (Slavin, 2014). Then the Black Death swept the land between 1346–1353 with estimates of the death rate as high as 50% of the population (Belich, 2016). Overlapping the plague, was the Hundred Years' War which ran from 1337–1453 and not only killed millions in England and France but had the effect of reorganizing those countries in the fashion of the military state (Fowler, 1973; Curry A., 2003).

Perhaps the most important consequence of these crises, and especially the war, was that after the long rise of the Church's supremacy in Western Europe, a nationalist tide weakened the legitimacy of the Church as a political authority, if not the totalizing logic of Christianity itself. Both sides in the war—England and France—appealed to the religious logic that held sway over their populaces to propagandize the war effort. So long as there was the religious narrative that promised an afterlife to adherents and linked support of the state to religious righteousness, anxiety over death in war was coupled with the anxiety of personal faith in the redemption narrative. As a logic that totalized the mind, the religious narrative was successfully wielded in this contradictory way by the profane forces of the state on both sides of the conflict to support their earthly squabble. Due to the success of mobilizing public support for the war using this tactic in France and England, after the war these states inserted a greater degree of national influence over the Church, further weakening the authority of the papal position (Green, 2014). In 1454, the summer after the war ended, the German papal legate, Aeneas Sylvius Piccolomini, wrote of his anxiety over the future of the Church to a friend:

> I prefer to be silent, and I could wish that my opinion prove entirely wrong and that I may be called a liar rather than a true prophet ... For I have no hope that what I should like to see will be realized; I cannot persuade myself that there is anything good in prospect ... Christianity has no head whom all will obey. Neither the pope nor the emperor is accorded his rights. There is no reverence and no obedience; we look on pope and emperor as figureheads and empty titles.
> PIUS II, 1571, p. 656[1]

The loss of power was not immediate in its consequences, but this anxiety was well founded as the future the Church faced was one of a steady, if only relative, decline in status and social identity, triggered by a series of interruptions in the social world that threatened the Church's seemingly impenetrable grip on the collective consciousness.

An example of this crack in the Church's domination was in the 1500s when Henry VIII broke free from the Church and anointed himself the head of the Church of England, while retaining the doctrinal positions of Catholicism. Not only did this challenge the Church's authority over spiritual matters, but Henry VIII seized Church land to pay for his wars (Bernard, 2005). This led

---

1 English translation of Latin original taken from Gilmore 1952, p. 1.

the Archbishop of Canterbury, John Whitgift, to write the Queen of England to try and recall some of the Church's authoritative power. He penned, many "have given to God, and to His Church, much land, and many immunities," reminding her that she was anointed "to maintain the Church-lands, and the rights belonging to it." He then added a threat for good measure, writing that he "denounced a curse upon those that break the *Magna Charta*, a curse like the leprosy that was entailed on the Jews." Finally, he demands on somewhat shaky ground: "Now, madam, what account can be given for the breach of this oath" (Hook, 1875, p. 134). Although the letter signals an attempt to maintain the state supported power structure that the Church had enjoyed for a millennium, they were facing the reality of all social power: it only holds sway and is maintained when people acquiesce to its will. Weakened as the Church was as a political institution, Henry VIII retained faith in the totalizing logic that they crafted, but not in the institution behind it, illustrating how totalizing logics take on an autopoietic life of their own that morph in shape while retaining their foundational power.

The static nature of traditional life did not afford many opportunities for recoding the collective consciousness or the minds of obedient individuals that the Church relied on to retain its unquestioned supremacy across the land. This challenge to their power did, however, foreshadow the impending social awakening in the West to the recognition that the world could be a dynamic reality and novel forces of change could emerge to challenge the reign of traditional thought. However, as this example illustrates, it was only those who held positions of social power who could challenge the Church directly, and even then, the challenges were only partial.

2   A Revolution in Space and of the Mind

In addition to the Hundred Years War, what changed in the 1400s that allowed fractures to appear in the Church's power in the 1500s was the acceleration of technological innovation, which in turn accelerated the pace of life by increasing the flow of information. In the mid to late 1440s, Johannes Gutenberg performed a groundbreaking feat of recombinant innovation. Taking existing technologies and repurposing them for novel uses, he developed the first system of movable type on which the printing press was based.[2] This technological

---

2   Printing had been developed in China some six centuries earlier but had not caught on as a commercial innovation in the West until Guttenberg's press (Pacey, 1991).

achievement ushered in a new era for Europe that laid the groundwork for the modern transformation of society. It enabled the dissemination, standardization, and reorganization of how information was transmitted and stored. Without needing to rely on hand-written copies—the skills for which many were lacking, and which were tedious and prone to human error—the scale of data collection with the invention of the printing press increased and allowed for the mass preservation of data.

The impact of this invention sent ripples throughout society. Economically, "between 1500 and 1600, European cities where printing presses were established in the 1400s grew 60% faster than otherwise similar cities" (Dittmar, 2011, p. 1133). Culturally, contained in this spread of information and knowledge was an implicit challenge to the Church, for until the invention and cultural adaptation of the printing press, they had control over the dissemination and interpretation of the Bible. On the one hand, this was because their work included a level of education and literacy denied to the masses, and on the other hand, because even for those who could read, the Bible was only available in Latin. Originally the Church blessed the invention as the work of God and "called on printers to help with the crusade against the Turks" (Eisenstein, 1979, p. 303). Expanded access to written materials, however, meant an increased demand for a literate public who could take advantage of these new cultural artifacts and use them to challenge hegemonic discourses in a public context. Just as the Church found the press to be a powerful tool to use against their enemies, so too did it serve as a weapon against their supremacy as demands for personal education that could take advantage of these new cultural artifacts arose.

Ideas came alive in those years with the invention of the printing press and Europe enjoyed a renaissance of innovation in culture and science that recalled the empires of old. In the Renaissance years thinkers reconceived and reconfigured ideas by performing an intellectual form of recombinant innovation. From Classical Greek philosophy and the Roman concept of *Humanitas*, a worldview developed that was based on an anthropocentric model of humanism (Zagorin, 2003). Owing to the totalizing logic of Christianity, this worldview in many ways reinforced the general doctrinal belief system of the Church that held the human at the center of creation. However, as commerce in ideas grew, many of the ideas, particularly in the natural sciences, were found wanting in explanatory power of observed phenomena. As observational data expanded, and could be recorded and disseminated through printing technologies, discrepancies in the theories could not be smoothed over without massive leaps in logic. The tendency to simply modify a theory to fit in new observations led to cumbersome and weak theories of the natural world. Reality and human understanding were put to the test in a fight between rational thought and empirical observations on the one hand, and on the other commonsense and traditional narratives; each of which was further complicated by internal tensions.

Copernicus is credited with starting the scientific revolution in 1543 due to his action, taken against the theories of the day which were in line with Church doctrine, to recognize that if the theories did not describe what was observed then they needed to be thrown out and reconceived (Kuhn, 1957; [1962] 2012). Based on his astronomical observations he concluded that the theory of geocentrism, which placed the Earth at the center of the universe, could not explain the data he collected. Either reality was wrong, or the theory was wrong. After repeated observations he opted for the latter interpretation, overturning some 1400 years of following Ptolemy's geocentric model, and proposed a heliocentric model which held the sun at the center and placed earth in a rotational pattern around it. This defied the commonsense human perspective which saw the sun cross the sky each day, as if it moved and the earth stood stationary, and threatened traditional systems of knowledge maintained by Church doctrine. As Sousanis (2015) tells it:

> By displacing the earth from the center and setting it spinning, [Copernicus] unwittingly sparked a revolution. *Nothing changed, except the point of view—which changed everything.* While others would expand upon this work, the fundamental shift of viewpoint irrevocably ruptured a stasis of thought, its implications rippled outward ... and a sun-centered outlook would fuel further revolutions. (p. 33; emphasis added)

What presents itself as the case, as the sociologist Niklas Luhmann (1994) cogently reminds us, can mask the true nature of the observed reality. Only by asking what lies behind the case can we reveal a higher or lower-order perspective that challenges the world of appearances, such as the earth rotating around the sun, and come to see that the reality of the world exists in fractal layers of perception (Crombez, 2015). Each layer that we peel away to reveal a new perspective can provoke anxiety as they challenge and complicate our worldview and sense of self, but doing so can also help us make meaning out of our reality and spark the quest for knowledge as we come to know ourselves in ways that were previously inaccessible. What became apparent in the years that followed Copernicus's discovery was that this quest for knowledge must be maintained in a permanent revolution of the mind, but it first begins with a revolution in the perception of material conditions which force a confrontation with this brand of anxiety in which the self is radically attached to the system of social domination.

Putting the case to the test meant challenging the world of appearances particularly as the world appeared to humans with the senses available to them given the totalizing logic that had trained them to think and adhere to a very specific worldview. The French Jesuit priest and philosopher, Pierre Teilhard de Chardin ([1959] 1964), describes the significance of this challenge raised by Copernicus:

> The conflict dates from the day when one man, flying from the face of appearance, perceived that the forces of nature are no more unalterably fixed in their orbits than the stars themselves, but that their serene arrangement around us depicts the flow of a tremendous tide—the day on which the first voice rang out, crying to Mankind peacefully slumbering on the raft of Earth, "We are moving! We are going forward!" ...
>
> It is a pleasant and dramatic spectacle, that of Mankind divided to its very depths into two irrevocably opposed camps—one looking toward the horizon and proclaiming with all its newfound faith, "We are moving," and the other, without shifting its position, obstinately maintaining, "Nothing changes. We are not moving at all." (p. 1)

At stake in this conflict is that if human senses could be wrong about the static nature of earth and suddenly awaken one day to a dynamic reality of motion and planetary rotation, then there could be other areas in which human perception had misguided the species to make false assumptions. As Teilhard de Chardin argues, this caused a rift in human thought with one side representing those who refused the implications of a dynamic reality and insisted on an immobile worldview and on the other side those who embraced the consequences of a world in motion.

Since the human scale is limited by the biological appendages we have for sensing our reality and these lead to commonsense observations deemed as 'natural,' the 'immobilists' were supported by the Church, who embraced this notion of a static human nature that could only change after physical death in a spiritual rebirth, which helped construct and maintain the system of tradition. Those who embraced the new worldview of motion and change had to develop new forms of seeing the world to access this reality hitherto unknown to them, and they had to do so against the authority of the Church, knowing that the anxiety they felt from challenging the Church's position was founded on the real knowledge of what happened to heretics and blasphemers. In other words, there were two branches of anxiety, the first from those who wanted to access this reality and challenge the system of domination because they could not locate their sense of self in the dominant narratives of reality, and the second from those who were so totalized by the logic of domination that they could not conceive of relocating their self in this changed reality. Most, therefore, were cautious and required ample empirical evidence before they committed to the new worldview and the risks its acceptance implied. To get this evidence and to see things beyond human capabilities required the development of new tools and technological appendages that could amplify and enhance human perspectives. These technologies were themselves the result of scientific inquiry and the marriage between science and technology was consummated in the Renaissance years from the 14th to the 17th centuries.

Of note was the development of the telescope, used to gain ever more precise observations of astronomical movements to gauge the feasibility of Copernicus's theory. Although there is much debate over who first invented the telescope as we now know it (Van Helden, Dupré, van Gent, & Zuidervaart, 2010), the Italian astronomer, Galileo Galilei, had an enormous impact on the success of the invention. Knowledge of optics and material processes were needed to develop the right lenses which would allow one to view magnified images of distant celestial objects (Dupré, 2002). This allowed Galileo to take even more precise astronomical measurements than Copernicus, and the evidence he gathered in support of heliocentrism strengthened his belief in Copernicus's theory. But it was not merely scientific and technical problems that controlled the view of material reality, there was also the social aspect of living under the totalizing logic of the Church and the anxieties of losing oneself to the power of its social control. "Copernicus, in fear of trial for heresy, long hesitated to announce his heliocentric view" (Polish American Journal, 1993), and he only did so in a book published the same year as his death, so he avoided the most serious social consequences of his blasphemy. The Church maintained that Copernicus's views were flawed and against holy writ and more than 70 years later, in 1616, when Galileo presented more evidence in support of these ideas he was ordered by the Church to abandon his defense of the heliocentric model (Mayer, 2010). Eventually, like Copernicus, when he had less to lose as an old man, Galileo confronted this anxiety when he threw caution to the wind in favor of advancing scientific knowledge. Against the orders of the Church he published a book that challenged the long held geocentric doctrines handed down since Ptolemy in antiquity and came out publicly in favor of the heliocentric model supported by logical argumentation and empirical evidence. This earned him a place before the inquisition, and he was condemned for holding heretical beliefs that ran contrary to the official position of the Church. Galileo was in his 70's by the time his sentence came down and he died a couple of years after, like Copernicus avoiding having to live much of a life branded as a heretic for challenging the totalizing logic of the Church. While Copernicus and Galileo escaped the worst punishments of the Church, others were not so lucky, such as Giordano Bruno, who refused to allow the anxiety of confronting a totalizing logic to sway the actions that he believed in and aligned with his sense of self. The result, however, of invoking his agency and channeling his anxiety into action against the Church's totalizing logic was that he received "condemnation to death for 'obstinate and pertinacious heresy' in 1600" (Rowland, 2013); a high price for maintaining his scientific views.

Something, however, had clearly changed in the way that people were responding to being dominated by the totalizing logic of the Church. For over 1000 years few had dared to publicly challenge its authority, but by the 1500's the evidence suggests that more and more people were drawn to look beneath the veil of domination that the Church had constructed to control the Western social world. As these challenges to the Church's authority played out in the natural sciences, fracturing its claims to epistemological and ontological truth, debates in philosophy began with a renewed sense of vigor. On the one hand there was the ever-present issue of logical consistency, while on the other, there was the question of whether the truth was contained in the faculties of the mind—which could be twisted to fit the totalizing narratives of social power in ways that felt perfectly normal and natural—or if truth could only be distilled from pure empirical observations—which could present contradictory information depending on the scale of observation. Whereas the Copernican scientific revolution represented a synthesis of the rational and the empirical, in philosophy these ideas were placed in tension with each other and were pushed to their radical limits in a revolutionary battle of the mind.

In this dominant Western philosophical tradition from which the critical method was birthed, two of the most influential representatives in this debate were the French philosopher, René Descartes, who in the 1600s took up the torch of rationalism, and the Scottish philosopher, David Hume, who in the early 1700s donned the mask of skepticism and assumed a stance of radical empiricism. The question at the center of the debate was ultimately related to an epistemological challenge over which source of knowledge was more trustworthy: that which was the product of rational thought or that which was the result of empirical sense data.

For Descartes the issue was that if commonsense knowledge, derived from human senses operating at the human scale, could be contradicted with observations made using those same senses at other scales, then how could sense data be trusted since it could ultimately deceive the observer? Rather than falling back on the totalizing logic of the Church to answer the question according to its narrative, Descartes's argument assumes the guise of an agnostic perspective that traces out the rationality of each belief system. However, Descartes remained under the influence of the Church and erred on the side of its system of thought, hoping to use rational thought to prove the existence of God and defend the Catholic faith from the challenges of empirical skepticism over objects of faith. Despite his religious aims, however, Descartes's advancing of rationalist thought further fractures the totalizing logic of the Church, even if unintentionally. Whereas the Church displaced anxiety from concerns with the material self to those associated with the potentiality of a spiritual self in the afterlife, the question of anxiety is reopened as a material question with Descartes.

Descartes probes the Church's epistemological and ontological logic raising the question of the self and its relation to, and place in, this world; which at the least is a recognition of the fact that the Church's narrative did not provide a satisfactory account for all and therefore required a philosophical defense. He writes that either "there is an omnipotent God who created us ... [or,] our existence derives not from a supremely powerful God but either from ourselves or from some other source" (Descartes, 1985, p. 194). Regardless of which answer is true, Descartes concludes that a God who has power over creation could also be a deceiver, "if he indeed wishes to, he can easily bring it about that I should be mistaken" or "God might have endowed me with such a nature that I could be deceived" (2008, p. 26). But even if the world is without God then it would be a world of imperfections and as a product of an imperfect reality it would mean that "I am so imperfect that I am perpetually deceived" (p. 16). Without being able to conclusively determine in the last instance that reality is itself not a deception, Descartes turns inward and concludes that all he can be sure of is that:

> certainly I did exist, if I convinced myself of something.—But there is some deceiver or other, supremely powerful and cunning, who is deliberately deceiving me all the time.—Beyond doubt then, I also exist, if he is deceiving me; and he can deceive me all he likes, but he will never bring it about that I should be nothing as long as I think I am something. So that, having weighed all these considerations sufficiently and more than sufficiently, I can finally decide that this proposition, 'I am, I exist', whenever it is uttered by me, or conceived in the mind, is necessarily true. (p. 18)

Descartes arrives at an epistemological and, therefore for him, an existential certainty, but as to what the "I" consists of in material reality, in which the guaranty of a future life in a spiritual reality is unknowable and unverifiable on purely rational grounds, remains a challenge in his framework.

What he concludes is that the "I" is a "thinking thing," one who must, therefore, create and make meaning for himself using his rational mind. At the same time, the "I" must navigate and sort through data provided by the senses that allow him to perceive that which is external to his thinking being, be those externalities real or imagined. Descartes, therefore, imagines a clean slate of the human as a thinking thing, a rational being, an "I" that interacts with reality without ever fully grasping and knowing the full impact of those interactions. This "I" can only answer the question of the self through meditations on the real and imagined external reality and must rationally determine its own ontology. However, this leaves an opening in Descartes's system because while he offers a certain kind of epistemological certainty, his rationalist system can never achieve the ontological *certainty* claimed and offered by the totalizing logic of the Church. If, therefore, the "I" must construct an ontology for her or himself, anxiety of the self must be linked to the way in which our ontological view is constructed. So, while the Church's ontological narrative is ultimately sufficient for Descartes to fill this hole, his probing of the epistemological side of the equation left the ontological side looking weak and ready to be exploited by this anxiety for any who did not share his unquestioned faith. Furthermore, the blank slate of an ahistorical rational being that Descartes constructs is only an imagined creature that exists as a metaphysical construct. To suppose its material existence is to ignore the fact that at the minute a thinking thing enters at birth into society the forces of history begin to shape not only the ontological perception of the "I" but also the ways that the self comes to employ rational thought.

This hole in the rationalist model is exploited and taken to the opposite extreme in the work of David Hume. He does this by explicitly challenging the totalizing logic of the Church, its epistemological and ontological claims, and those who attempt to defend it under the banner of rationalism:

> Examine the religious principles which have, in fact, prevailed in the world. You will scarcely be persuaded that they are other than sick men's dreams; or perhaps will regard them more as the playsome whimsies of monkeys in human shape than the serious, positive, dogmatical asseverations of a being who dignifies himself with the name of rational.
> HUME, [1757] 1889, p. 74

Hume's distaste for religion is based precisely on this masquerading of opinion in the guise of rational thought. Rather than seek to find the truth on its own terms, the attitude that dominates the philosophy of his day is to use philosophical tools to find arguments in favor of the Church's totalizing logic.

Philosophy at this stage is generally not a critical enterprise, rather it is one that is fully in the service of power and the system of domination and used to justify and maintain that system. Natural science and technological advances, although being wielded by the structure to maintain the system of domination are, by their ability to force shifts in material perspective, far more radical in their societal impact than the metaphysics of the day.

Although he advanced the doctrine of rational thought, philosophical work in the vein of Descartes succumbs to the totalizing logic of Christianity rather than providing a system of thought that runs counter to that logic by penetrating the veil of domination. That we can see in our scenic landscape that it helped to further fracture the totalizing logic of the Church should rather be understood as an unintended consequence. Descartes's position only masquerades as objective, when in fact it represents a subjective rationalization of the dominant system of belief, reflected by a mind that was socialized to think that way under a specific form of social domination. Rather than default to the Church's totalizing logic to gain freedom from the anxiety of ontological uncertainty, Hume spins the narrative around and enters the realm of critical thought. It is not the Church's logic that relieves anxiety, rather it produces anxiety because those who adhere to its logic see reasons all around them to doubt their own faith and therefore their self-image. Believers risk living a life in which they cannot embrace their authentic self because they have sacrificed it to this totalizing logic and are controlled by its demands for accepting its narratives of epistemological and ontological certainty, and unbelievers are generally not free to explore other narratives without fear of immediate social sanctions. For Hume, however, there is comfort in asking the questions themselves, even if the answers are elusive, and he contents himself with the philosophical inquiry. He writes:

> The whole is a riddle, an enigma, an inexplicable mystery. Doubt, uncertainty, suspense of judgment, appear the only result of our most accurate scrutiny concerning this subject. But such is the frailty of human reason, and such the irresistible contagion of opinion, that even this deliberate doubt could scarcely be upheld, did we not enlarge our view, and, opposing one species of superstition to another, set them a quarrelling; while we ourselves, during their fury and contention, happily make our escape into the calm, though obscure, regions of philosophy. (p. 75)

Therefore, if rational thinking can be so easily swayed by the power and influence of the institutions of domination, then, for Hume, it is not empirical reality where we are deceived, but it is in those thoughts we have which we so easily mistake as being our own, rather than recognizing their origin in the totalizing logic of a system of domination.

While the imagination can be given to flights of fancy and can conjure images of a dreamlike quality, Hume ([1748] 2007) insists that this is the work of the mind, and that it is dependent on material conditions. He writes:

> though our thought seems to possess this unbounded liberty, we shall find, upon a nearer examination, that it is really confined within very narrow limits, and that all this creative power of the mind amounts to no more than the faculty of compounding, transposing, augmenting, or diminishing the materials afforded us by the senses and experience. (p. 13)

It is not a giant leap from this empiricist position, which recognizes the value of perspective on material reality in shaping thoughts, to link the Copernican revolution in our understanding of our positionality in the universe to the revolution of the mind in the Enlightenment that lays the groundwork for modern thought. Even through threats of excommunication, Hume approached death in 1776 free of anxiety and at peace with his worldview that stood as a challenge to the totalizing logic of the Church with an unbridled skepticism of opinion disguised as rational thought (Rasmussen, 2017).

3 Enlightenment and the Birth of Modernity

Two events happened around the time of Hume's death that further developed the foundation upon which the modern world is built and changed the shape of anxiety. The first is America's Declaration of Independence in 1776 from the British empire. The second is the philosophical work of Immanuel Kant.

The Declaration of Independence established a new nation built on modern principles that came to be codified in its constitution. What makes it historically unique is that it represented the founding of a nation state that was actively attempting to free itself from a large part of the history of tradition found in the cultural artifacts (books, art, architecture, etc.) and deep embeddedness of the Church coupled to the State in Western Europe, without simply seeking to eradicate those traditions using authoritarian and militaristic means. It was not, however, an anti-religious enterprise. Rather it built off the Protestant challenge to Catholic domination, which during the Protestant Reformation (which I will consider in more detail below as we confront the work of Max Weber) in the 1500s contained a kernel of the modernist spirit.

Despite its fundamentalist approach to the Bible, the Protestant vision is one that is highly individualistic. To end the mediated relationship of the priestly class between God and human that the Catholic Church relied on as a means for exercising their power and maintaining social control, the Protestant vision imagined a direct relationship between the individual and God that limited the mediated nature of the relationship. This relationship could only be nurtured if the notion of individualism was sufficiently developed so that people could rely on and trust in their own abilities on the one hand, and on the other, if there was an open and available means of cultivating a spiritual identity that was not dependent on priestly mediation. The burden of anxiety is therefore shifted in large part onto the individual in Protestantism. They are responsible for developing their own coping mechanisms to reconcile their newly developed sense of self with a social milieu that they must judge according to principles divined through the development of a personal relationship with God.

The cause of increased individualism will be explored in greater detail as it relates to Durkheim's theories and the transition to modern society below, but the historian Reinhart Koselleck ([1959] 1988) simply attributes the rise of individualism to the cultural influence of Enlightenment thinkers, such as Thomas Hobbes and John Locke, whose ideas are implicit in the Declaration of Independence and the American Constitution (Dienstag, 1996; Coleman, 1977). According to Koselleck ([1959] 1988) this spirit of individualism was found in the seeds of "the Enlightenment, that is, the criticism of Church and State that furnished the dualistic counterpart for the development of the bourgeois sense of self openly and plainly threaten[ing] the existing State" (p. 172). In England, for example, the union of Church and State was used as a means of oppressing individualistic expression that ran counter to the structure of power. However, once the challenge of Protestantism enters the mainstream discourse in the mid-1500s with Edward VI in England and Henry IV in France, the monarchies had to worry about the bipolar power structure of the Catholic Church and the State. From the perspective of the State, the problem was whether people would in the final instance submit to the commandments of the State or those of the Church when their interests did not align. For example, will the people listen to their king if he commands a war against a Catholic country that has the Pope's support, or will they follow their spiritual leader and refuse the king the authority of the crown? For Koselleck, this question leads to the division of the public and the private spheres and, for our purposes, signals one of the earliest shifts toward the social cultivation of anxiety. The new-found individualism that spread to the masses meets its challenge in the schizophrenic split between the public commands of the king and the private dictates of religion. Torn between competing value spheres, the self that identifies with a religious identity that is not shared by the king, is subject to a loss of that identity in the public sphere where it must be repressed by the demands of the social milieu cultivated by the monarchy and vice versa. This opening salvo of splitting the modern self into a fragmented identity is the beginning of a permanent crisis that is mirrored in the psychosocial dynamic of self and society in conditions of modernity where demands made of us in our private and public lives often result in role strain and role conflict (Creary & Gordon, 2006).

Beyond the political ideologies operating at the macro level, cultivating the spiritual identities of Protestant individualism relied on two intermingled advances in material conditions already hinted at above. The first was a means of mass communication as embodied in the printing press which enabled the distribution of textual information for mass consumption; in other words, it allowed the average person to have greater access to biblical and other religious texts. This in turn increased the demand for personal education so that people could read for themselves what these religious texts—which prior to the 1500s were largely only available to the priestly class—said about spiritual matters and decide for themselves as a matter of personal conscience how they were to behave and adhere to the commandments contained therein. While many of the signers of the Declaration of Independence were admittedly deists, rather than subscribing to a particular form of organized religion, they fell on this individualistic side of Protestantism in their rejection of authoritarian means of control evidenced by their embrace of democratic principles (at least insofar as they applied to white, property-owning, men; for women, enslaved peoples, and racial and ethnic minorities authoritarian control remained). Therefore, it was upon that religious revolution that the philosophical and scientific revolutionary spirit followed, culminating in the modern political revolutions, as in the case of the American Revolution of 1775–83 and the French Revolution of 1789–99.

What these two wars shared in common was a commitment to the idea of modern individualism enshrined in their enlightenment slogans: (America) "Life, Liberty, and the pursuit of Happiness" and (France) "Liberté, égalité, fraternité."[3] This translated into a modern democratic political system in America built on freedom of religion and a ban on the establishment of a state religion, effectively overturning the reliance of the state on the totalizing logic of religion as the primary mechanism of social control which it would need to locate elsewhere. In France although feudalism was abolished, and it moved in the direction of liberal political values, politically the social control question was

---

3 Liberty, Equality, Fraternity.

answered by it becoming a dictatorship under Napoleon in 1804. The effects on religion were, however, no less dramatic as "the revolution abolished the tithe, then nationalized and sold off church property" (Blackbourn, 1991, p. 779). However, rather than embracing Protestantism, the moves in France were taken to challenge and reshape the power of Catholicism, demonstrating that its totalizing logic was not immune to the impact of the rising tide of the modernist spirit and the normative values that underpinned these revolutions (Gibson, 1989; McManners, 1969).

These normative principles of the French and American revolutions formed the bedrock of the critical method that emerged during those revolutionary years and advanced in a systematic fashion as a method in the work of the German enlightenment philosopher, Immanuel Kant. During Kant's most productive years, beginning in the late 1770s and stretching through the 1780s, he enjoyed the rule of Fredrick II (also known as Fredrick the Great) in Prussia. Although Prussia did not experience a political revolution akin to that of France and America in those years, as a young man Fredrick "explained to his uncle that he was not surprised "by how many philosophers declare that they have no religion," but he "could not imagine that a king would ever speak like this," because of how "religion supports royal authority"" (Kloes, 2016, p. 103). Fredrick maintained the monarchy, but he opted to rule in a manner that privileged rational thought, religious freedom, and tolerance—key components in how the enlightenment thinkers conceived of their revolution of the mind—over authoritarian modes of coerced obedience. This ruling style was successful, and he enjoyed a long reign, but by maintaining the system of monarchy that was so closely tied, as noted above, to the system of religious authority and its totalizing logic, and by not applying the principles of enlightenment to the political system as such, after his death there was a conservative backlash in Prussia (Lestition, 1993). By that time, however, the key enlightenment texts had already unleashed their ideas and the seeds of change were sown in the consciousness of the age.

In philosophy as opposed to the natural sciences, Hume's work was a major spark for the development of the critical method, but it is perhaps best remembered not on its own terms and rather for the impact that it had on Kant during those years when there was less political support for religious resistance to critiques of systems of thought. Kant says that Hume "interrupted my dogmatic slumber" (Kant, [1783] 2001, p. 5) and while not accepting Hume's solution to the epistemological problem that placed empiricism in an antagonistic relationship to rationalism, Hume's skeptical challenge to rational philosophy—with its agenda largely dictated by the desire to prove the existence of God as a counter to increasing knowledge of the natural world that challenged such beliefs—and his insistence on the need to respond to external reality *as well as* the inner workings of thought, set Kant's philosophical agenda. Kant, in turn, set the agenda for modern Western philosophy.

Kant's major contribution to the philosophical debate was to synthesize the perspectives of rational thought and empirical reality by determining the proper domain of each while uncovering when and how they must overlap. He accomplished this by developing a critical method of critique through which he took up the question of epistemology, on its own terms, that is, on its philosophical merit, rather than in relation to the personal belief system of socially constructed knowledge to which he ascribed. In other words, Kant aimed, to the extent possible, to do what Descartes could not by putting his own beliefs in a box to the side while he tackled the problems on their own terms. However, Kant too, sought, if not to prove the existence of God using the tools of philosophy, then at least to save a place for God that existed independently of validation by the philosophical method so as to end the tradition of rational philosophy that was primarily dedicated to theological concerns. In this, we see an agreement between Kant and Hume that such tasks did more to demean the output of philosophy by locking it to a static issue than it did to expand the foundation upon which new knowledge could be produced or uncovered. Kant's project is therefore one that sought progress by modernizing philosophy, pulling it from the concerns of the traditional world to those of a modern world system.

Kant accomplishes this by introducing the concept of critique as "the whole plan of science, both in regards its limits and as regards its entire internal structure" (Kant, [1781/1787] 1998, p. B xxii). As McCarthy (1985) states, this early attempt by Kant to locate critique methodologically, refers "to the examination of the roots and structure underlying the very possibility of cognitive experience" (p. 16). In this way, the critical method which progresses via critique is precisely geared toward tasks such as ours: to scientifically analyze the concept of anxiety—which is a cognitive process of interpreting the self in the social milieu and the reconciliation of their incompatibility—and to diagnose the structures that produce it and the forms in which it emerges. However, this method is still not in its fully developed form in Kant, and rather than producing a system of science for the production of knowledge, Kant's version of critique is set to the more philosophically oriented task of clarifying knowledge by testing the boundaries of reason.

Kant's first major work in this vein is *Critique of Pure Reason* ([1781/1787] 1998), which performs the analytic and boundary marking work of outlining the means of attaining knowledge, the categories of knowledge, and the distinction between that which can absolutely be known by means of pure reason (*a priori*) and that which can only be known as a result of interactions with empirical reality (*a posteriori*). For Kant, it is not worth totally doubting our senses, which would lead one into a philosophic black hole of unchecked skepticism and would severely limit any such philosophic enterprises by reducing them to the study of subjectivity and nothingness. Rather, he determines that there must be some assumptions on which we inherently rely and to which those of our species generally agree. In this way Kant opens the *Critique* by admitting that "all our knowledge begins with experience" (p. 41/B1), and such experience triggers in us a sense of understanding which must then be subjected to a rational grounding so that we can differentiate between those things which are knowable and those which are not.

In contrast to empirically derived knowledge, for Kant *a priori* knowledge is foundational and underlies all other knowledge; that is, the assumptions which we must know to be true in order to understand and know the rest of the reality of which we are a part. *A priori* knowledge is thus the knowledge which is divorced from the senses. It represents that which is left over once we have removed all things given to us from our senses, and yet, something remains. That something is part of the foundational makeup on which we can then build upon by assigning meaning to it which allows us to understand the data received by our senses. *A priori* knowledge carries with it the further requirements that it relates to things which are both necessary and universal, because through reason, and verified in experience, we learn that they must be so.

To these Kant adds the further definitions of analytic and synthetic. Analytic, on the one hand, is that which is understood in and of itself from the object of study, that is, by mere analysis of a concept the things which we can know about it but not remove from it while still maintaining the concept. Analytic can thus be understood as tautological, in that all that we gain from analytic exercises are predicates which are contained in the subject concept and which cannot be removed without destroying the concept itself. Synthetic, on the other hand, is where advancements in knowledge are made and it applies to the combination of concepts by which we advance our knowledge of things as they relate to other ideas and other things of which we have prior knowledge. The synthetic is thus a descriptive way of obtaining knowledge through the combination of concepts, in which we learn something new about our concept by applying knowledge gained elsewhere to it.

Prior to Kant's *Critique* these classifications of knowledge were much debated by the rationalist and empiricist traditions. On the one hand, rationalists, building off the Cartesian model, believed that metaphysical truths could be reached through mere analytics of *a priori* concepts. On the other hand, the empiricists, like David Hume, followed a skeptical model which, through a reduction of metaphysical concepts, found many wanting because they required synthesis, which they determined to only be possible with *a posteriori* concepts, thus they dismissed the possibility of many *a priori* claims as altogether unknowable. Kant establishes that the limits of knowledge are "space and time" because they "are [the] only forms of sensible intuition" ([1781/1787] 1998, p. B xxv) and "our mode of intuition is dependent on the existence of the object, and is therefore possible only if the subject's faculty of representation is affected by that object" (B72). In other words, Kant cannot limit himself to the radical rationalism of Descartes in which all that is knowable is our "self" as a thinking thing, nor can he follow the path of Hume to its extreme in which empirical reality is left untouched by the faculties of rational thought and vice versa, which is to say that for Kant reason and empirical reality are co-constructed by an exchange between the sensing subject, the object as appearance, and the faculties of reason that distinguish between the two and make judgements about them. Kant's novel and controversial solution, therefore, was to claim demonstrations of *a priori* knowledge which are also synthetic, thereby finding the ground upon which his stronger claim to the interconnectedness of *a priori* and *a posteriori* knowledge rests.

Kant thinks that at least in intention metaphysics must contain *a priori* synthetic propositions, because without this form of knowledge metaphysics would be too severely limited by the analytic criteria others held for *a priori* propositions. Without *a priori* synthetic propositions, metaphysics could make no claims about the outside world and could only analyze ideas as such, with little hope of claiming legitimate value as a truth-seeking science that could make judgements and have practical value for everyday life. These are considered in his next two tomes: *Critique of Practical Reason* written in 1788 and *Critique of Judgement* in 1790.

With Kant's claims on what constitutes knowledge, the boundaries of "knowing," and therefore, the way in which we think, the claim that our senses are limited to the world of appearances means that spiritual questions, such as those concerning the existence of God on which the totalizing logic of the Church depended, suffer a setback given the modern tools of analytical thought that he refined. Kant concludes that God must exist outside of space and time, which is in line with the Church's view that God is the alpha and the omega (the beginning and the end), therefore that which exists beyond or before/after space and time, so his work offers a critique of the view while at the same time he does not dismantle the claim of God's existence by directly challenging it. However, given that we can only perceive within space and time, our senses and our faculties of reason are not capable of defending, using either the philosophical or the scientific method, the existence of God. While Kant reserves a place outside of our means of knowing for God, he essentially puts the nail in the coffin on the question of God for the agenda of modern philosophy and science, regardless of whether one wishes to answer the question in the affirmative or the negative. Without having the agenda of answering the question, Kant's attempt to save God from philosophy instead ushers in the death of God by removing the question from enlightened thought, at least from the perspective of modern metaphysics and science. Anxiety forks again at this moment, for now anxiety couples with a more powerful form of doubt. Those who were previously anxious over their spiritual existence after their material death are now confronted with a system of thought that challenges the validity of their beliefs not only on empirical grounds, but also on rational grounds, for their belief lies beyond the realm of appearances in space and time.

Experiencing the success of his method—insofar as Kant is convinced that no one can again raise the question of epistemology without confronting his system—which does not place the belief in front of the study, but rather attempts to create a value-free mode of thinking through problems for what they are in their essence, Kant comes to recognize that this practice requires at a minimum a level of freedom so that it can proceed according to its own claims in its own domain. In other words, the philosopher must at the very least be allowed to operate free from domination of the mind and the influence of others who would demand that she must adopt a specific position and defend it prior to the start of the investigation. However, Kant is frequently blind to the implications of his own assumptions, in that he maintains a racist and sexist attitude in his work, implicitly suggesting, by his dismissal of 'others' (women, Jews, etc.), that it is only white male Europeans who possess the faculties needed to engage in this kind of intellectual activity. Even when he turns his attention to social issues, Kant does not directly challenge power, he instead makes the argument that the philosophical method and the advance of enlightenment thought must be completed beyond the reach of power so as to achieve truth rather than opinion. Kant's claim is that this will also benefit those in power because it will produce knowledge for the benefit of all. And although he does not extend this beyond the privileged status position that he and others like him share, the application of the critical method to his work pushes the logic beyond its roots by opening a new door in the history of thought through which even those he denied as 'others' could follow.

In his famous essay, *Was ist Aufklärung?* (What is Enlightenment?, written in 1784), Kant (2007) invites a new social reality that lies beyond systems of totalizing logic so that progress and advancement can be made through critique. He writes, "*Sapere aude!* "Have the courage to use your own reason!"—that is the motto of enlightenment" (p. 29). The goal of enlightenment is not to have people who always rely only on their own thoughts to the abandon of accumulated knowledge and wisdom, rather the goal is that in thinking for ourselves we will work "gradually out of barbarity if only intentional artifices are not made to hold [us] in it" (p. 36). These artifices are precisely the totalizing logics that tell us: "I need not think, if I can only pay—others will readily undertake the irksome work" (p. 29) of confronting the anxieties of our reality for us, be they of material or spiritual origin. For Kant, giving into the temptation to have others simply tell us what to do, rather than engaging in the intellectual challenge of confronting our anxieties, be they individual or social, is an act of cowardice. The only condition that is needed for critique to flourish is the removal of all social barriers to it, to allow people the freedom to think and challenge the systems of power and domination, and, in the end, to have the courage to channel anxiety at personal cost without knowing the outcome in advance. This is to risk the self willingly for the sake of progress.

This promise of progress was the carrot that overturned the traditional world and birthed the modern world. But just as Kant hangs onto his prejudices and fails to confront the totality of his society by contenting himself with the abstractions of philosophical universals, the progress of the modern world developed unevenly, and the system was built on a new kind of exploitation that developed within a new system of dominance that emerged at precisely the right time to replace the crumbling social power of the Church. Voltaire emphasized this uneven development when he wrote in 1771 that "more than half of the habitable earth is still inhabited by two-legged animals, living in that horrible state which approaches nature, who are scarcely alive or clothed; barely enjoying the gift of speech; hardly perceiving that they are unhappy; living and dying, almost without knowing it" (1771, p. 91). Thus, the modern world that Kant is striving for in his work came to be conceived of as a project that had to be spread out to the world in a process of modernization. This process and the consequences of it served as the foundation for the next leap in critical thought, as the new technologies, systems of politics, economics, and shifting cultural norms reproduced the very systems they tore down.

CHAPTER 2

Critical Methods 1

1   From Philosophy to Social Theory: (Hegel, Feuerbach and) Marx's
        Critical Method and the Totalizing Logic of Capital

Georg Wilhelm Friedrich Hegel was the next major figure in Western philosophy to advance the critical method in the social milieu of a bourgeoning modern society with an eroding traditional world. It was Hegel's work that had the most direct and profound effect on Marx's rigorous development of a system of critical thought for the diagnosis of the psychosocial effects arising from the conditions of life in that new social order. Hegel, unlike Kant, had a greater focus on the effect that the immediate social landscape of one's development has on the questions and methods posed in philosophical discourse. Hegel's youth overlapped Kant's old age (as Marx's youth overlapped Hegel's old age), but the political scenario of Prussia in which Hegel came of age morphed in the late 18th century under the rule of the anti-enlightenment reactionary Fredrick William II. Fredrick William II was a poor economic planner and squandered the massive financial reserves left by Fredrick the Great which led to national economic woes. This was coupled with a turn to an "authoritarian cultural policy whose objective was to curb the supposedly corrosive effects of scepticism on the moral fabric of school, church and university" (sic) (Clark, 2006, Ch. 8). When Fredrick William III assumed power in 1797, he not only had to rebuild the financial reserves of the state, but he had to spend his energy for many years focused on foreign affairs that arose from the Napoleonic Wars. By the time those were settled in 1815, he turned his attention domestically to a conservative mode of governance that sought, much like France had in controlling the power of the Catholic Church, to control the Protestant sects. This was attempted through an ambitious plan to consolidate those sects into a singular Prussian Church Union (Clark, 1996). So, while Kant ([1784] 2007) enjoyed "an age of enlightenment, if not an enlightened age" in Prussia, the situation in which Hegel developed as a thinker and wrote was one of confrontation between enlightenment thought and counter-enlightenment principles.

As a result of the changed historical circumstances of Hegel's psychosocial development, he finds in Kant a far more radical and underdeveloped thesis than Kant seemingly recognized. Whereas Kant placed space and time as the boundaries and conditions of human knowledge, his philosophical intervention stayed true to tradition by maintaining a resolute focus on the quest for universal and absolute truths, which were presumed to be static and unchangeable across space and time. This mode of thinking ignored the ramifications of time as a boundary and condition of human knowledge which is definitionally linked to change and dynamism. Kant's mode of thinking, therefore, while attempting to modernize philosophical discourse was still rooted in traditional thought patterns, which given the glacial pace of change under that mode of social organization appeared as largely static and unchanging. Tying concepts to the absolute gave less treatment to the dynamic implications of the temporal dimension as a condition of knowing while overemphasizing a dependence on the spatial restriction as a boundary line of truth. In other words, Kant, as a representative of philosophy, locked onto a one-dimensional method of critique that conceived of systems of thought which were presumed to maintain their logical consistency throughout space *as if* time, as a secondary boundary, was itself always already a spatially anchored and defined constant that could not, or at least rarely did, impact the conditions of knowing. This allowed an easier justification for the belief of unfettered societal progress on a grand scale, as if history followed a unidirectional development that progressed not only in the domains of technology and economics, but followed a political trajectory that was tied to the discovery (natural process), as opposed to invention (artificial/social process), of philosophical and social thought; as if they could be reduced to rules of nature that existed independent of human artifice.

That is, in Kant's work the attitude is one that still imagined knowledge as a staircase. Each step would then represent another achievement in knowledge which, once uncovered, would become the foundation for the next step. This logical progression of knowledge had the appearance of being not just an ideal but rather the natural process of advancing thought. What it betrays is Kant's placement of rationality and reason as ideal modes of thinking above the material actuality of how people think, which does not have to match up to the ideal. The pursuit of knowledge, for Kant, was therefore a unidirectional climb and was dependent on the paths set forth by the previous generations which progressed steadily toward truth. This constraint on thought, while representative of the ideal of the scientific enterprise, had no power to bind those who engaged in philosophical discourse for purposes of advancing ideology as opposed to pure knowledge. The thread that Hegel pulls on is precisely that time by its very nature implies change, but not directionality, and when it is framed in terms of human knowledge and understanding it takes on the more specific form of human history as the artificial construct of free-willed human decision making about how they see, interact with, and comprehend their socially constructed reality. This accounts for the anti-enlightenment backlash experienced under Frederick William II. History, then, rather than being unidirectional, was for Hegel "constituted by the *Geist* as it unfolds dialectically" (Postone, 2009, p. 67) and the whole of this process, which progresses by means of negation, is conceived of as the totality.

At a certain scale we can witness a dialectic unfolding in the thought patterns sketched out in the scenic landscape above that led to Hegel's work, which only appears as progressive and unidirectional if it is tracked in thought alone and material reality is ignored. Moving from the natural sciences to philosophy, and from debates between the rationalists and the empiricists, the scene in which modern thought began to challenge the totalizing logic of religion was one that progressed by means of negative thought. Copernicus and Galileo's observations negated the logic of celestial perfection that placed the earth and humanity at the center of the universe. Descartes's attempt to save the knowledge of God from its empirical negation, does so at the cost of deflating empirical reality to mere illusion, stripping God of the qualities of creation to the essence of abstract perfection. Hume negates the purely rational mode of epistemology by critiquing it through the lens of empirical skepticism. And Kant negates both extremes by offering a synthesis out of their seemingly contradictory perspectives. Attempts to reinforce religious doctrine through the scientific and philosophical method exposed gaps—no less than attempts to refute these perspectives—in the logic which propelled the critique of the totalizing system of thought and deconstructed its truth claims. Although the advance of this dialectic in philosophical discourse weakened the totalizing logic of religion, it did not, and could not, deliver a death blow to that logic among the material masses whose adherence to religious tradition continued to hobble along in a fractured, but no longer totalized, form, which in many ways preempts Marx's criticism of Hegel's model.

Hegel ([1807] 1977) writes that "the whole is nothing other than the essence consummating itself through its development" (p. 11), establishing time as a central category through which we must understand the development of thought and therefore the effect that the social dimension through history has on modes of thought. Hegel points to an internal, rather than an external, causal mechanism inherent to the totality through which it develops historically as a subject. That is, nothing acts on the totality, but rather it acts through and within the totality, allowing history to be read as the story of the totality in the process of becoming itself. In other words, the totality does not become itself until its story, and our history, ends. Philosophy's claims to universal and absolute truths are ones which imagine that the temporal condition of certain forms of knowledge have already, at least in part, reached their end and exist in a quasi-static state rather than as part of a socio-historical dynamic. Under the framework laid out by traditional philosophy, all possible thought is always already present, unchanging at the level of Truth, and is simply waiting to be uncovered. Attributing internal over external causes for empirical transformations of the totality shifted the problem of conceptualizing it from a paradox—where any externality to the totality negates the concept of the totality—to a contradiction in terms of the positive knowledge one may have about the totality—the internal movement changes the totality from within, allowing it to be historicized but making all positive assertions historically contingent, or under certain conditions, impossible.

Unlike Kant's system of thought which still wishes to maintain allegiance to the idea that some universal truths exist prior to experience in and of the world, Hegel's solution dialecticizes the empirical (the totality exists, and it changes over time as it becomes itself) and the rational (the logical consistency of the totality in thought), reading them in tension with each other to a far higher degree than Kant's system allows. Hegel accomplishes this by recognizing that questions of beginning are impossible to verify, because as creatures of history we are bound to always start in the middle with our present circumstances serving as the given case. Rather than being content to clarify knowledge as the object of critique, as in Kant's system, "Hegel's philosophical "construction" is amenable to "social reality"" (Vouros, 2014, p. 176), to that middle ground of history, and provides a model and method that allows for the making of meaning out of the connections between the surface level of appearances and the depth of material reality. Therefore, Hegel's system ushers in a new mode of thought and knowledge construction properly called social theory and offers the first social analytic system, insofar as it recognizes the socio-historical impact on thought.

Hegel's system becomes the model for thinking the totality as a critical concept because it "perfectly complements and harmonizes with modern political economy" (Vouros, 2014, p. 175), that which becomes the force that learns from the totalizing logic of religion and replaces it as the primary totalizing logic of modernity. Modern political economy—represented by the logic of capital first in the West and then eventually spreading to the whole of humanity—extended its gaze to the whole material universe by transforming nature into its artificial other: the commodity. By the time Marx comes of age in the mid-1800s, this new system of political economy based on the logic of capital is radically transforming the world and it is doing so at pace that has greatly accelerated from the time of Hegel's writing, which provides an opening for Marx to push Hegel beyond Hegel. For example, James Watt's rotary steam engine was patented in 1781 and was quickly adapted to the railroad system, to the extent that "[b]y 1840, some 4,600 kilometers of line had been built in America, compared with 2,400 in Britain and 1,500 on the continent of Europe" (Pacey, 1991,

p. 138). Like transport, communications were also accelerating thanks to the invention of the electric telegraph which had near instantaneous political and economic implications. In 1858 the telegraph allowed Queen Victoria to communicate across the Atlantic with President James Buchanan, however, this pales in comparison to the economic implications of the telegraph which allowed for the creation of "11 commodity exchanges in major US cities between 1845 and 1871" (Sussman, 2016, p. 44) in support of capital development.

In the 19th century material reality underwent a transformation fueled by industrial speed, which had the effect of shrinking space in time. By Marx's time, it was no longer the social conservatives of the church or the state who were the primary guides of society, but the machinic logic of capital that was reaching to every corner of the globe and overturning the static order of life. This is not to say that religion and the state lost their power, rather it is to imply that they had to adapt their power to the totalizing logic of capital and work within a changed set of social assumptions to retain their positions of dominance in the social order. It was the structure of capital, not that of religion, that had the most direct impact on the shape of Marx's adult life and which led to his successive exile from Germany, France, and Belgium, and eventual life in the United Kingdom (see McLellan, 2006; Gabriel, 2012; Sperber, 2014; Jones, 2016; and Liedman, 2018). Just as those who challenged the logic of religion faced forms of social and physical death, Marx faced similar obstacles from the forces of power when he publicly challenged the marriage of the logic of capital to the system of politics. This is the reality that Marx contended with in his writings and is what motivated his challenge to the Hegelian model.

By placing the empirical and the rational in a dialectical relationship it allows for the overcoming of contradictions that arise from a reality in which we must account for both, even, or especially, when they appear at odds with each other; as increasingly became the case in modern societies oriented to political economy as their central organizing principle after the weakening of religious logic. Furthermore, overcoming these contradictions is what, for Hegel, internally fuels the totality as it realizes itself in a dynamic reality. This process is what constitutes the totality as the subject of what needs and has yet to be explained in history, thereby setting the agenda for a science to be practiced on negative foundations that cannot positively assert what the totality is, but can logically determine what it is not and what prevents it from assuming its ideal guise. However, Hegel's concept is not without its problems, retaining as it does "a residue of false positivity" (Adorno, [1963] 1993, p. 43) in his insistence that "[t]he True is the whole" (Hegel, [1807] 1977, p. 11) which places his gaze toward positive assertions of the end of history (the point when the True would become whole or total), meaning that Hegel, like Kant, does not fully develop the most radical implications that emerge from his critical method.

The problem with Hegel's critical method arises from the fact that he stopped short of applying the method to itself to root out those traces of positivity which are incompatible with thinking the totality as a negative concept, and it was Marx who discovered that new contradictions emerged when he applied pressure and pushed on the logic of Hegel's system. Although Marx owes an enormous intellectual debt to Hegel, he reproaches Hegel's philosophy because when it "has sealed itself off to form a consummate, total world, the determination of this totality is conditioned by the general development of philosophy" (Marx, [1841] 1978, p. 11) rather than the development of material (i.e. real) historical conditions. In other words, this means that Hegel's philosophy in its ideal formulation dialectically progresses philosophy following the model of ideal-material-ideal', but it does not provide a model for the progress of material reality as such. Therefore, for Hegel, although the totality is an ideal that emerges out of the relationship between thought and empirically observable transformations of material reality, it forms something that is distinct from material reality as such, and its dialectical becoming is only resolved once material reality is left behind. Because this totality, as the 'True' totality, is only an ideal, it can be erected and perfected in thought without the baggage of material reality that endlessly drags contradictions along with it. However, in equating the ideal with the totality, even when done in the abstract, it enters contradiction with the concept yet again by ignoring the historical potential that the transformation of material conditions has, and will continue to have, on thoughts that think the ideal in material reality. In other words, the ideal is a positive attempt to determine the truth content of the totality because it only resolves materiality by eliminating it in the final instance; that is, once history is over, time as a condition of knowledge would presumably cease its function while space would remain static and unchanging thus vindicating the old mode of philosophy that thinks in terms of universal and absolute truths. The essence, then, of the totality as ideal is that it is an eventual replacement for material reality that can persist as Geist (Spirit) alone.

From Marx's perspective by limiting the totality to the philosophically ideal, at the cost of excluding the material from its completion, Hegel failed to genuinely think in a manner that was consistent with the totality as a critical concept. The result of Hegel's system, per Marx, was that "the world confronting a philosophy total in itself is ... a world torn apart" (Marx, [1841] 1978, p. 11). Rather than leading philosophical thought and material reality into an eventual synthesis as the totality becomes itself, Hegel ripped the two apart as competing totalities with a winner and a loser. There are two ways of reading this, either

Hegel leads us to a contradiction in his concept of the totality in which he mistakes what is only partial (the ideal) for what is whole and total (the ideal[non-material] + the material[non-ideal]), or he assigns a value to the ideal(+) that supersedes the material(-), eliminating the importance that materiality has on and in the hierarchy of thought. Although Hegel can account for the dynamism of empirical reality in the present with his dialectical method, this final move of a dialectic that can only resolve itself as an ideal debases the material and is reminiscent of the ancient Greek philosopher, Parmenides, whose early thinking on the totality elevated the rational over the empirical.

Marx, however, sees within Hegel's system a methodological brilliance that by means of its critical power grants it the dynamic ability to work through contradictions, so he sets himself to "a ruthless criticism of everything existing ([i.e. the totality])" (Marx, [1843] 1978, p. 13). Marx ([1867] 1990) explains his method:

> My dialectic method is, in its foundations, not only different from the Hegelian, but exactly opposite to it. For Hegel, the process of thinking, which he even transforms into an independent subject, under the name of 'the Idea', is the creator of the real world, and the real world is only the external appearance of the idea. With me the reverse is true: the ideal is nothing but the material world reflected in the mind of man, and translated into forms of thought. (p. 102)

Following Hegel, by looking at the totality in the process of becoming, Marx concluded that thought and material reality were so dialectically entwined that one could not advance without the other or proceed in a lopsided fashion with the goal of eventually sacrificing one for the other. With the revolutionary transformations brought on by the process of industrialization, Marx was sensitive to how thought was itself linked to material reality and the ways that politics, economics and technology revealed new modes of thought while working to conceal others. How could Hegel think through the complexities of a world guided by the logic of capital if he lived in a world that had not been touched by the acceleration of reality through transportation and communication technologies operating under the logic of capital? Modern reality was for Marx far more dynamic and subject to change than Hegel's system assumed, because the material circumstances in which he developed intellectually contained far more dynamism in everyday life due to the industrial revolution. For Marx ([1845] 1992), the goal was not merely to advance thought, as in Hegel's formulation, but to transform material reality, and a material system that transforms itself continuously must be met with a methodological system that is itself also subject to transformation so as to be able to adapt to the changed circumstances of that reality. To accomplish this, he inverts Hegel's dialectical method to a version that is geared toward historical transformation (material-ideal-material'), keeping the goal of changing the world in mind. Whereas Hegel's dialectic resolves itself in philosophy, Marx's version can only be resolved in history outside of the confines of thought. Proceeding in this manner Marx developed a system that challenged the power of the philosophical titans of thought by pushing Hegel beyond Hegel in the new science of social theory.

Although Marx is primarily concerned with material reality and the political economic forces that had taken the reigns in driving the transformation of reality, he is also tuned into the historical dimension of thought and grounds his critique in the history of critique born of and sharpened against the totalizing logic of religion. Marx asserts that "religion is the general theory of this world ... it's logic in popular form" and "the criticism of religion is the prerequisite of all criticism" ([1844a] 1992, pp. 243–244). He recognizes that the critical method was historically developed in philosophy as a means for overcoming the totalizing logic of religion that dominated the minds of humanity. The reason that it is important to overcome this logic is because religion only offers an "*illusory* happiness" (p. 244). If humanity hopes to attain "*real* happiness," then the task of critique is to "*call on them to give up a condition that requires illusions*" (p. 244). The problem is that humanity does not recognize the illusory nature of their predicament when they are subjects of totalizing (alienating) logics. Turning back to Hegel through Feuerbach, Marx sees that there is a structural logic that maintains this condition of illusion beyond the realm of religion where it is birthed, and it is this logic that pulls the production of alienation as the general condition of the world from the power sphere of religion to that of political economy (Fischbach, 2008).

Hegel ([1807] 1977) framed this condition in positive terms, believing that self-consciousness—that is, "what consciousness knows in knowing itself" (p. 103) as a sober process of seeing the truth of oneself free from illusion—is only possible once self-consciousness is recognized in the other. Here self-consciousness refers to seeing oneself with sober eyes beyond the veil of illusion produced by totalizing logics. On the one hand, Hegel is issuing a direct challenge to the Cartesian model that insists that we can know nothing except our subjective "self" as a thinking thing. On the other hand, because self-consciousness requires that we treat and study the self as an object, from the position of a subject, we must undergo a process of objectification, which occurs when we recognize that others possess self-consciousness as well. Therefore, self-consciousness at the moment it is recognized as a quality of the other "has lost itself, for it finds itself as an *other* being," meaning that it cannot be recognized as an object simply by turning inward through a study of the self, nor can self-consciousness become an object until it is experienced as something that also exists outside of the subject: "in doing so it has superseded the other, for it does not see the other as an essential being, but in the other it sees its own self" (p. 111). What this entails, then, is that to escape illusions about one's self, one must first recognize the condition of self-consciousness in others leading to a situation in which "[t]hey *recognize* themselves as *mutually recognizing* one another" (p. 112). The brilliance of Hegel's thought here is that he grounds the state of stripping away illusions about ourselves and our reality in the social, for it is only in the social setting where mutual recognition occurs so that self-consciousness can be developed. This places Hegel's theory of history on the shoulders of the human, for it is only upon the basis of the human essence as a social creature that thought and history can be understood without becoming lost in a quest for the origin story or a beginning of history (as thought was oriented under the totalizing logic of religion). However, he ultimately ignores the full implications of what it means to ground recognition in a social act because in a material social system that is dominated by a totalizing logic there is a structural facet of reality that prevents the philosophically pure form of recognition and human essence (as an ideal) from developing and this failure can only be understood through the negative conceptualization of recognition: alienation. Here, again, the trace of false positivity infects the Hegelian dialectical account by resolving in thought (recognition) that which could only be resolved materially in history (alienation).

Alienation is the state in which we are totalized by a dominating logic to the extent that we can neither recognize in ourselves or others the essence of self-consciousness as discrete and whole thinking subjects apart from that logic because we have come to treat ourselves and others as objects of that logic rather than as the embodiment of the human essence. In other words, anxiety is precisely one of the "symptoms" of alienation in modern society, because in an alienated state, human subjects are structured in such a way that they are not able to locate their "self" in their social milieu because their "self" is dialectically entwined with that milieu which produces an alienated distortion of the "self." Although this implies a form of objectification of the self, which for Hegel is a requirement in achieving a state of recognition, alienation is a perversion of objectification for Marx. Marx develops this line of thought from Hegel through Feuerbach, who pinpoints the foundational logic of alienation in the totalizing logic of religion and plays another major role in the development of the critical method (see Deranty, 2014).

Feuerbach ([1842] 2012, p. 156) writes that "theology [i.e. religious thought] *dichotomizes* and *externalizes* man in order to then identify his externalized essence with him" (p. 156). What he is doing here is making an anthropological argument that sees religion as the externalized essence of the human, with God being the symbolic representation of the human ideal ("*man* is the *truth and reality* of God—for all predicates that realize God as God ... are posited first in and with man" (p. 159)). The consequence of this is that humans are alienated under the totalizing logic of religion because they come to associate the characteristics that first originated in the species, with a transcendental other whose characteristics are held up as unobtainable in material life. Historically, then, these characteristics are no longer seen as being the essence of humanity *as is*, but are treated as alien to the human, as otherworldly and beyond our ken.

Religion solves this problem with the illusory narrative of the afterlife in which humans, once their material existence is over, achieve the promise of a second life through death, a spiritual existence in the presence of God where these characteristics are made available to them. The implications of this are that humanity under the dominating and totalizing logic of religion, are unable to see their material lives for what they really are. Their material lives become alien to them, as they come to believe that their *real* and *true* authentic life will not begin until after their material life has ended. As mentioned above, this totalizing and alienating logic is made manifest particularly in the monotheism of Christianity which concentrates the ideals of the human in a single God. As Sloterdijk ([1983] 1987) has argued, God—that is, God as the ideal of humanity—had to learn how to become God, and this complete "God as ideal" is unique to Christianity because in polytheistic religions, and even in the monotheism of Judaism that predated it, the gods retain negative human characteristics (for example, Zeus is a serial rapist and the Old Testament God in Genesis regrets creation, nearly wipes it out with a flood, and then promises never to do that again). Feuerbach ([1842] 2012) explains: "The essence [or, ideal] of the Christian religion is, in truth, human essence" (Hegel's recognition thesis is a parallel argument to the biblical commandment to love your neighbor as yourself); "in the consciousness of Christians it is [materially], however, a *different*, a *non*-human essence" (p. 158), meaning that what was the essence of humanity is through Christianity rediscovered in its alien form in the ideality of God. It is for this reason that Feuerbach makes the provocative claim that "Hegelian philosophy has alienated man *from himself*" (p. 157) because Hegel's positive philosophy of recognition shares the theological logic of the God narrative, in that it also posits an unrealized and displaced ideal of human essence that lacks the immediacy of material conditions. In other words, Hegel's thesis retains the logic of religious alienation in the positive philosophy of recognition, it does not negate it.

Hume ([1738] 1965) had already concluded that "errors in religion are dangerous; those in philosophy only ridiculous" (p. 272) and although he was satisfied with keeping philosophy in this detached realm, for Marx, traditional philosophy was an insufficient vehicle for the critical method. This idea is illustrated in his famous thesis on Feuerbach: "The philosophers have only *interpreted* the world, in various ways; the point is to *change* it" (Marx, [1845] 1992, p. 423). The dangers of religion were made evident in the critique of religion by exposing the ways that its totalizing logic alienated humanity from locating their essence in their material conditions. Hegel's system recreated this problem because he stopped short of applying his critical method to itself, in the form of auto-critique, which would have aided in the recognition that his own thought was structured by the system to think in the manner it did. Marx's move of inverting the Hegelian method and engaging in a ruthless critique of everything was intended as a way to avoid that problem by allowing the critic to understand not only the object of critique, but what effect that object has on structuring the thought of the critic to perceive it the way she does so that she may change her mode of thinking to one that challenges the structuring logic. In other words, Marx's system was a refined method of social analysis because it accounted for how totalizing logics reproduce themselves in our psychosocial makeup, in a way that Hegel's philosophy outlined but ultimately failed to fully do. Only by understanding how our own thoughts succumb to illusion could it be possible to turn the new philosophical science of social theory into a dangerous method that could challenge dominating logics materially.

By the time that Marx developed his critical method, he had concluded that "the criticism of religion has been essentially completed" ([1844a] 1992, p. 243), i.e. in the modern societies of his day, the illusion of god had ruptured, and as a result, so too had the totalizing aspect of religious logic. This does not mean that the religious logic disappeared and that all were free of its alienating effects; in fact, this only represented the reality of a tiny fraction of the population. What it meant was that religion had in those societies lost the ability to dominate the social totality, because its logic had been exposed, as is evident by the number of texts appearing in those societies that were critical of religion and were at least partially, in some areas, allowed to be debated publicly (such as Feuerbach's critiques). The work for the negation of religious alienation had been completed *in thought*, by Marx's time, but it was the task of history to see if it would ever be completed *materially*. In other words, the critical method had demonstrated its value by providing the tools for enlightened individuals to root out the ways that they were dominated, and this influenced the social milieu by opening new modes of thought. But it was up to individuals to engage in the socioanalytic work necessary to break free of its influence, which was only possible once religion lost its totalizing power; this could not occur from the top down perspective of the social without reproducing the very system of domination that such critical work sought to dismantle. Meaning that domination, illusion, and alienation persisted in modern society even though religion no longer held the same power that it once did. Versed in recognizing the patterns of this alienating logic, Marx ([1844a] 1992) sets his agenda when he writes: "[i]t is the immediate *task of philosophy*"—now on a materialist track we can call this social theory—"which is in the service of history, to unmask self-estrangement in its *unholy forms* once the *holy form* of human self-estrangement [i.e. alienation] has been unmasked" (p. 244). He uncovers these unholy forms in a duplicated and recoded totalizing logic, now hard at work on the minds of a species fueling its new source of power: capital.

Of the material transformations that enabled capital to replace religion as the dominant totalizing logic of modern society, the downfall of feudalism (Bloch, [1939] 2014; Markoff, 2004) and the development of the factory (Freeman, 2018) were among the most profound. Turning out the serfs from the feudal system that dominated the Middle Ages was a necessary step because it "comprised the abolition of all types of personal dependence" (Katz, 1993, p. 366) which was a necessary step for stripping people of any basis for reproducing their lives beyond the sale of their labor. This provided a readymade pool of laborers for the emergent capitalist class to put to work in the pursuit of profit. In 18th century England, the development of the factory provided the surest means of mobilizing the working class to that end. Although there was little to praise about the life of serfs in the feudal system, their transformation and that of the peasantry as a whole, into a laboring class was not without its friction (Thompson, 1966; 1967). Culturally, attacks on the feudal system persisted long after it had been abolished, evidenced, for example, in the song "Downfall of Feudalism" published in the working-class newspaper *Young America* in 1845.

> Base Feudalism has foundered,
>   The demon grasps for breath,
> His rapid march is downward,
>   To everlasting death ...
> Democracy untiring,
>   Strikes at the monster's heart ...
> And how we fought for freedom,
>   Let future ages tell. (p. 4)

Although this was in a "working man's" newspaper, the romantic appeal to freedom hardly reflected the material reality captured by the poet William Blake, who had in 1804 referred to the reality of the factory system in his designation of them as "dark Satanic Mills." Freedom from spatial bondage, as occurred with the overthrowing of the feudal system in which serfs were no longer bound to the land, did not translate into freedom from temporal bondage, as is the case in the capitalist system where workers are bound to the sale of their time. Neither did it improve the working conditions, hours of toil, and quality of life. It is telling that from its onset the capitalist system has had to be propped up with cultural propaganda like the above song to obfuscate the issue of domination by stressing the "freedom" that this system provides. But it is Marx who exposes in his critique just how illusory this brand of freedom is in the system of capital, which he does by illustrating how it carries the same alienating logic as religion had before.

Marx outlines his concept of alienation in the *Economic and Philosophic Manuscripts* ([1844b] 1992) where he takes it from its immaterial roots in religious logic and demonstrates how it has been woven into the material fabric of modern society through the logic of capital. In the system of capital, people are split into two camps or classes. These camps are established by the doctrine of private property which seals off the natural world through artificial means and divides up the world into those who have legal claim (or the strength of force and/or capital to establish claim) to property and those who have none. Rather than recognize that private property revolutionizes the world and splits humanity from its natural roots into a system of artifice, those in power and mainstream political economists insist that this system of domination is a new state of nature and is the normal condition of humanity. Marx's starting point, however, is that private property is not a natural phenomenon and has yet to be explained. Building off this premise in the modern system of political economy, the first camp who Marx identifies is the bourgeoisie made up of those who make money off the rent of their property and those who have a reserve of capital at their disposal; which is distinct from the money form in this system because capital is specifically dedicated to the reproduction and growth of capital rather than to the exchange of goods for personal consumption. The second camp is the workers, or proletarians, who, because they have no reserve of capital or rights to property, are compelled to sell the only thing available to them in order to survive: their labor-power.

Wielding their capital, the capitalists develop the means of production, the factory system and the industrial machinery used in the production of goods. Denied land and means of production, the workers must sell their labor-power to the capitalists so that they can make enough money to purchase and rent the goods needed to reproduce labor-power (both their own and, through their children, the class of laborers). This requires an enormous amount of flexibility on the part of the workers, who in traditional settings developed skill sets appropriate to the labor of their immediate needs, but in modern settings they must develop a variety of skill sets, including many new ones, that are primarily used to meet the needs of others. Adam Smith's early work in political economy points to how the division of labor played a central role in structuring this new system by taking a task and dividing it into "distinct operations ... all performed by distinct hands ... [which] occasions, in every art, a proportionable increase of the productive powers of labor" ([1776] 1981, p. 15). For the capitalist, this translates into increased productivity and mass production of goods, but as Marx ([1844b] 1992) points out, "it impoverishes the worker and reduces him to a machine" (p. 287). It does this by splitting the laborer from the object of labor. No longer is the laborer involved in an intimate relationship with nature in which they transform it as an extension of their human essence made material. Instead, they produce objects that are sold as commodities, ones that they cannot possess even if they are in need of the object unless they approach it first on the market in its commodity form.[1] Labor, therefore, is no longer an activity of personal fulfillment in modern society, it "appears only in the form of *wage-earning activity*" (p. 289). Just as the manufactured object is reduced to its "price" so too is the worker reduced to their "price," to the extent that "Human life is a piece of capital" (p. 306).

A cascading effect was triggered by this structure of labor under the logic of capital, and it totalized the world anew by extending an epistemology and an ontology to cannibalize the mind and body of the human species. The effect is that under the totalizing logic of capital, "individuals experience ... an alienation from the product of their labor, from themselves, from nature, from each other, and from the species" (Dahms, 2008a, p. 40). The first consequence, as outlined above, is "that the worker is related to the *product of his labor* as to an *alien* object" (Marx, [1844b] 1992, p. 324). The fruits of the workers' labor are not theirs to enjoy, they belong to the capitalist who has paid the laborer for the consumption of their labor-power on the grounds that they forgo claim

---

1 Jonathan Lethem parodies the absurdity of this in his post-apocalyptic novel *Amnesia Moon* (1995), where in one scene we find that the McDonalds corporation continues to operate but in a horrific manner with employees who are never relieved for their shift change. Slowly starving to death and unable to clock out, they refuse to eat the food that they have prepared for customers, even to the point that they throw it away when it no longer meets the company's freshness standards, because they have no money to purchase it and are prohibited from eating on the clock.

to the object of their labor. Because the product of the worker's labor-power assumes a material form that is external to the worker, "this realization appears as a *loss of reality* for the worker" (p. 324), or the loss of their individuality and creative potential. Their time and how they choose to spend it is no longer their own. It now belongs to the capitalist as does the object of their labor, so they are alienated from themselves and from the objects that they use to materially define themselves.

Nature, which Marx calls "the *sensuous external world*" (p. 325), is the reservoir of raw materials and assumes a price in the market, as do the laborers who are themselves objectified as labor-power. Denied access to the raw materials of nature and without the capital needed to invest in raw materials themselves, the laborer "deprives himself of the means of life" (p. 325) by transforming nature into commodities for the capitalist. From an environmental perspective this also means that as nature is transformed under the logic of capital, the laborer is effectively destroying the very basis of life itself, both from the immediate perspective of needing the goods nature provides to sustain life and in the long run by depleting the natural and finite store of raw materials that provide an object for the laborer to define herself through labor and to sustain that labor. Being entirely dependent on the sale of their labor to the capitalist for their survival, workers assume a precarious position. On the one hand, their time and the fruit of their labor are alien from them, so they do not get the feeling of satisfaction in having expressed their essence through their labor (it is the capitalist who enjoys this feeling by syphoning off the surplus enjoyment of the laborers). On the other hand, being dependent on selling their labor power to survive places them in competition with other workers. This alienates the worker from other workers, who they see not as other humans suffering under the same dominating system, but as competition in the capitalist system pitted against each other in the struggle for survival.

The result of this alienating process is as Marx writes, "It estranges man from his body, from nature as it exists outside of him, from his spiritual essence, his human essence ... An immediate consequence ... is the *estrangement of man from man*" (p. 329). By alienating "man from man" the totalizing logic of capital effectively dismantles the power of the social as a sphere of solidarity, and being a social species, life under this logic finally alienates us from our species-being and begins the mechanthropomorphic process of our becoming something other, something more machine-like than human-like in modern society. All of this is the consequence of private property, which is maintained by establishing a totalizing logic that is grounded in a material system which allows it to appear as natural, and the result is a more total alienation than the religious mode that predates it.

Alienation is a symptom of life in a modern society that is under the totalizing spell of the logic of capital. The central problem, however, remains the fact that just as those who lived in the traditional world under the domination of religion, "we are naturally positioned neither to recognize alienation as a byproduct of the pursuit of prosperity nor to conceive of the detrimental impact it has on our ability to acknowledge and make explicit the dynamics that are at the core of modern society" (Dahms, 2008a, p. 41). A socioanalytic intervention is an artificial interruption in our everyday lives that creates the space and time needed to cultivate our ability to recognize how our alienation is affecting our subjective and our intersubjective lives. Short of this, in advanced modern societies where anxiety has reached its social maturity, the affect is bound to persist in its coagulated state and increase the likelihood of our turning inward to escape from the places where the presence of the social is most strongly felt. All the more so when there is no concretely discernible object that discrete subjects can point to in their immediate vicinity that is identifiable as the object cause of anxiety since it is woven into the very fabric of our lives. The outcome is that the subject will come to believe that the problem is theirs alone and not a result of social pathologies. Without the development of a critical method that can point out this condition, there is no hope of addressing the issues that stem from it.

Development of a critical method, does not, however, imply that the logic for which it offers a critique can be broken. Unlike religious alienation, which was real in terms of the emotional energy invested into its logic and had material consequences as a result of that investment, this form of alienation under the logic of capital is weaved into the fabric of our modern lives. For religious alienation the promise was of a reward in the afterlife when the individual would be made whole again. For alienation in the system of capital, there is no promise of an eventual reward, there is only threat of penalty (again social and/or physical death) for those who refuse to acquiesce. The totalizing logic of capital took what was transcendent in religion and made it material. This is why the positive concept of recognition, proposed by Hegel, is of no real benefit beyond philosophy under the system of capital, because it reproduces religious alienation by promising a better life in the future while ignoring the material reality of life in the now. Alienation, as a form of negative critique, allows us to uncover its logic at work in our daily lives, and yet, it does not promise us refuge from the structure of the reality that reproduces the condition. In gaining certainty over why we are the way we are, in how our identity is linked to our social milieu, and in how it is eroded and prevented from developing by the structure of material alienation in which we live, we also must be prepared for encounters and confrontations with new forms of uncertainty that shroud the vision of our shared future. This means that as we address the problem of anxiety, even by approaching it in the setting of critical socioanalysis, we must expect that our anxiety will be heightened and agitated by the process. Critical socioanalysis does not promise a "cure" it promises us a means of accessing the roots of our problems and doing so can be an emotionally provocative exercise. In this sense, the critical method as outlined by Marx, is a diagnostic tool, not a prescriptive one.

For Marx alienation was the material starting point that needed to be confronted, but the quest to uncover it in all its unholy forms provides the agenda for the critical method which progresses dialectically and can only be resolved historically. In his own work he went on to develop a means of understanding how alienation could possibly be negated through a dialectical movement in the class structure. However, true to the method he outlined in this early work, Marx refused to give into the temptation of resolving in thought what can only be resolved historically. And as capital evolved historically, his later work returned to the structure of alienation by exploring how this logic cannibalized ever more of our minds and bodies in the cancerous form of the commodity and our fetishization of it, and therefore, how it continuously prevents the realization of a working-class consciousness and political solutions to the problem. This later work is crucial to the diagnosis of our current political economic worries, recognizing that this logic leads to a mechanthropomorphic system in which "machinery does not just act as a superior competitor to the worker"—a situation where workers become more alienated from their fellow humans by the force of competition in both the biological and the technological realms—it is "always on the point of making [the human] superfluous" (Marx, [1867] 1990, p. 562). Marx forecasts the amplification of the posthuman reality that is fueling many present anxieties, but he is cognizant of how these advances teeter on the brink of anxiety and ecstasy. Anxiety, because without the capitalist system to provide the means of sustenance in exchange for labor where are we to turn? Ecstasy, because the machinic division of labor contains at a minimum the possibility of eradicating human labor; which is the precondition for meeting the needs of survival.

In critical socioanalysis, alienation is the foundational symptom that we must look for and encourage the analysand to confront because it represents the de facto state of being for modern subjects. It serves as the guiding key that unlocks the logic of domination in all its unholy forms. While the logic of alienation that Marx uncovered works as a starting point, it does not exhaust the ways that we are alienated in modern society. The socioanalytic process, while buttressed by social theory, touches on our individualized psychosocial landscape and it is the task of the individual analysand, who, in the process of auto-critique, must come to uncover the logic of their own alienation in its myriad forms. As alienation, like the logic of capital that has embedded it in our material existence, demonstrates its resiliency and plasticity in contemporary societies, many of the subjects of capital are so alienated by the totalizing logic of capital that they may strongly resist the suggestion that the system of capital is a primary cause of their everyday anxieties. Just as they may resist how they are alienated in modern society by the hierarchies of race, ethnicity, gender, and sexuality. Instead, operating at a lower level of abstraction, they may locate their anxiety in the ways that capital, or modern society as such, has rejected, rather than accepted, them. This rejection feels personal, it attacks and threatens our sense of individuality, and it does so by at times masking and at other times amplifying the social causes of our experience. It is these categories of individual and social that we must now clarify if we are to sort out this tangled web of modern anxiety.

2   The Individual, the Social, and the Knot: Toward the Durkheimian Symptom of Anomie

Durkheim begins his work with the same starting point as Marx: the division of labor. While acknowledging that "[w]e can no longer be under any illusion about the trends in modern industry ... powerful mechanisms, large scale groupings of power and capital, and consequently an extreme division of labor" (Durkheim, [1893] 2013, p. 33), for Durkheim, this "law" has consequences that extend far beyond its reduction to economic issues. As Marx turns to the division of labor to explain private property, Durkheim turns to it to explain the birth of the social and the individual as powerful forces of influence on the development of modern society and of how we come to think and experience ourselves and life in that society. Durkheim finds within the emergence of the division of labor a historical account of these forces appearing as distinct categories in the development of our species and, as societies reach the extreme point of this division, he uncovers a modern pathology, which he names anomie, that can amplify our anxiety to devastating effect. Furthermore, he demonstrates that the social and the individual cannot be captured in their concrete specificity by relying on the existing methods and theoretical perspectives of psychology and biology. A sociological explanation is needed to explain how the individual and the social appear as contradictory objects that offer competing perspectives of analysis and it must account for the fact that they emerge as part of one and the same process; a process that creates a new object of scientific inquiry that is distinct from psychological and biological processes. There are two reasons that we can point to for why Durkheim diverged from the purely political economic explanations made popular by Marx and set himself the goal of developing a sociological method that is proper for the scientific illumination of these objects. Before digging into Durkheim's theory let us briefly examine those reasons by way of considering his psychosocial milieu.

First, Durkheim's France was less developed and dominated economically by the logic of capital than England was in Marx's time. From 1815–1848, France experienced a slow wave of industrial development. This was opposed to the rapid British industrialization, aided by the growth of the rail system and the subsequent market for financial speculation that came with it, as witnessed by Marx. According to the historian Arthur Louis Dunham (1955), industrialism did not take off as quickly in France due in part to the enlightenment belief in individualism that had so captured French culture and cultivated an air of pride in their labor that was incompatible with the growing logic of capital. From a cultural perspective, this individualistic attitude in France led to a spirit of innovation in technology and science; as evidenced in a variety of areas, such as the Niépces' invention of the internal combustion engine in 1807 and brother Nicéphore's invention of the photograph in 1822 (Rosen, 1987), Gerhardt's development of aspirin in 1853 (Levesque & Lafont, 2000), Pasteur and Bernard's method of pasteurization in 1862 (Latour, [1988] 1993), Michaux and Lallement's first mechanical pedal powered bicycle in 1864 (patented in 1868) (Bijker, 1995), and the many imaginary devices and fictional explorations of these new technologies throughout the scientific romances of novelist Jules Verne beginning in 1848 (Evans, 2013). Much of this innovation occurred in small scale artisan shops that were not geared toward mass production for commercial application, so very few brought their inventors monetary success in their lifetime.

Generally, the economic condition was one where French workers resisted the mass production of cheap low-quality goods for mass consumption such as those that were thriving in the factory system in England, but this put them at a competitive disadvantage in foreign trade. Despite France's reputation for producing high quality goods and for innovating technology and scientific techniques, England's economic system undercut them on quantity and price, effectively stunting the growth of the French economy. So, while Marx was busy critiquing the emergent and thriving logic of bourgeois political economy that had chased him across Europe to exile in England, when Durkheim was born in 1858, France was dealing with issues that related to a higher degree to the politically tumultuous century it was enduring, which included economic issues but could not be completely reduced to them.

Without enough foreign trade to sustain the national costs of industrialization—a problem exacerbated in the mid-1800s by a financial crisis in England caused by volatility in the railroad speculation market—unemployment levels rose in France and led to increased social instability (see Hobsbawm, [1962] 1996, especially Ch. 13; 1975). Another revolution was on its way that pitted "class against class" and invoked the idea of "permanent revolution" as the limits of the bourgeois project became more apparent in each short lived French political experiment since the French Revolution of 1789–99 (Calhoun, 1989, p. 210). The revolution of 1848 ousted the conservative monarchy of Louis Philippe I in France, and Louis-Napoléon Bonaparte III—the nephew of Napoleon I—was elected by popular vote. It was, however, a short-lived triumph for liberal politics, as by 1851 Napoleon III staged a coup d'état and named himself the new emperor of France while not abandoning many of the goals of liberalism. Even with the financial crisis in England, "Napoleon III was a great friend to the railroad" and followed the English model in many ways, helping to modernize the French economy and usher in an international age of capitalism built on investments in communication and transportation technologies (Calhoun 1989, p. 224). In large part, it was thinks to the policies initiated by Napoleon III that Paris eventually flourished as a center of technological development; expressed by the many expositions to showcase these technologies to the world which continued the successful tradition that was first put on display at the French Industrial Exposition of 1844 (Curmer, 1843–44).

The totalizing logic of capital worked its way into French life, but the religious logic it sought to replace was still strongly felt in the predominantly Catholic France and it competed with the logic of capital over the directionality of the society. Other than social domination, what these logics shared was that neither was fated to a single state or mode of governance, although both relied in various ways in part on the state to enforce their brand of domination for social control of the masses. In France this was made evident yet again with the fall of the Second Empire when Napoleon III was defeated in the Franco-Prussian War of 1870 and the Third Republic of France was established as "the first stable electoral democracy ... on the European continent" (Hanson, 2010, p. 1024). The introduction of a democratic mode of governance did not immediately translate into a more robust system of sustained capitalist development and economic growth, nor the downfall of religiously oriented worldviews, rather progress in both of those directions slowed. Shifting from an authoritarian empire—albeit one cloaked in a liberal veneer that could dictate the direction of the economy and exert influence over national religious practices largely at the emperor's whim—to a democratic mode of governance that allowed differing political ideologies the chance to compete for power meant that the direction of French society was up for debate. Unsurprisingly, given the erosion of religious power under the modernist system and the loss of workers' freedom in the capitalist economy, the peasants and the Catholic church dug in and supported a traditional approach with a conservative agenda, while the bourgeoisie and the liberal republicans supported a progressive agenda. Politically, this attempt at a pluralistic society, rather than leading to a deliberative politic that advanced a progressive agenda via compromise or a system of unfettered capitalist growth, created what the historian Stanley Hoffman (1963) called a "stalemate society."

Modern society as a whole was a manifestly dynamic system, but France at the time when Durkheim came of age was a far more static society economically and religiously than the England of Marx's time. This did not mean that capitalist logic ceased to have an impact on French society; following the success of the Industrial Exhibition in 1844, the Exposition Universelle received international attention and continued to put French innovations on display for the world, boasting over 13,000,000 visitors in 1878 and over 32,000,000 in 1889 (Bureau International des Expositions, 2018). Capital demonstrated its totalizing effects as it continued to slowly penetrate and transform the economic mode of life, but the more prominent dynamism that Durkheim experienced in France at the time was in the secular and intellectual culture of urban existence where the modernist dream still imagined an open future, both economically and politically.

The enlightenment ideals that underpinned the socio-cultural setting of France are what allowed Durkheim the freedom to make the move from a rabbinic education, as was his familial tradition, to a secular education and pursue an academic career, but it also speaks to why his interest lay beyond a pure focus on economics. As Jeffrey Alexander (1986) contends,

> Durkheim came to maturity in the late 1870s and 1880s, in the crucible of the formation of the Third Republic in France. From the very beginning of his identification as a sociologist—which Mauss dates from 1881—he linked his intellectual vocation to certain normative or ideological goals: first, French society must be changed so that it could become stable; second, this stability could be achieved only if there were justice, particularly justice in economic distribution; third, the increased state organization necessary to create justice should never occur at the expense of individual freedom. (p. 94)

Although it is common to look back on these early classics as coming from positions of individual privilege due to their being white, male, Europeans, it is important to remember that the issue of social justice was of paramount importance to them, especially given their marginal positions in the social hierarchy of the day. Durkheim, like Marx, faced a variety of challenges and uncertainty in his life, economically as well as politically, due to the openly anti-Semitic prejudice commonly demonstrated against Jews in their societies. So, the question of domination was both a personal and a social question for Durkheim that had high stakes for both his and France's future (see Lukes, [1973] 1985 and Fournier, [2007] 2013). This may also explain why these classics of modern social thought developed their methods with a critical perspective in mind, and why they held views on the issues of race, gender, and religious tolerance that stood in contrast to much thought, even intellectual and sociological thought, of the time. Critical thought is that which runs counter to the systems of totalizing logic that shape our thoughts, and Durkheim makes it very clear that his project follows this principle when he insists that his brand of "science presupposes the entire freedom of the mind" (Durkheim, [1893] 2013, p. 6). But this requires freedom at two interlinked levels because

> Two consciousnesses exist within us: the one comprises only states that are personal to each one of us, characteristic of us as individuals, whilst the other comprises states that are common to the whole of society ... Now, although distinct, these two consciousnesses are linked to each other, since in the end they constitute only one entity, for both have one and the same organic basis, thus they are interdependent. (p. 81)

Durkheim's reasoning here, explains why socioanalysis is a necessary development and outgrowth of psychoanalysis for the treatment of anxiety. Once anxiety's origin is interwoven with our social fabric it becomes a part of our individual constitutions and impacts the construction of our identity. It was on that understanding of the social scientific enterprise, at once free from totalizing modes of thought and grounded in a set of normative principles, that Durkheim insisted on a sociological approach that explored the tensions between individuality and social life when he established the first sociology department at the University of Bordeaux in 1895 and the first sociology journal, *L'Année sociologique*, in 1896.

Durkheim ([1893] 2013) maintained an allegiance to enlightenment principles when he expanded on his methodological approach, arguing that "[w]e must rid ourselves of those ways of perceiving and judging that long habit has implanted within us. We must rigorously subject ourselves to the discipline of methodical doubt" (p. 6). This call for autocritique of our methods meant that even something as seemingly natural a concept as society required a thorough examination to understand how it is held together and of what it is composed. He recognizes that is not an easy process, though, because "[p]hilosophy is only possible when religion has lost some of its sway. This new way of representing things shocks collective opinion, which resists it" (p. 224). Meaning that as the totalizing system of religion ruptures, there is need for a new force that provides the stability and control over moral action, but any such change is bound to be radical, meaning that it will necessarily have at least a temporary destabilizing effect. An age of "permanent revolution" and political instability in France illustrates that effect and made the necessity of this approach clear to Durkheim. Since the material reality of his "society" was dynamic, that dynamic reality implied that society was not necessarily a permanent state for humanity. While Marx was suspicious of sociology because of its roots in the positivism of Comte and Saint-Simone, he used the notion of the "social" to ground many of his arguments. In Durkheim's work, however, this could be taken as another example of the social being presupposed but having yet to be explained. Investigations that failed to adequately establish the social as a historical phenomenon are why, for Durkheim, the sociological tradition as outlined by the early French positivist thinkers lacked a sufficient foundation to ground sociology as a scientific enterprise. Durkheim, therefore, demonstrates at least a partial agreement with Marx's position that a failure to account for the dynamism of modern society blinded those early sociological investigations from seeing the actual and potential dark sides of societal transformation.

Society implies a bond of solidarity between individuals, and since this bond can erode and weaken over time, it must be understood historically how it came to be as a precondition to our reinforcement of it or our thwarting of processes, deliberate or otherwise, that weaken it or seek to wield power over it in a totalizing way. Durkheim explains how he will proceed to explore this topic with the following guiding questions:

> How does it come about that the individual, whilst becoming more autonomous, depends ever more closely upon society? How can he become at the same time more of an individual and yet more linked to society? For it is indisputable that these two movements, however contradictory they appear to be, are carried on in tandem. ([1893] 2013, p. 7)

Due to the dynamic nature of the questions posed, Durkheim can only answer them by means of a thorough examination of how these changes came about.

He does this by comparing traditional lives to modern lives and locating the major differences and ruptures between the two. Only upon this knowledge serving as the foundation can we glean the effects that the process has had on our modern existence and how it has entangled us in this contradictory knot of self and society. Durkheim locates the division of labor as a necessary precondition for solidarity, because "political societies cannot sustain their equilibrium save by the specialization of tasks" ([1893] 2013, p. 50) and society cannot exist without a bond that holds it together. Here, he credits Comte for being the first to recognize that the division of labor has consequences that extend beyond the economic realm, but it is Durkheim who realizes that if solidarity is a requirement for the function of society, then the division of labor must produce a new force that binds people together despite its differentiating effects. The way to discover this force is to proceed by way of a comparative analysis; a method that is central to these early modern critical thinkers. If modern society is not a state of nature, and the social is a power that arises from the ability of humans to artificially alter their reality and separate themselves from nature, then in the course of the development of the human species a causal mechanism must exist that explains how this division of labor lent itself to the construction of a modern society that broke with traditional forms of life and yet managed to avoid a descent into anarchy.

Durkheim's first step is to develop a theory of mechanical solidarity, or solidarity by similarities, to explain the bonds that existed in pre-modern forms of human organization. "What characterizes [mechanical solidarity] is that it comprises a system of homogeneous segments similar to one another" ([1893] 2013, p. 142). As explained in the scenic landscape above that outlined the rise of modern society, the vast majority of people in traditional settings lived a static life in which they all shared the same occupation, thought patterns, and ideas about their reality, in large part due to the necessities for survival and geographical limitations that nature imposed upon them. A static life implies that there is little potential for variation in terms of available life paths, and beyond this Durkheim demonstrates that it also implies a generally static set of cultural and moral codes which regulate behaviors.

Although early anthropological accounts are at best logically derived sketches of how life is presumed to have existed, due to the limited availability of recorded data on these modes of life, gleanings from the archeological record suggest that early humans lived in nomadic hunter/gatherer groupings and eventually settled down only once techniques were developed that made it easier to work the land (i.e. transform nature) and coerce it to provide its bounty in specific geographic locations. Karatani (2014) argues that this settling was due to the development of technological appendages (especially those needed for fishing) that anchored people to a particular space because the technical apparatuses they needed for certain activities were either too cumbersome to move with the tribe or took so long to build that to abandon them and start over at the next camp made little sense. Once these sedentary lifestyles became the norm, those born to this life specialized in whatever form of food cultivation was customary to the community in their geographic locale and were indoctrinated into the set of moral standards enforced by that community. This process developed as the tribal unit formed out of groupings made of close familial bonds in small segmented settings. But as many groups settled in a similar way, although there were variations in belief systems, the overarching logic persisted as to how people within any given group developed.

Since there were so few alternatives available within the community to pursue other occupations or opportunities to explore other modes of thought and systems of morality, there was little variation, other than in the form of personality traits, between the kinds of people that lived in the community. Life for men and women was generally dictated by a gendered division of labor, but as discrete persons there was little variation between one man or woman and another man or woman in the community. Durkheim concludes that, "this particular structure enables society to hold the individual more tightly in its grip, making him more strongly attached to his domestic environment, and consequently to tradition" (p. 236).

The structure of life under conditions of mechanical solidarity required a tight bond between its component parts that was reinforced by the limited availability of alternative modes of thinking and living. Failure to attach oneself to those communal bonds was to risk social death and exile. The bonds were held together by the glue of sameness: same occupations, same religion, same family structures, same opportunities for personal development, same geography, same material circumstances, same thought patterns, etc. However, once the division of labor was applied to the community in the pursuit of wealth extraction and efficiency of labor, different minds began to develop in their inhabitants as a result. The farmer and the factory worker no longer live under the same conditions in their rural and urban settings, share the same concerns, or are limited by the same grouping of available marriage partners. Even among the factory workers, each is assigned a different task and becomes preoccupied with a different set of concerns focused on their own tasks in the process. Although several of these differences are of degree, rather than kind, these shades of difference produce revolutionary effects on their minds and the makeup of society.

Demand for options rises, including, for example, among places of worship and choice of vocation. Communities that only had one place of worship eventually developed a variety of options for their inhabitants who no longer shared a single vision of their shared reality. This also opened the door for those who chose to forego religious worship altogether. Whereas previously everyone in the community attended the same church and it was obvious who had missed the service, now with competing services, those who chose not to attend could not as easily be found out and ostracized for their private choices. Being able to either choose a vocation or being forced by necessity to choose a new one, rather than assume the only one available by means of inheritance, put pressure not only on the community, but also on the more intimate family unit. As children were able to pursue alternative lifestyles from their parents, the differences between them came to be felt generationally as gaps in ways of thinking, that widened with each generational advance. Some preferred the new way of life while others held on desperately to the traditional way of life that they rightly understood to be slipping from their grasp.

In these newly developing modern societies, the form of solidarity that bound people together was transformed from that system of rigid and mechanical tradition—defined by lack of alternatives—to one that proceeded in a more organic manner. The structure of societies held together by organic solidarity "are constituted, not by the replication of similar homogenous elements, but by a system of different organs, each one of which has a special role and which themselves are formed by different parts" (p. 143). The advantage in the previous model was that people could easily fill in for each other and each familial unit was largely self-sufficient, growing enough food to reproduce the labor power of the family and create enough surplus for the family to grow. The advantage in the new model, is that people are allowed greater freedom to explore their interests and develop as unique individuals. The downside is that they are no longer self-sufficient and so they cannot be self-reliant. Exchanging labor geared toward the immediate production of needs for wage-labor meant that those whose lives were touched by the division of labor now had to rely on others to produce the goods that they needed to survive. Whereas under conditions of mechanical solidarity the need to rely on the services of others was limited to the occasional instance (e.g. going to the doctor when sick) became a daily act under conditions of organic solidarity (e.g. going to the grocer for food, the tailor for clothes, the carpenter for furniture, etc.). Reliance on others to meet every day needs in a system of exchange leads to a "slow task of consolidation ... [a] network of ties that gradually becomes woven of its own accord and that makes organic solidarity more permanent" (p. 286). Under these conditions it becomes increasingly impossible to be self-reliant because the structure of the communal grouping of likewise self-reliant members withers away, as do the conditions needed to sustain such a model.

The outcome is two-fold. On the one hand, due to the application of the division of labor across all spheres of societal life, individuality emerges as a new feature in the human species because the experiences of each member now vary in countless ways, with each variation contributing to the building of unique realities and worldviews in each novel life experience. On the other hand, these discrete individuals are required by the organization of society under the logic of the division of labor to rely on others to a higher degree in a series of everyday interactions which, taken as a whole, comprise a new objective force that compels their behavior to a set of organically arising norms. It is this force that earns the name social and comprises the object of study for sociology.

With the social identified as an object of historical origin, Durkheim turns his attention once more to the question of methodology but runs into difficulty when he tries to operationalize his definition of the concept as a foundation for sociology as a scientific practice. How, if the social is an intangible but objective force, can the scientist locate it and study it as a concrete object? For Durkheim, the answer lies in the "social fact," which "is identifiable through the power of external coercion which it exerts or is capable of exerting upon individuals" ([1895] 1982, p. 56). There is much debate over whether Durkheim's proposed method and description of social facts is in fact logically consistent. Prominent social theorists have dismissed this aspect of his contribution to methodology, saying that it "made little to no sense, not even to Durkheim" (Lemert, 2017, p. 24) and that it is "the weakest of Durkheim's major works" (Giddens, 1977, p. 292). The dismissal arises in part because of confusion over the contradictory nature of Durkheim's statements on how the individual connects to the social and how the social in turn penetrates the individual through the force of social facts. How, after all, can a "fact" exert a force that coerces behavior?

The problem is that thinking of the social in the coagulated form of a fact, implies a positivist orientation that ruptures the dynamism of the object in the very attempt to operationalize in this static form what can only be demonstrated through an analysis of its dynamic nature. On the one hand, Durkheim argues that the social is the result of the same process that produces individuality (and the condition of organic solidarity) and it is in the necessity of interconnections between discrete individuals that this force called the social comes into existence. On the other hand, Durkheim ([1897] 2002) argues that "there can be no sociology unless societies exist, and that societies cannot exist if there are only individuals" (p. xxxvi), meaning that the social cannot simply be the aggregate of individuals. What the social "is" can only be made clear in the historical dynamic of its process of becoming, which emerges only in the examination of Durkheim's sociological practice. Lukes ([1973] 1985), while following the criticism that Durkheim is not entirely clear on this issue, correctly identifies that this tension between precisely what constitutes the individual and the social is "the keystone of Durkheim's entire system of thought" (p. 22). This keystone is evident in Durkheim's assertion that social facts "constitute a reality *sui generis* vastly different from the individual facts which manifest that reality" ([1895] 1982, p. 54). The social is greater than the sum of its parts, but it emerges from those parts in a historical dynamic whereby it eventually comes to eclipse and orbit those parts and exert an influence over them. This is not an altogether clear process for the very reason that this only makes sense when one performs historically grounded sociological analyses, meaning that the social is only made visible by sociological analyses, which in turn cannot be performed unless there is a social force present that is visible in its exertion of control over individual behaviors. The contradiction of sociology is a mirror of the contradiction of the social and individual tied in a knot, yet having separate and distinct functions, limitations, and abilities. Their mutual dependence on one another is self-reinforcing and rests on the central hypothesis of sociology, that the social exists. Due, however, to the dynamic nature of modern society and the historical dynamic that gave birth to the social, sociology can never assume that the social is a permanent force, which means that every sociological investigation must uncover the social as its foundation anew. This negates arguments for sociology as a positive science built on a static foundation and places Durkheim firmly in the critical tradition.

Mainstream sociology has, however, often ignored Durkheim's nuanced argument. Rather than accepting that the social is engaged in a historical process of becoming that is both tied to the individual and orbits individuals as a force unto its own, they opt instead for a positivist reading of him that reifies the social as a static concept that exists trans-historically and they imagine it as an isolatable concept that can be applied at will to various empirical phenomena. Durkheim is partially to blame for this misrepresentation, because he leaves the door open to this reading in his attempt to operationalize the social for a *sociological* theory—that is, a theory of practice for sociology—which does not demonstrate the social anew, as in social theory, but deals with it abstractly in this misleading manner. By saying that "[t]he first and most basic fact is *to consider social facts as things*" (Durkheim, [1895] 1982, p. 60), sociologists have reified the social and taken for granted its dynamic nature. But Durkheim is not telling us to treat the *social* as a lifeless and static *thing*, rather he is saying that when we observe the effects of the social, we must treat those effects as the object, *the thing*, of our study, and there we will find the social at work. Failure to treat the social as a dynamic and historically specific force, has weakened mainstream sociology as it has lost sight of its most radical implications to the point where it does little more than empirical taxonomies without connecting them back to the social dynamic of self and society. Critical socioanalysis recalls these classic methodologies to interrogate the dynamic interconnection of the individual and the social and put them to the test historically, which, following Durkheim, can only happen in the very practice of performing sociological analyses. In the socioanalytic session, then, as the analysand is engaged in the performance of their analysis, the social, if it exists, will manifest itself anew in their narrative as a dynamic force interacting with and shaping their individuality, if that too, still exists.

Beyond the groundwork that Durkheim lays, both for establishing the historical dynamism of the social and the method for uncovering its influence, his most important application of this was to the diagnosis of anomie as a pathological symptom of modern society. Anomie is a condition of rootlessness, or a lack of being restrained in our individual passions by the social structure, because that structure does not provide a sufficient mechanism for meeting individual passions while it simultaneously produces them. "Anomie, therefore … results from man's activity lacking regulation and his consequent suffering" (Durkheim, [1897] 2002, p. 219). Which pushed to its extreme point can be a cause of suicide. In his early work on the division of labor, Durkheim identified anomie as the source of "the continually recurring conflicts and disorders of every kind of which the economic world affords such a sorry spectacle" ([1893] 2013, p. 9). Recognizing that the modern world was organized according to occupational success and opportunities, meant that people came to craft their identities in an intimate manner with their occupation. This attitude persists in the common American social ritual of meeting someone new and immediately asking the question: What do you do? It is a seemingly innocuous question, but it can often fill one with dread as it can enhance the feeling of anomie by tying one's self-worth as an individual to the social status and success that their occupation may or may not afford.

Anomie, therefore, first arises as a result of the division of labor which in the process of differentiating individuals, creates a situation in which the means of satisfying individual wants proliferates at an accelerating rate but is tied to the success one has in their occupation and the social forces that dictate whether or not society finds value in that occupation. This means that the wants that modern society meets do not have to correspond to those that the individual has, nor does it mean that the social structure will provide a guaranty of success in the pursuit of individuals' occupational or general life course aspirations. Rather, society cannot form or maintain itself without requiring of us perpetual sacrifices that are costly to us. For the sole reason that it goes beyond us, it obliges us to go beyond ourselves; and to go beyond itself is, for a being, in some measure to emerge from its own nature, something which does not happen without a more or less painful tension.

DURKHEIM, 2005, p. 44

The Luddite rebellion of 1811–12 provides an example of this process in its acute form and its interconnection between both political economy and technological development. The Luddites were not opposed to the new machinery that increased the productivity of their industry as such, rather what led them to smash the industrial machinery was the "wage reduction and unemployment" (Clancy, 2017, p. 393) that the machinery enabled and which negatively impacted their ability to function and survive as individuals pursuing their occupation within that society. The subsequent anxiety is a direct result of encountering an anomic condition that destabilizes the sense of self as society attempts to achieve stability without accounting for human needs and values.

Frequently the presence of anomie triggers a call to conservative action that puts the brakes on innovation just to retain labor for the sake of labor. As was the case with the arguments put forth by American conservatives against the Energy Independence and Security Act of 2007, which sought a ban on certain household incandescent lightbulbs (One Hundred Tenth Congress of the United States of America, 2007). The goal of the legislation was to improve energy efficiency to address environmental concerns by mandating a more efficient technology to replace the older less efficient one, but this meant that the factories that produced those incandescent bulbs would either need to produce something else or close their doors, leaving the workers without a job. As in the case of Appalachian coal miners who have faced similar social constraints on their occupation for years now, the underlying social problem cannot be resolved by simply maintaining labor for labor's sake. This is further complicated by existential threats, such as climate change, which produce moments when the social good must outweigh individuals' wants. The social problematic in modern society is that this process has too often come at the cost of sacrificing not only wants but it also produces a more intense form of anomie by simultaneously sacrificing their needs. On the one hand, this points to one of the effects of the modernization process that elevates the goals of the logic of capital over those of the individual, while on the other hand, it illustrates the failure of social forces to adequately shape and meet the needs of individuals by providing viable alternatives for them to continue to succeed according to the totalizing logic of capital that dominates them. The contradiction exposed by Durkheim's development of the diagnosis of anomie is that it is the same process that produces individualism that also fails to meet the needs of the individuals it produces, because the goals of the social are not identical to the goals of the individual. The result of ignoring anomie is, as Durkheim pointed out, a rise in a pathological state of modern society that can lead to an increase of anxiety, feelings of worthlessness as our identities fail to find purchase in the social milieu, and ultimately a rise in suicides by those who in the final analysis find no alternatives available to them—because they receive no structural support and are left to fend for themselves in a society that no longer provides the necessary conditions for the self-reliance that it now demands of those it has rejected.

While alienation provides a symptom of a totalizing experience of those living in modern societies oriented to the logic of capital and is the key symptom that must be brought to the fore in the socioanalytic session, anomie represents an acute from of this pathology that varies in time and place and can lead to the most extreme solution to anxiety when left unchecked. The socioanalyst must be especially sensitive to issues of anomie and ever vigilant to its unchecked creep, as it touches on the most sensitive core of this knot among those populations most vulnerable to the erosion of modern solidarity. A further question must now rear its head, why, if alienation is the de facto condition of life in modern societies and the proliferation of anomie is tolerated as the modernization process accelerates, are so few individuals able to identify these pathologies when we have known of their diagnosis since Marx and Durkheim exposed them so long ago? We must turn now to the work of Max Weber to find an answer that we can add to our critical socioanalytic toolbox.

3 Totalizing the Mind and Spirit: Weber and the Unholy Union of Religion and Capital

Born in 1864, to the same Prussia as Marx and only six years Durkheim's junior, Max Weber's life began in definitively different circumstances than either. Weber was not an outsider to his society in the way that Marx and Durkheim were due to their Jewish ancestry, economic situations, and in the former case, political radicalism. In contrast, Weber's family was solidly bourgeois and ingrained as members of the status quo in their society (Radkau, [2005] 2009). Whereas the motivational force behind Marx and Durkheim's work can be linked to their positions as outsiders looking in on the forces that were responsible for shaping the overarching logic of modern society that denied them social security and individual acceptance, Weber's motivation is more easily linked to his intimate personal biography and the ways that the dominating logic of his society directly shaped the psychosocial milieu, the mind and spirit, of its core members.

His father, Max Sr., was a lawyer engaged in civil service and a politician in the National Liberal Party. Despite keeping their distance politically from Bismarck's conservatism, they came to dominance in the 1870s after Napoleon III's defeat in the Franco-Prussian war when they supported Bismarck's unification of the German empire. At the age of five, the Weber's moved to Charlottenburg so that Max Sr. could serve on the Berlin city council. Due to his father's social standing, and located in a robust urban environment, the Weber's home was busy with visits from prominent politicians and academics. His mother, Helene, was a committed Protestant and a demanding woman whose "capacity to keep going from morning till night" was described by Max's wife, Marianne, as "*shaming*" (Radkau, [2005] 2009, p. 16). With the drive to success of his father and the unrelenting work ethic of his mother, Weber's parents commanded much of him.

Intellectually he proved up to the task. As a young man he excelled at school but was bored with it, so he set himself intellectual challenges such as reading all 40 volumes of Goethe's work (Käsler, [1979] 1988) and writing critical historical essays (Sica, [2004] 2017). His parent's dominating personalities fed into the young Max and he assumed the typical "eldest child" personality trait of wielding authority over his younger siblings. This personality trait led to a lifelong rivalry of sorts with his brother, Alfred, who became a well-known economist in his own right. In 1893, Max married his cousin, the accomplished feminist Marianne Schnitger, but had an unconventional marriage that is generally accepted to have never been consummated. The following year, at the young age of 30, he was appointed to a professorship of political economy in Freiburg, and in 1896 he took up a similar position at the University of Heidelberg. Being a professor provided Weber with the money needed to escape the dominance of his parent's influence, but their impact on his psychosocial development ran deep. Although Weber demonstrated his intellectual prowess as both a child and an adult, in 1897 he had a violent altercation with Max Sr.—over his treatment of Helene—and kicked him out of the family home only to see his father die less than two months later with the altercation still unresolved. The effect on Weber's mental state and intellectual output was devastating, eventually leading to the point that he required a several months stay in a sanitarium and an extended leave from academia (Weber did not begin teaching again until 1918/9).

There is unfortunately not enough detailed information on Weber's mental illness, except for a few brief accounts such as the following provided by his contemporaries, Karl Jaspers and Karl Lowenstein, who visited him during this troubling time (Dreijmanis, 2008). Jaspers wrote:

> Only one thing causes a little anxiety. Often you notice an aroused expression pass across his face, his eyes become peculiarly piercing and you fear that at any moment he might become nervously ill just like he was for almost two years. It is as though a mighty will is constantly wrestling to control a nervous system that is going to become agitated. The battle is not to give a trace of this away.
>   quoted in KIRKBRIGHT, 2004, p. 77

Lowenstein (1966) recalled that Weber had "a daemonic personality. Even in routine matters there was something incalculable, explosive about him. You never knew when the inner volcano would erupt" (p. 101). The descriptions here serve us well as illuminations on the paralyzing effects of anxiety in which the self is locked in an internal battle against the external but internalized agitating force of a social that refuses to give respite. Torn between the bureaucratic rationalism of his father, the unrelenting work ethic of his Protestant mother, and the knowledge he gained through his sociological investigations on how modes of domination effect the psychosocial development of self and society, Weber was a man who struggled with the contradictions of an identity that was not wholly shaped of his own accord. These personality traits, and the social conditions in which they flourished, motivated Weber's intellectual agenda and led to his producing critical intellectual inquiries on their historical development of the highest order. Although it is intellectually fashionable to avoid making such direct claims between a person's private life and their texts, imagining them as fully separate entities, Weber in true critical fashion recognized how the psychosocial milieu directly impacts our modes of thought when he claimed that "a *learned inner quality* decides a person's choice of occupation and further course of occupational development. And this inner quality is influenced by the direction of one's upbringing which in turn is influenced by the religious climate in one's native town and one's parental home" ([1904–05] 2011, p. 70). Before turning to see how he developed his critical method to examine these phenomena, let us briefly examine the broader social context in a little more depth so as to better understand how Weber's and his parent's traits exemplified in concrete form the burgeoning social characteristics of a rapidly modernizing German society.

Durkheim's France had moved in the direction of democracy after the Franco-Prussian war which allowed conservative and liberal forces to compete for power, but in Germany, although there were political parties of varying persuasions and degrees of influence, the political trajectory remained firmly planted in the conservative tradition due to the powerful influence that Bismarck wielded. As the historian Jonathan Steinberg (2011) argues, "A 'Liberal Era' under Emperor/King Fredrick III ... might have begun ... [except] King William I ... did not die at 70, nor at 80, nor at 90 ... [and] had become desperate" (pp. 6–7) in 1862 when the parliament clashed with the crown over military reform. Fearful of losing the authority of the crown, King William chose to rely on Bismarck instead of his own power or that of the other royals to dictate German affairs, which allowed free reign of Bismarck's reactionary policies to shape the empire.

Economically, prior to 1850 the German empire was less industrialized than France and England, but Bismarck's policies had some success in changing the economic fortunes of the German peoples:

> The German states varied a good deal: some were very rich, others much poorer. By 1850, the average German GDP per capita was very close to the French level, and it remained so for the next quarter of a century. By the mid-1880s, the German advantage had become noticeable and remained so until 1913.
> CARRERAS & JOSEPHSON, 2010, p. 44

Rapid industrialization between 1871 and 1914, led to estimated "growth rates in real per capita income of 14.6 percent per decade! The percentage of the German labor force employed in agriculture dropped from 54.6 percent in 1849–55, to 35.1 percent in 1910–13" (Twarog, 1997, pp. 286–7), with measures surpassing the rates in France and England over the same period. One reason for these transformations was that, prior to Bismarck's rise to power, railroad development largely depended on private enterprise. Whereas after he inserted himself in the process in 1870, he was able to negotiate with the various German states and secure funding for rail development "that would help sustain a German economic take-off in the decades ahead" (Mitchell, 2000, p. 60). Railroads were fundamental to the industrialization process because they linked cities that were previously isolated geographically, stimulated economic and social change, and allowed for the transportation of raw materials to production facilities and of commercial goods to a much wider market than previously possible (Wolmar, 2010).

The technology on which this transportation system depended was not the result of German ingenuity, rather it was aided by imports of machinery that were already well developed in Britain. Germany, now known for its superior engineering capabilities, did not have the social or institutional structure at the time to develop these technologies on their own. They recognized that direct competition on this front would only reinforce how far behind Britain, France, and America they were in the modernization process. To gain a competitive advantage in the expanding world market they turned their attention to the development of necessary social and institutional structures. This began with the channeling of investments into the modernization of technique through education and the refinement of scientific processes in research.

Technique and technology flow from the logic of systemic rationalism as interlinked processes, but they often develop unevenly, in different contexts, and with uses that are socially constructed after the fact; even in ways that are unintended by their designers. Germany was "[t]he first polity to realize ... [that] for techno-industrial supremacy, what mattered most was science and technology ... the research laboratory thus became as important as the colony [was to Britain for agriculture and extraction of raw materials] ... more so" (Hugill & Bachmann, 2005, p. 160). With modern technologies available for purchase from British engineering firms to get the ball rolling on the modernization process, Germany turned its attention in the mid to late 1800s to educating their population so that they could perform the labor enabled and necessitated by these new industrial technologies. Furthermore, Germany aimed to develop a competitive advantage over the modernized economies of France, Britain, and America, by focusing on the discovery of novel techniques that would boost efficiency and productivity. In hindsight, this shift of focus away from the immediate development of practical technology to basic research on rational techniques and bureaucratic action greatly sped up the German race to industrial modernization and would soon be copied by competitor nations.

Universities in Germany already had a long history, with several being established in the 14th and 15th centuries, but because of those long histories they were viewed in the 1800s as potential grounds for resistance to modernization, so they became the target of reformist policies. The general logic of these reformations was to orient the universities around the concept of *Wissenschaft*—the systemic pursuit of knowledge—and an orientation toward basic research. The effect of this reorientation on the German university system was that "[t]o gain acceptance in the inner sanctum of one's field, one had to demonstrate that one's work had both mastered previous scholarship and superseded it—a skill that the seminar system exquisitely fostered. This logic accelerated tendencies of specialization" (Howard, 2006, p. 277) which in turn accelerated the development of specialized German techniques with a variety of applications. Practically, as the ideas of *Wissenschaft* and basic research were adopted by the natural sciences, the use of these new techniques began to set Germany on the same modern footing as France, Britain, and America, and gave them a new product for export: specialized, or expert, knowledge. For example, German developments in organic chemistry for industrial agriculture became a 'carrier technology' (Hall & Preston, 1988) "allowing renewed investment and return to profitability in the world economy along the lines suggested by Schumpeter [([1943] 2003, p. 83)] ... spawning many associated technologies along the way" (Hugill & Bachmann, 2005, p. 163). The development of these techniques accounted for a decline in the percentage of those working in agriculture.

The German university system's focus on basic research provides two important takeaways from the scenic landscape of Weber's psychosocial development. First, it accounts for the nationalist pride nurtured in German citizens as they succeeded in forging their own path to modernity. Second, as Howard (2006) argues, the transformation of the university into the paradigmatic model of basic research was aided not only by the state but by Protestant theologians. Recognizing that the totalizing logic of religion was weakened and was being replaced by the logic of capital, Protestant theologians in the German university system pushed for these reforms. They believed that the rational and systematic study of theology would help to legitimate Protestantism in modern society by aligning it with the new-found successes of, and trust in, science. German Protestants, although conservative at the time, aligned themselves with the modern world, while in France, Catholics formed a conservative resistance that was aligned with the traditional world.[2]

In 1904—against the social backdrop of a rapidly modernizing Germany and the psychological backdrop of a personally debilitating mental illness— Max and Marianne took an extended trip to the United States. Like Hume who awoke Kant from his dogmatic slumber, this visit to America appears to have had a comparable effect on Weber. The stubborn contradiction that

---

[2] It is worth making a point of distinction that is generally forgotten today in contemporary America, that the terms conservative and liberal are not synonymous with traditional and modern, neither do they carry the same meaning that they do in American politics espoused by the Republicans and the Democrats in either a historical or a global context. Therefore, in France, one could say that the alignment between the Catholics and the peasantry was both traditional and conservative while in Bismarck's Germany Protestants were both modern and conservative.

must have weighed on Weber's mind, as to how the demands of individualism could be reconciled with the social forces that so tightly shape even our most intimate thoughts, broke open in a detailed historical narrative. This seminal work of Weber's, *The Protestant Ethic and the Spirit of Capital* ([1904–05] 2011), uncovers an astounding and highly compatible synthesis between the totalizing logic of religion and that of capital, explaining how there was a transfer from the one to the other of totalizing effects on the minds of individuals and the social spirit of the age. The central thesis of the book follows from Weber's observation that "people who own capital, employers, more highly educated skilled workers, and more highly trained technical or business personnel in modern companies, tend to be, with striking frequency, overwhelmingly *Protestant*" (p. 67). How then did Protestantism emerge as a branch of the totalizing logic of religion that was not only sympathetic to the modern vision but reinforced it, and why did it appear as if those who adhered to the Protestant, rather than the Catholic branch of this divide, reap greater rewards from the system of capital? Radkau ([2005] 2009, p. 111) downplays the impact that the Webers' trip to America had on how he developed an answer to this question because Weber began outlining the project prior to the trip, but Ringer (2004) asserts that the text "provided a major theme of Weber's observations during his visit to the United States" (p. 137). While these claims do not directly contradict each other, the direction of Ringer's claim is the stronger one for understanding how Weber thought through and developed his thesis.

When moving from a profoundly religious mode of existence found in the Protestant ethic to the profoundly secular spirit of capital, what survived "in contemporary America," according to Weber's observations, "are the derivatives of a religious regulation of life which once worked with penetrating efficiency" ([1904–05] 2011, p. 219). As the former ever more gave way to the latter "[t]he American who is "modern," or wants to be regarded as modern, becomes increasingly embarrassed when … the ecclesiastical character of his country is discussed" (Weber, [1985] 2011, p. 227). What Weber experienced in America was a more mature material example of the social phenomenon—this synthesis between Protestant rationalism and the logic of capital—than what he experienced in its infant form in Germany and witnessed first-hand in the personality traits of his parents. While in America, he not only attended a variety of Protestant rituals and meetings but, he was able to see the effects that industrialization and modernization had on life in both urban and rural settings; including the cultural effects that were, according to his thesis, the result of this synthesis and the advancement of modern societal norms. That he was thinking of America, and that it influenced the formulation of his argument is also apparent by the fact that the example he chooses as representative of the 'spirit of capital' is taken from the writings of Benjamin Franklin.

The modern world is not oriented toward a willingness to engage with critique because critique, at a minimum, exposes the internal dynamic of the dominant logic which hints at the possibility of rupture in its contradictions. These logics build cultural mechanisms within their constructed realities that reinforce the adoption of their catechism and place a very high cost on obtaining the time needed for critique to take root. Furthermore, they create a system of social organization that is actively opposed to the cultivation of patient and willing audiences by demonizing activities that are not directly tied into the logic. Weber's example of Benjamin Franklin is a nod in recognition of this fact and serves as a sample of the 'spirit of capital' bleeding through the common parlance of colloquial metaphors presented as traditional folk wisdom. The spirit of capital is transmitted in Franklin's words: "*time* is *money*", "*credit* is *money*", "The good *paymaster* is lord of another man's purse", and moral lessons such as "He that idly loses five shillings' worth of time, loses five shillings and might as prudently throw five shillings into the sea" (Weber, [1904–05] 2011, pp. 77–78). These folk wisdoms are precisely what is taken as self-evident under the domination of the logic of capital, but like Marx and Durkheim, Weber believes that they are presupposed and have to be explained historically in a way that shows how they came to embody what he calls "a peculiar ethic" whose "violation is treated not simply as foolishness but as a sort of forgetfulness of *duty*" (p. 79). Remembering back to the claim of Ellul ([1954] 1964) that in premodern times hard work was not idealized as an end unto itself, Weber correctly identifies that this mode of thinking about time as monetized is something that differentiates modern society from traditional modes of life. Therefore, he correctly identifies there must be a historical narrative that can account for this coding of the mind and why it makes modern subjects think in a way that precisely reinforces the new totalizing logic of capital and its mode of domination. Because he witnessed the fact that Protestant countries were adapting to this new totalizing logic in an easier manner than Catholic countries in the West (he also produced several negative case-studies looking at other religious systems around the globe to reinforce his thesis), Weber set himself the task of tracing out the development of Protestantism so as to uncover how the totalizing logic of religion along this Protestant branching became sympathetic to a synthesis with the totalizing logic of capital in modern societies.

Historically the process follows a contradictory narrative from the causes of the Protestant Reformation inaugurated in 1517 by Martin Luther with his ninety-five theses posted as an ethical challenge to Catholic doctrine, to where it culminates in that spirit of capital found in Franklin's maxims. At its core, Weber's thesis is that this is a process of rationalization that is undertaken to justify dominant worldviews and correspondingly appropriate social and individual actions to ends which reinforce that mode of domination. However, "'Rationalism,' is a historical concept that contains within itself a world of contradictions," and these contradictions make up the *"irrational* element" hidden within the narrative that must be exposed by employing the critical method (Weber, [1904–05] 2011, p. 98). The first contradiction uncovered by his application of this method relates to the orientation of one's life toward the material and the spiritual world, and the second relates to the conflict between the individual and the social. These relationships and the thought justifying them go through several reversals in a narrative that begins with the Catholic tradition, is transformed by the Protestant Reformation into a new ethic, and then evolves into the spirit of capital.

The totalizing logic of religion, which prior to the Reformation is represented by Catholicism, alienates people by having them focus on a spiritual afterlife as the beginning of their true and authentic existence at the cost of devaluing the material existence of their everyday life. On the one hand, the Church supports the most extreme version of this alienation as a path to salvation in the form of monastic asceticism: a highly individualistic practice in which the spiritually devout turn inward and shun the material world. On the other hand, there are the common worshipers and the priestly class whose existence is rife with the pangs (anxieties) and temptations (ecstasies) of material existence; the extent of their religious alienation varies by degree according to Marx's maxim that "the more man puts into God, the less he retains within himself" ([1844b] 1992, p. 324). The role of the priestly class is to serve as mediators between the common worshipers and God, making the path to salvation dependent on maintaining that social relationship. Luther's own background was of the monastic variety, but his years there "were haunted by a dark shadow of acute anxiety as he sought, without any sense of success, to win God's favor and forgiveness for his (largely imagined) sins through many acts of self-mortification" (Mullett, 2003, p. 47). Following the Catholic logic, the young Luther believed that to achieve salvation and gain a fully justified sense that God accepted someone despite their sins required that the sinner actively work for it. This logic of salvation being linked to action or work, is visible in the requirement for priestly mediation between God and worshiper in the act of confession. It was also present in the practice of seeking indulgences, which were granted to a sinner to reduce the punishment of their sin in exchange for completing an action or working on some task. In the Middle Ages, however, this practice led "to abuses rooted in institutional cupidity and theological distortion" (Newsom,

2010, p. 370) as indulgences became monetized and were sold to petitioners by members of the Catholic clergy.

The practice of indulgences illustrated two contradictions: first, while encouraging religious alienation the Church was itself pushing a program that was in violation of the biblical commandment to build up treasures in heaven over those on earth (Matthew 6:19–20) and second, the program favored the rich over the poor, allowing those with money to purchase their forgiveness, which violated the biblical principle that wealth is an obstacle to attaining salvation (see Proverbs 11:28, Matthew 19:23–24, Mark 10:23–25, Luke 18:24–25). Luther's training gave him the knowledge needed to understand this as a material corruption of Church doctrine. But because Catholicism is built on a mediated relationship to God, in which the petitioner must go through a priest to gain access to God, the priestly class had the ability to wield their power over those who feared the spiritual implications for their immortal soul and coerce obedience to them as the representatives of the system of totalizing logic, no matter how that logic was warped to the individualistic ends of those representatives. These contradictions—that *irrational* element—in the belief system flourished because the Bible was unavailable in the common tongue and common worshipers were dependent on the priestly class as mediators of the biblical message upon which the Church's dogma supposedly rested. Without Luther's critique to serve as a catalyst they had no solid basis for believing that these practices were irrational distortions of the teachings they claimed to follow.

Just as the natural scientists of the 15th century risked their lives when they went against Church doctrine and published their theories of heliocentrism, Luther faced enormous risks in publishing his critiques of Catholicism. Even though he was ultimately a dedicated fundamentalist to the totalizing logic of religion, his challenge of the most prominent institutional representative of that logic made him little more than a David going up against a Goliath. But Luther was above all a rational thinker and he maximized his impact by spreading his message as quickly and as widely as he could to preempt the Church's ability to counter his protest. This was accomplished with the aid of the recently commercialized printing press, which enabled him to publish numerous sermons in low-cost pamphlets using the everyday vernacular of the people to spread his critique. Although the Catholic Church enjoyed immense power and the ability to silence dissenting voices, they underestimated the revolutionary potential of this technology and "between 1521 and 1525, when the pamphlet war was at its height, Luther and his supporters out published their opponents by a margin of nine to one" (Pettegree, 2015, p. 210). Luther's use of this novel technology to further his own ends demonstrates a masterful understanding of rationalism, in both thought and material application, something not lost on Western capitalism which is "strongly influenced above all by advances in the realm of *technology*. The nature of the rationality of modern Western capitalism is today determined by the calculability of factors that are technically decisive" (Weber, [1920] 2011, p. 244). The success of Luther's strategy to use technology as an aid to his protest was made evident by a public who endorsed and was emboldened by Luther's critique. That the Church underestimated the power of his critique and this new mode of information distribution is found in the words of frustration written by the papal legate Aleander to the Cardinal de' Medici:

> A shower of Lutheran writings in German and Latin comes out daily. There is even a press maintained here, where hitherto this art has been unknown. Nothing else is bought here except Luther's books even in the imperial court, for the people stick together remarkably and have lots of money. Until the edicts shall have been promulgated, we are helpless … Another recent annoyance is that those who return from Rome tell everyone that the Lutheran affair is considered a joke and a matter of no importance.
>
> SMITH, P., 1913, p. 456

Technology helped Luther's ideas to spread like a virus and once they were in the hands of an eager public they were picked up by others who, now emboldened by Luther's example, further refined them and continued the project of rationalizing the Protestant belief system.

Rather than abandon the logic of Christianity, Luther's protest instead aimed to reform the religious logic by ending the practice of mediation. In his vision, individuals were responsible for developing a personal and direct relationship with God rather than depending on other sinful humans for salvation. Although on the surface this would appear to encourage the life of monastic asceticism, the closest Catholic practice that could be justified along these lines, Luther rejects this interpretation. To cultivate a personal relationship with God, people needed personal access to the Bible and texts to help them interpret it. Luther provided both: first in the form of religious pamphlets and then in a translation of the Bible into the common tongue. Weber zeros in on Luther's choosing the German word "Beruf" in his translation of the Bible, which in English means "**calling**: one's *task* is given by God" (Weber, [1904–05] 2011, p. 99), as central to the new dogma of Protestantism. This calling is seen as a spiritual duty entrusted to individuals in the material realm because, for Luther, "the only way to please God … [is] the fulfillment of one's duties

... under all circumstances" (p. 101). Rather than linking salvation to the performance of specific actions, Luther now comes to understand salvation in a passive light, whereby it only requires faith in the power of Jesus's sacrifice on behalf of all sinners. Sinners then demonstrate their allegiance to that faith by following their vocational calling in the material world as an activity pleasing to God.

The emphasis on being called to a form of labor or an occupation, for Luther, is absent from "the monastic *organization of life*" whose "ascetic withdrawal from the world" (p. 100) abandons social salvation for personal salvation, benefiting only the individual practitioner. The vocational calling instead becomes a duty that is socially good because "the division of labor forces every person to work for *others*" (p. 101). Whereas the life of monastic asceticism followed Jesus first commandment to love God above all else (Mark 12:28–30), it ignored the second commandment to love one's neighbor as one's self (Mark 12:31), which in modern society is, at least in principle, a requirement of the forced life experience of those who are compelled to pursue secular labor and must rely on others in the social relation of the exchange market because conditions for self-reliance are absent. Living in a system of organic solidarity, to borrow Durkheim's phrase, thereby necessitated and reinforced the justification that it was a religious and spiritual, not merely a material, good for people to be compelled to avoid base individualism in favor of social deeds that were in line with biblical commandments. Therefore, although Luther relies on an individualist argument for the spiritual relationship between a person and God, he relies on a social argument for expressing faith in salvation by focusing on how one lives their material life among others.

The work left in refining Luther's system was to deal with the burden of anxiety that persisted when one could no longer perform a specific action to resolve their consciousness of guilt and now had to rely on the belief that all could be saved simply by exercising faith in Christ's sacrifice and working hard at one's vocation. Additional rationalizations were provided by John Calvin's answer to this problem with the introduction of the doctrine of predestination. According to this narrative, if God is omniscient and omnipotent, then it would be impossible for a human to change God's mind. Therefore, God must know beforehand whether someone is saved or damned, because no human action could possibly sway God from condemnation at one moment, to salvation the next. And if they were saved at one moment, only to be damned the next, then it would imply that God had granted their salvation in error. This reinforced the notion that salvation was passively rather than actively obtained, but according to Fromm ([1941] 1969) it did nothing for anxiety, it merely made people double down on the only action they could take: work.

> The state of anxiety, the feeling of powerlessness and insignificance, and especially doubt concerning one's future after death, represent a state of mind which is practically unbearable for anybody. Almost no one stricken with this fear would be able to relax, enjoy life, and be indifferent as to what happened afterward. One possible way to escape this unbearable state of uncertainty and the paralyzing feeling of one's own insignificance is the very trait which became so prominent in Calvinism: the development of a frantic activity and a striving to do *something*. Activity in this sense assumes a compulsory quality: *the individual has to be active in order to overcome his feeling of doubt and powerlessness*. This kind of effort and activity is not the result of inner strength and self-confidence; it is a desperate escape from anxiety.
>
> FROMM [1941] 1969, p. 91

Rather than alleviating anxiety this system had the potential to pile sin on top of sin, and therefore anxiety on top of anxiety, with no clear way of obtaining external validation of God's forgiveness. Lacking the appearance of certainty that the mediated relationship in Catholicism provided, Protestantism was prone to the production of anxiety. As Kierkegaard ([1844] 2014) saw it "*anxiety about sin produces sin*" (p. 89), meaning that if one became uncertain as to whether or not their sin was forgiven, then that was itself a sin and through the compounding nature of these sins their salvation was further called into question. Weber summarized that since salvation was no longer doled out by subjective human priests who could be bought, "the salvation destiny of every person," in Protestantism, "must be exclusively attributed to the hand of an objective power—and one's own influence has not the slightest effect" ([1904–05] 2011, p. 117). This did not free people to live any way they wanted, but rather firmly cemented the Protestant ethic on the doctrine of the calling as the only manner through which one could divine their salvation. If they were saved, then it made sense that God would bless their hard work, and if they were damned then he would not.

The problem is that rewards for hard work from worldly vocations in modern societies take the form of money and wealth, bringing us right back to the biblical contradictions that Luther saw in the Catholic practice of selling indulgences, but one more easily solved in thought if not in practice. Through a detailed analysis of the many different Protestant sects that arose in the centuries after Luther and Calvin, Weber explained how this was rationalized through a return to asceticism. Wealth as such could not be bad on its own as it was a reward for hard work, it only becomes "suspect when it tempts the devout in the direction of lazy restfulness and a sinful enjoyment of life" (p. 164). Unlike monastic asceticism which involved a vow of poverty that helped serve as an external constraint of the passions, Protestant asceticism required an internal constraint to curb the temptations of wealth leading one either away from their calling or toward "the irrational use of possessions" (p. 169). In other words, the system had to maintain that making money was good, while using that money to pass the time in idle behaviors or in ways that exceeded one's needs, through consumption in excess or of luxury goods, was wrong. This rationalization of "ascetic Protestantism shattered the bonds restricting all striving for gain—not only by legalizing profit but also by perceiving it as desired by God" (p. 169). The goal was not to punish the wealthy for being blessed in their labors, rather it was to rationally orchestrate life around the use of wealth "for necessary, *practical,* and *useful*, endeavors" (p. 170).

Given that the totalizing logic of capital is based on the unending production of wealth, the Protestant ethic, by orienting itself toward material life in a way that Catholicism did not, reinforced this new totalizing of the mind and spirit by directing people to use their wealth in ways that perpetuated the logic of capital. If gaining profit is good, and one cannot spend their money on goods for consumption beyond their needs, then they can take that money and turn it into capital. Investing capital in the production process therefore allows one to continue to practice their calling, seek God's blessing in the form of more profit and, as a form of social good, create new vocations that allow others to do the same. However, once it is detached from Protestantism and society is imbued with the spirit of capital, without a moral system that treats labor in rational terms that are focused on the ends, rather than the means, labor becomes an irrational enterprise that is undertaken for its own sake with the production of wealth as its only visible objective. Wealth grows as abstract tallies on a score board, and people labor on without spiritual, and often without material, reward. The result is that the supposed social good of the Protestant ethic is corrupted by the spirit of capital in an irrational way that no longer carries any social thrust for the use of wealth in ways that are practical or useful for the social whole. Instead the spirit of capital only maintains the ethic that work for its own sake is a necessity, and in fetishizing labor as a rational use of human power it has irrationally ignored the ends of this process.

As Weber concludes, "[t]he Puritan *wanted* to be a person with a vocational calling; we *must* be" (p. 177). The totalizing logic of capital conditions all of its subjects as carriers of this 'spirit,' "*not* only those directly engaged in economically producing activity, but of all born into this grinding mechanism. It does so with overwhelming force, and perhaps it will continue to do so until the last ton of fossil fuel has burnt to ash" (p. 177). This is our answer for why people, as individuals possessing agency, despite being aware of our alienated condition and that intense bouts of anomie will arise in society, are unable to successfully combat either on their own as individuals. While the spirit of capital shapes our psychological dispositions, the social structures of modern society prevent our ability to rationally organize our lives in ways that are conducive to an authentically fulfilling material life. Anxiety has even less potential to be relieved under the totalizing logic of capital, than it did under the totalizing logic of religion because of this unholy alliance. At least in the religious mode, there was a belief in the promise of salvation waiting at the end of the life spent laboring. A life of labor and the end of material life were co-constructed as reinforcing systems for both material and spiritual existence. In this mode of capital, labor is the means to its own end and we are trapped in this logic as if in "a steel-hard casing," with the destruction of our planet and resources as the only constant material obstacle in the path of this process.

Here is the biggest difference between these two totalizing logics. The religious version was predominantly one of the mind, and a revolution in mind was necessary and sufficient for its dismantling as a totalizing force (but not its dissolution). But, having learned the lessons from the decline of religious power in modern societies, the totalizing logic of capital is woven into the material fabric of our lives to such an extent that it controls our minds, spirits, and bodies, and "even with the best will, the modern person seems generally unable to imagine *how* large of a significance these components of our consciousness rooted in religious beliefs have actually had upon culture, national character, and the organization of life" (p. 178). In other words, what was carried over from religion and embedded in the material order of capital is a more powerful force than it was in its religious stage. What this means is that the material conditions created by the totalizing logic of capital cannot simply be overturned by critique, and critique is often, at best, viewed with suspicion as this organizational mode of life and its supporting attitudes are deemed true and natural. Weber's critical method demonstrates the irrationality of this system, but such irrationality is built into the system itself as a means of rationalizing its own ends; that is, the ends of capital, not those of humanity. To rationally organize society in a different manner would require nothing less than crossing the psychosocial divide and orchestrating a complete revolution of the mind and the complete revolution of material life. For Weber, however, such a process was unlikely, and perhaps impossible, given the way that rational thought became instrumental in modern societies, with the means available in modern society becoming an end unto themselves as rationality at the level of immediacy appears as something irrational at the level of long term goals.

Weber was aware that because the critical method is historical, the concepts that we develop to aid with our analysis, ultimately for the express purpose of diagnosing modern society, "must be gradually put together from its single component parts, each of which is taken out of historical reality. Therefore, the final formation of the concept cannot appear at the beginning of the investigation: rather it must stand at its *conclusion*" ([1904–05] 2011, p. 76). "Historical truth, however, is served equally little if either of these analyses claims to be the conclusion of an investigation rather than its preparatory stage" (p. 179). What Weber means by this reversal—where the concept does not emerge until the end, and the end is really only the beginning—is that the critical method, now, under the totalizing logic of capital, must always be engaged in the preparatory stage. That is its task. It uncovers how these logics of domination function, how they come to totalize our minds and spirits, and root out the irrational elements as the foundation in thought for a revolution of the mind which could serve as a precondition for the revolution of material reality, unlikely as it is for the latter to happen given the dynamic nature of modern societies under the logic of capital. The task of the critical method is to illuminate what the preconditions would be if modern society—as a whole—was determined to engage in a rational accounting of how it structurally reproduces social pathologies. However, critical narratives must be wary of becoming irrational narratives that promise salvation when there is none. Until history demonstrates that the material conditions have changed and there is a material basis for salvation, neither theory, critical or otherwise, nor individual forms of praxis can provide it. Our task, as critical socioanalysts, is to diagnose domination, but we do not have the power to end that domination. Only the recoding of our psychosocial dynamic and its historical synthesis with material reality can accomplish the necessary work of orchestrating our escape from the anxiety produced by life under the logic of capital.

Again, this critical method, central to the development of a practice of critical socioanalysis, stands apart from the mainstream practice of sociology today, which Dahms (2008b) argues: "work[s] from the implicit assumption that efforts to illuminate economic, political, cultural, and psychological features of social life do not require *rigorous reflection on how the immersion of social research in space and time, i.e., within concrete socio-historical circumstances, impacts on our ability to truly illuminate social life*" (p. 36). The mainstream approach therefore assumes that because our work is empirical and deals with actually existing reality, the concepts and conclusions of our investigations are self-evident and true so long as the rules of research methods have been stringently adhered to in the collection of data and the results have been presented in simplified 'everyday' language. Furthermore, it is precisely this misunderstanding of social reality as a historically dynamic concept that conditions mainstream sociology to offer positive solutions for change that are little more than irrational platitudes which ultimately serve the totalizing logic of capital, reinforcing a state of material alienation by refusing to confront with sober senses the reality of our current social state bound as it is in a casing as hard as steel. However, the critical tradition is geared toward exposing that which is not self-evident of our condition precisely because critical thought retains a firm cognitive grasp on the fact that the systems of domination imposed on us by the totalizing logics are the very forces that dictate what is self-evident about our condition and how we represent it linguistically. It is this grasp which was so brutally fought for in the long historical struggle against the totalizing logic of religion that cannot be relinquished by a critical social science under the totalizing logic of capital, even if the historical conditions that logic constructs reveal that a wide-spread acceptance and recoding is at best unlikely or at worst impossible at the present moment.

If the goal is to understand how our thoughts are constructed by those logics and shaped to understand our reality, then the task of sociology is exactly the opposite of what the mainstream approaches offer: it is not to mirror reality as a perfect reflection by conforming to the rules of the dominant language game, but to challenge that reality by exposing the latent social forces that bend our minds to see reality in the ways that it socializes us to do, and this often requires that the reader take the time to learn the rules of new language games that break free from the protocols of the dominant narratives. In an accelerated society that cannibalizes our time, people are coded to demand quick fixes, summary reports, and easily digestible sound bites that do not challenge the dominant mode of thinking because doing so is a painful task that demands we take the time to cultivate the mentality needed to see beneath the veil of domination by challenging and actively changing our cognitive functions. This task, however, cannot be rushed with the main points of empirical analysis summarized for quick consumption because it requires the ability to see the way our minds are coded by modern society to interpret the empirical reality those societies reinforce. If our thoughts are the result of the coding of the logic of capital, then the ability to recognize its effect on us requires at the very least a partial re-coding of our minds. In this sense the socioanalyst is an experimental vanguard who is committed to the recoding of their own mind and an acceptance that they must embody the painful contradiction of living in a society to which their thoughts do not align; while simultaneously doing a deep dive into the internal logic of that society to better understand its coding and look for possible vulnerabilities where viral recoding strategies could potentially be employed. Although this is a necessary task for all modern individuals who are committed to the pursuit of 'the good life,' present circumstances suggest that most will either ignore the call or act in hostile ways when their modes of thought are challenged and the irrational forces that guide their lives are exposed, for that reason it is all the more important that critical theorists engage in this practice to keep it alive and out of the dustbin of history.

With the layered psychosocial structuring of our minds this task demands nothing short of continuously deconstructing our minds to strip away and reveal each foundation upon which we have built our thought while, at the same time, confronting the fact that the construction of distorted thoughts will continue with no end in sight. Given that we cannot escape the bombardment of the dominant coding process, which never rests because it is powered by the social dynamic, the socioanalytic task works against the grain of society and at present can best hope to bend our code enough so as to see how the dominant logics have warped our views of reality. This process is one that demands the embrace of pure anxiety—a pure mode of radical uncertainty—because that is precisely what our societies produce. The problem that we must next confront is that in the process of confronting anxiety we are conditioned to repress consciousness of the very things that provoke it. Our final foray into the classics will therefore examine this issue of repression through the work of Sigmund Freud and his critical method upon which socioanalysis is based for challenging it.

4   From Psychoanalysis to Socioanalysis: Anxiety, Repression, and Talk Therapy in Freud

Like the other classics of critical modern thought, Sigmund Freud was a product of the newly modernized European world, enlightenment thought, and the shift from the totalizing logic of religion to that of capital. His intellectual project was radical and ambitious. Not content to simply add another series of texts to the growing archive of psychology, Freud sought to establish an entirely new subdivision of that archive in the library of science that would be built upon a revolutionary practice that placed his theoretical discoveries at the core. The ambition was for this practice to rival establishment psychology by offering a critical methodological approach for therapy built upon a psychodynamic framework that not only treated mental ailments but also formed a material basis upon which to build a critical theory of the subject matter that psychology claimed as its domain. Freud used the term psychoanalysis to describe both the practice and the corelated body of theory. Because his work was grounded in the material reality of his patients' psychic lives and guided by critical thought, it was above all a self-critical enterprise that continuously subjected its own assumptions to critique and refinement, and thereby provides an example for tracing out how thought transforms over time when considering psychosocial developments and the continuous application of the critical method. As with the other classics above, there is not sufficient space here to trace all these transformations and the full development of Freud's thought. I will limit the following to a few key aspects of his thought that will, on the one hand, serve as building blocks for the recombinant innovation that is the practice of socioanalysis and, on the other hand, link to the theoretical contributions of the classics examined above. Following that, I will examine how his theory of anxiety transformed over time as he continuously subjected it to the critical method. In keeping with the depth hermeneutic approach, Freud's psychosocial background likewise plays an important role in understanding the sociohistorical conditions in which psychoanalysis developed as it did, which will foreshadow the later contextualization for why socioanalysis is needed as a challenge to mainstream sociology that can serve as the critical social scientific practice of the 21st century.

The transition from traditional to modern sensibilities played out in rapid succession in Freud's family. His paternal grandfather was a rabbi in the *Hassidic* tradition, but his father, Jacob, pursued a rather unsuccessful career as a merchant. Jacob, while ultimately leaning closer to an adherence with religious logic, assimilated to his cultural surroundings as a Jew in the *Haskalah* tradition "which extolled the virtues of enlightenment and, while retaining some traditional practices, rejected rabbinic Judaism with its emphasis on Talmudic law as being narrow, backward and largely antithetical to the modern world and to scientific progress" (Aberbach, 2003, pp. 122–123). It was through this *Haskalah* tradition that Jacob exposed Sigmund to the Bible. That tradition placed the text in its historical context and included the study of other cultural artifacts and modes of exegesis that went beyond religion into the fields of secular scholarship to better understand the wider cultural relevance of the Bible as a written text. Jacob wanted to instill in the young Sigmund an appreciation for the Jewish religious tradition and modern thought but, by subjecting Jewish orthodoxy to the enlightenment tools of rational thought as encouraged by the *Haskalah* tradition, Sigmund came to have a greater appreciation for the critical methods he was exposed to than for the object of study (Rizzuto, 1998). Put differently, Sigmund gained an appreciation for the Bible as a literary text that had great cultural influence rather than as a sacred object that was inspired in a realm beyond that of profane life. This was clearly evident in his 1939 book *Moses and Monotheism* ([1939] 1967) wherein he argues that Moses was an Egyptian who developed the Jewish religion of the Israelites. Throughout his life he demonstrated a certain ambivalence toward Judaism, writing in *An Autobiographical Study* ([1935] 1952) that "My parents were Jews, and I have remained a Jew myself" (p. 6), but thought of himself as "a completely godless Jew" (1963, p. 63). Although he distanced himself from the religious life of Jewish belief, identifying as an atheist, his sense of self was developed in a psychosocial milieu that was dominated by the Jewish question and whether modern society could overcome the prejudiced barriers of religious and ethnic intolerance that persisted despite societal claims of enlightened progress.[3] Like Durkheim, Freud's ties to Judaism are therefore best thought of in the ethnic/cultural, rather than in the religious, sense; but like Marx, he was deeply influenced by the writings of Feuerbach and assumed a critical view of religion as an institution that dominated the mind in an irrational manner (Levitt, 2009).

Similarly, the transition from traditional to modern life played out rapidly in Freud's larger social milieu of Austria with developments in technology and political economy paving the road to modernity. Life was not easy for Jews in Austria, but in the late 1800s Vienna developed into an urban hub built on the principles of a liberal approach to modernity that stood in direct opposition to Bismarck's politics that guided the fate of Prussia. Supported by Wilhelm I's desire to retain power whatever the domestic cost, the historian Wehler ([1973] 1985) argues that Bismarck's unification of the German empire and the wars fought for Prussian hegemony "were used as devices to legitimize the prevailing political system against the striving for social and political emancipation of the middle classes, or even the proletariat" (p. 26). But that system was challenged when Franz Joseph I assumed the role of the Emperor of Austria in 1848 as he increasingly responded to the desires of the middle and lower classes that were in part stoked by economic change, caused by a nearly complete agrarian revolution and budding industrial revolution. From the years 1848 to 1873, Austria rapidly built a railroad system and by 1913 it boasted the third largest track in Europe, behind only Germany and Russia (Broadberry, Federico, & Klien, 2010).

The desires of the middle and lower classes were also linked to religion, with the Jews, in many ways, being the freest to explore the desires enabled by these economic revolutions. They were neither suspicious of wealth in the Catholic sense, nor were they constrained by the Protestant ethic and its brand of asceticism. This allowed Jewish desires to grow in ways more closely aligned with the development and appreciation of cultural work in music, art, theater, and literature, as well as scientific development in chemistry, medicine, physics, economics, philosophy, psychology, and history to name but a few of the areas that expanded and benefited from Austrian Jewish contributions.

---

3   On the influence of Judaism in Freud's intellectual development, see Fuks, 2008.

Therefore, the economic revolutions were reinforced by intellectual, cultural, and pragmatic advances that earned wide popular support in urban hubs, like Berlin and Vienna, which fed into the growing demands of the lower classes for greater individual autonomy and social participation. "In 1860," the year the Freud's moved to Vienna when Sigmund was almost four years old, "the liberals of Austria ... assumed power over the city of Vienna ... and transformed the institutions of the state in accordance with the principles of constitutionalism and the cultural values of the middle class" (Schorshe, [1961] 1981, p. 24). Practically, the liberals developed Vienna into the model of modern urbanization with a public support structure that included a variety of social services, such as, a flood control system, a water supply, and a public health system that overturned the traditional church-controlled charity-based healthcare system into a municipality-controlled system grounded in the developing medical sciences. The first telephone was installed in Vienna in 1881 and by 1913 their use was so widespread that Austria-Hungary had the fourth highest number of annual telephone calls in Europe, numbering around 568 million (Broadberry, Federico, & Klien, 2010).

Internationally Austria and Prussia were aligned with common interests, but domestically Austrian influence was seen as leading the German peoples in a way that ran counter to Bismarck's Prussian conservatism (Bueno de Mesquita, 1990). In 1866, with social, economic, and political crises threatening Austria, a brief war broke out between the two that split German support. In spite of the liberal reformist policies guiding Austria, a large part of the blame for the economic crises that led to the war were thrown at the feet of the Jews, as, for example, "[i]n April 1866, Vienna's official *Militärzeitung*—[the military newspaper]—blamed ... Jewish speculators: 'The Jews are withholding their money; the Jews are mocking us'" (Wawro, 1995, p. 239). Rather than change Austrian politics and bring the country into the fold of German unification, Bismarck won the war and opted to exclude the Austrians from his project, thereby negating any influence they had previously wielded over the future of German affairs. Ultimately, free from Bismarck's influence, Vienna continued its trend toward liberal principles and Austria continued the modernization process.

Up until "World War I, Austria-Hungary's machine-building industry ranked amongst the leading producers of the world in terms of total output and employment, surpassed only by the United States, Britain, and Germany" (Schulze, 1996, p. 15). However, despite the success in the machine-building sector, and the industrialization project in general, this divorce from the German empire did lead to a slowdown in the overall rates of Austrian industrialization (2.8% growth per annum from 1870–1913), compared to that of the

German empire (4.1%) over the same period of time (Broadberry, Federico, & Klien, 2010). However, escaping Bismarck's politics was not altogether undesirable from the Jewish perspective. In 1857, Vienna only had around 6,000 Jews, but "by 1880," when Freud was 24 years old, the Jewish population "had grown to over 72,000, one in every ten inhabitants of Vienna was a Jew" (Gay, 1998, p. 20). Austrians were not particularly welcome to this growth in the Jewish population and continued to publicly espouse anti-Semitic views, but the sheer number of the Jewish population lent itself to a social support system in Vienna which enabled opportunities that were hitherto foreclosed to those of Jewish descent. This transformation of the Jewish experience was also legally supported by Franz Joseph 1 who by 1867 granted Jews equal rights (Wistrich, 2006). By the late 1800s, Vienna had become a capital of Jewish emancipation and enlightenment, with progress made not only in economic affairs but in the development of a rich cultural and scientific climate.

As a young man Sigmund was torn between embracing the bourgeoning Jewish climate in Vienna and his father's inability to provide the financial means that would allow him full participation in that world. His attitude toward the temptations of bourgeois life developed while he was a child in large part as reflections of his parents' behaviors. The frustration of his father's financial situation was amplified by his mother, Amalie, and her "difficult" personality. She experienced a major depression in Sigmund's youth in which her sense of loss was compounded. On the one hand, because she was many years Jacob's junior—closer in age to his sons from his previous marriage—and felt betrayed by his passing himself off to her father as a more successful merchant than he was, coupled with the reminder that she had given him her youth in this lopsided exchange. And on the other hand, her brother Julius died a month before her son of the same name was born, who in turn died less than a year later. In those early formative years, Amalie was emotionally unavailable to the young Sigmund as she withdrew into an inner world constructed by her grief and general unhappiness. The move to a crowded Vienna where the Jews had not yet left the segregation of their 'wretched quarter,' with a depressed mother and a financially downtrodden father, weighed on the young Freud. "At the same time as the grown-ups were transmitting their agitation to the young boy, they were also unavailable to help him contain his own fear, sorrow, and anger" (Whitebook, 2017, p. 47). As Freud developed as a child he turned inward focusing on his own interests, sharing many of the same intellectual traits as Weber, which included an early appreciation for literature and history as a young boy, and top marks in his classes as a student (Collins & Makowski, 1993). His proclivity to and success in intellectual affairs drew his mother's praise and a reversal in her attitude from withdrawal to doting. She came to believe that "Sigmund had been given unusual gifts, that he was destined to become famous. For him, therefore, no sacrifice was too great" (Freud, M., 1957, p. 19) even if it meant placing his desires and ambitions above those of his siblings. Raised in circumstances that would amplify his narcissism, Freud's early life demonstrated both the increased sense of individuality afforded by modern society, as well as the cruel truths of life under the power of nature and its artificial other, the social. While there were certain anxieties present in the young Freud's life, his journey started with an idealistic belief in the Enlightenment project and the cause of progress which were reinforced by the material milieu of Vienna.

The contradictions of modern society that guided Freud's fate were visible in his father, a man caught between the currents of religious and enlightenment thought, and his mother, a woman torn between sorrow for the lives and paths shuttered in the past and hope for those yet to be traveled in the future by her son. In an environment that produced the genuine dilemma of life caught between the past and the future, Freud had to select a course of study and a vocation that was amenable to his talents and his intellectual and class-mobility ambitions while being constrained by his Jewish and financial background. Capitalist industry was growing in Austria, as were business opportunities for Jews, but his curiosity drove him toward the intellectual avenues open to someone of his background that would provide the financial security he desired: law and medicine. For a man of Freud's talents there was nothing in law that could provide such a tempting ground for making new scientific discoveries that would leave his mark upon the modern project. It was the riddles of human nature yet to be solved which were exacerbated by the contradictions of modern life that pushed him toward medicine, which was still in the exciting process of incorporating the revolutionary insights of Charles Darwin who, in 1859 when Freud was three, published *On the Origin of Species* as a scientific challenge to the traditional biological narratives. As a student, and in the years prior to the establishment of his own practice, he pursued a course of study in the fields of physiology and neurology and was employed by several well-regarded scientists to assist in their medical research (Gay, [1950] 1989). By the 1890s he established himself as an expert with a specialty in nervous ailments.

Freud came to reject the mainstream explanations of neurotic behavior and pathological symptoms, finding that one of the key problems in the psychological profession was the hierarchical approach to the doctor/patient relationship that imagined the doctor in a privileged position of knowledge and the patient as nothing more than the object of study (Freud, 1910). In this position, the doctor, as the holder of expert knowledge, was "in-the-know" while the patient maintained a position of ignorance, even and perhaps especially, about their own ailments. To the dismay of the medical profession, Freud claimed that doctors, although possessing expert medical knowledge, "cannot understand hysteria. [They are] in the same position before it as the layman" (p. 183). While trained in the medical and psychobiological approach, which were developing as laboratory sciences, over the course of his career Freud increasingly moved in the direction of a psychosocial approach that attempted to engage with the dynamic nature of human behavior rather than treating it as a static phenomenon that could be isolated in a lab. Freud arrived at this conclusion by way of his direct interactions with patients who—once engaged in the talk therapy he developed to better understand hysteria from the patient's perspective—demonstrated that they knew far more about the root causes of their symptoms than the doctors who were merely treating their symptoms and failing to listen to, or account for, those roots. This was not, however, to suggest that the patients were conscious of the reasons for their suffering. Rather, because the root causes of these nervous ailments often came from traumatic events, they were repressed by the conscious mind. So, Freud concluded that this knowledge must reside elsewhere and that it would require a novel method to uncover it. Locating where that knowledge resides and how to access it was Freud's ultimate and largest contribution to the critical method. It is precisely because Freud's concepts have entered everyday speech and popular culture that they are often riddled with conflicting, confused, or flat out wrong interpretations. As such, it is worth briefly revisiting his foundational ideas.

Freud developed two tripartite classifications of the mind to help aid the understanding of how, on the one hand, one could know and at the same time not know, and on the other, employ rational thought while simultaneously having those thoughts thwarted by irrational impulses and actions. In one of these classifications Freud provides a structural theory that divides the mind into the id, the ego, and the superego, which link up to the drives (libidinal and death). These drives "are two essentially different classes of instincts: the sexual instincts, understood in the wildest sense—Eros, if you prefer that name—and the aggressive instincts, whose aim is destruction" (Freud, S., [1933] 1989, pp. 128–129), which Freud comes to call Thanatos ([1920] 1990). The id is "a chaos, a cauldron of seething excitations" (Freud, S., [1933] 1989, p. 91) and is nothing more than "[i]nstinctual cathexes seeking discharge" (p. 93). The id is ruled by the pleasure principle. It is the most animalistic part of our mind, the one that is closest to nature, and is the first to develop. Like a small child, the id is irrational and demands constant pleasure (libidinal satisfaction), or at least the avoidance of unpleasure (which at the extreme point accounts for the death drive as the state of absence of unpleasure). The ego "is the sense-organ of the entire apparatus" (p. 94) and mediates between the stimuli provided by the external and the internal world. It is upon the ego that the demands of action are made, and which carries the burden for satisfying the sources of those demands. Finally, there is the super-ego which is the internalization of prohibitions on actions, that "which lays down definite standards for its conduct, without taking any account of its difficulties from the direction of the id and the external world, and which, if those standards are not obeyed, punishes it with tense feelings of inferiority and of guilt" (p. 97). In other words, the super-ego is like an authority figure made up of the internalized composite of parents and familial values, religious leaders and moral codes, teachers and social norms, government and the legal code, etc., which is unbendable, unbreakable, and unreasonable in its commands to obey. In this sense, the super-ego is the most socially controlled part of the mind, the one that is closest to the artifice of modern society.[4] Caught between the id's pleasure seeking and the super-ego's control mechanisms, the ego must satisfy both while also contending with the reality principle, or, the external world and its natural as well as social constraints.

Complementing that model, Freud's other classification of the mind attributes "three qualities to psychical processes: they are either conscious, preconscious or unconscious" (Freud, S., [1940] 1989, p. 32). What is conscious is the thought content that is currently loaded in the mind, the active known knowns. What is preconscious are the passive known knowns, that is, the knowledge content of our minds that we have the ability recall and make conscious by means of our mental efforts and energies. What is unconscious are the unknown knowns, the knowledge content of our mind that is repressed or inaccessible to our conscious recall. It is this last quality of the psyche, the unconscious, upon which the entirety of Freud's psychoanalytic enterprise hinges, for that is where the latent knowledge resides, where repressed thoughts, actions, and events evade our conscious mind but bubble up in unexpected ways that shape our life course and our actions regardless of our rational ideas and plans. Freud was not, however, the first to suggest that the unconscious was a component of the mind (Nicholls & Liebscher, 2010). F. W. J. Schelling ([1800] 1978) introduced the concept to philosophical audiences and Eduard Von Hartmann ([1869] 1884) greatly developed and expanded upon it, but it was Freud who brought the concept to the masses as an object of scientific inquiry and made it a part of Western culture's understanding of the mind (Ellenberger, 1970). We may call the unconscious Freud's discovery,

---

4 On the concept of artifice, see Dahms, 2017a; *forthcoming*.

insofar as "the discovery of the unconscious, such as it appears at the moment of its historical emergence [in Freud], with its own dimension, is that the full significance of meaning far surpasses the signs manipulated by the individual. Man is always creating a great many more signs than he thinks. That's what the Freudian discovery is about—a new attitude to man" (Lacan, [1978] 1991, p. 122), which can only be fully understood by turning the critical gaze on the unconscious as Freud presents it. This new attitude is historically linked to the modernization process and arises once the individual becomes historically possible in modernity as Durkheim demonstrates, but it is Freud's discovery of the unconscious that then allows us to organize the mind "within a dialectic in which the *I* is distinct from the ego," thereby warning against the conflation of the subject and "the individual" (Lacan, [1978] 1991, p. 8), or put differently, from the mind and the conscious. Furthermore, this attitude is precisely what was cultivated in Freud's Vienna, and is what motivated his early work to be critical of psychology, which following the biological perspective denied that these were "new" minds. He bought into the mentality of modern humans and critically evaluated it, but by ignoring the social dimension in his early work his view was only partially conceived. As Weber had demonstrated, as the Protestant mode of asceticism came to spread through modern minds like a virus in its secular form as the spirit of capital, modern people increasingly found themselves in social positions that were at odds with their individual desires, which they had to repress to function in the new social order. Early on, then, psychoanalysis did more to help adjust people to the conditions of bourgeois society than it did to challenge the reasons why repression was increasing in these societies.

Because there are common misconceptions about these two tripartite divisions of the mind, they demand some further clarification. These are separate models that do not perfectly align and although there is some overlap, they are not descriptive of identical mental phenomena. Freud writes that "id and unconscious are as intimately linked as ego and the preconscious" and while the id remains inaccessible to conscious change, its content can be "transformed into the preconscious state and so taken into the ego" (Freud, S., [1940] 1989, pp. 35–36). The ego, as the mediating force between external pressures and the internal pressures of the id and the super-ego, is where most of the conscious activity takes place, but "we may say that repression is the work of this super-ego and that it is carried out either by itself or by the ego in obedience to its orders" (Freud, S., [1933] 1989, p. 86), meaning that the super-ego is also partly conscious (and that repression is linked to the social structure of modern society). Freud also claims that "large portions of the ego and super-ego can remain unconscious and are normally unconscious" (p. 87). Therefore, while the id is fully unconscious, the ego is primarily preconscious, partly conscious and partly unconscious, and the super-ego is also partly conscious and partly unconscious.

There are several conclusions we can draw from this that challenge not only cultural usage of these terms, but also the prevailing philosophy of science that guides much mainstream research and which has coopted psychoanalytic research in haphazard fashion, particularly in the United States and Britain, where it has pursued a greater focus on the ego at the cost of downplaying the role of the unconscious.[5] First, Freud's concept of the ego is not interchangeable with the common usage of it to refer to the sense-of-self that one has, as in the saying that one has an 'over-inflated ego' or a 'big ego,' since this would be incompatible with the fact that the ego is primarily preconscious and partly unconscious, therefore a large portion of it is hidden from conscious thought. Second, the term subconscious—which is commonly substituted for the preconscious (and at times is used interchangeably with the unconscious)—is a misnomer that implies a hierarchy of the mind that is not present in Freud's model. Third, since most of the mind is unconscious, it does not easily fit into the positivist notions of the scientific method. Freud's models contend that most of the mind's content is not manifest and easily accessible, rather the greatest part of the mind's content is latent. This means that the largest share of the mind's content cannot be directly observed and studied but must be understood through the effects that it produces, the access and understanding of which requires a critical approach to the subject as an object of scientific inquiry. Furthermore, this also suggests that the conscious mind is not master of its own house, which means that if we restrict ourselves to studying only the manifest content, as in mainstream psychological approaches, we conflate the subject and the individual thereby distorting our view of both.

---

5 In the years following Freud's death, mainstream psychoanalytic discourse moved away from many of his theories, assuming, along the lines of positivism—laid out by Auguste Comte ([1896] 2000)—that the birth of the discipline was wrapped up in the Theological and Metaphysical stages. In other words, it was too steeped in the mythological, i.e. the Oedipus complex, and in concepts, like the unconscious, that were more metaphysical than the empirically minded, and increasingly positivistic, sciences were comfortable with. In the post-World War II years, under the leadership and research of Anna Freud and Heinz Hartmann many began to place their psychoanalytic roots in Freud's text *The Ego and the Id* ([1923] 1990) signaling a shift toward a more empirically grounded ego psychology, which in particular came to dominate the American psychoanalytic scene (Wallerstein, 2002). The Neo-Freudian, Lacan ([1966] 2006), however, viewed this as the abandonment of "Freud's discovery [that] calls truth into question"; this discovery being the "unconscious" (pp. 337–8).

Freud's solution for accessing the content of the unconscious was to locate those places where it bubbles forth, circumnavigating the conscious mind to do so. For Freud ([1899] 2010) the unconscious made itself visible in dreams, slips of the tongue, and in repetitive behaviors that did not coincide with the rational, consciously laid, plans the individual claimed to be following. Because the unconscious cannot be accessed directly, talk therapy is a necessary component of psychoanalysis, for it is only within the psychoanalytic session that the analysand can dedicate themselves to the analysis of their dreams and that the slips and repetitions can be studied in the dynamic form through which they are revealed. Rather than see the conscious as the most trustworthy source of information on the mind, and therefore the most deserving of scientific inquiry, for Freud the conscious was untrustworthy because it was not 'in-the-know' either; it was precisely this difficult to access realm of the unconscious that touches most closely upon the truth content of the psyche and therefore serves as the necessary object of study.

Freud's guiding hypothesis for psychoanalysis is that some of this unconscious knowledge can be brought to the conscious mind through dedicated work by the analysand where they examine their dreams, slips, and repetitions, so that the content of the unconscious can be better understood by the conscious and so that the analysand can better bear witness to how it affects their life. Therefore, it is correct to say that the patient possesses the knowledge of the root causes of their symptoms but, at the same time knows nothing of them because they reside in their unconscious. This is why Freud ([1958] 2001) concluded that "we may say that the patient does not *remember* anything of what he has forgotten and repressed, but *acts* it out. He reproduces it not as a memory but as an action; he *repeats* it, without, of course, knowing that he is repeating it" (p. 150). Rather than tracing the causal logic to find the latent sources of neurotic ailments, the mainstream psychological approach focused on the manifest content of these ailments and developed treatment plans geared toward the elimination of the symptom, but they left those latent wounds festering beneath the surface with a predictable outcome: the symptoms would eventually manifest anew. In other words, they focused on changing the patient's future behavior while leaving the past largely unexamined. While it is true that the past cannot be changed after the fact, the ways that those past events come to program the mind to certain responses could not be overcome if the goal was simply to attempt to place a more powerful code on top of the problematic code running in the patient's unconscious.

This process created a dependency on the medical professional that aligned with the logic of modern societies, in that it reproduced the system of means (treatments) by sacrificing the goal of a rational end (cure), because the new code would be thwarted time and again by the original code whose function was left unchallenged. For Freudian psychoanalysis, however, the therapeutic session progressed by means of a division of labor: the patient talks, while the psychoanalyst "employs the art of interpretation mainly for the purpose of recognizing the resistances which appear ... on the surface of the patient's mind ... and making them conscious to the patient" (p. 147). In other words, the psychoanalyst "uncovers the resistances which are unknown to the patient; when these have gotten the better of, the patient often relates the forgotten situations and connections without any difficulty" (p. 147). For Freud, it was precisely the original coding of the mind which was hidden in the unconscious that had to be understood if the patient was to have any hope of escaping its domination over their future behaviors. By bringing it to the fore in the psychoanalytic session, the patient is able to analyze the interdependence of their actions and their coding in a space and time dedicated to this task, so as to gain some level of awareness of when the code would activate in their everyday life in ways that ran counter to their conscious ambitions.

Summarizing the psychoanalytic session Bruce Fink (2007) says: "The psychoanalyst's first task is to listen and listen carefully" (p. 1) looking for those slips and moments of repetition, followed by "asking [the analysand] questions so that she will fill in missing details, finish sentences that have trailed off, and explain what she means by certain things she says" (p. 24). The point of analysis is to put the analysand to work on their own mind, it is not to create a scenario for the analyst to resume and reinstate the hierarchal relationship of doctor→patient with the doctor having the knowledge that the patient lacks, as the mainstream approach does. In other words, "the goal is not to get [the analysand] to substitute the analyst's understandings for his own understandings (that is, to internalize [the analyst's] point of view) but rather to get him to become suspicious of all meanings and understandings insofar as they partake of rationalization and fantasy" (p. 81). Stated yet another way, the goal is not to teach them what to think, but how to think, and specifically, how to think critically using the critical method. For the psychoanalytic treatment to bring about a "cure" of the nervous ailments, the analysand must interpret the frothy parts that they can skim off the surface of their unconscious and recognize the partiality of what is accessible to conscious thought, so that they come to know the limits of thought not only in their own mind but also in recognizing the limits of others' thoughts that are subject to these same mental limitations. Only then can they come to recognize that all meaning is necessarily partial, so, no authority, be it the internal authorities of the id and the super-ego or the external authorities of society, possesses pure understanding that would justify either external or internal forms of authoritarianism. The authoritarian model corresponds only to a static reality, but society and the individual's mind are dynamically bound in a process that links the possibility of thought to the historical material reality in which they live. This is why the analysand must learn to work through their thoughts to see how each domain of their mind influences them, locate where their thought has calcified, and where it is in need of agitation. If curing nervous aliments, like anxiety, is the goal, then this critical approach that accounts for how people are socialized to think and act, how behaviors come to be sorted as either civilized or backward, and how individuals repress certain tendencies and thoughts while never fully exorcizing their influence, is required if the contradictory riddles of rational thought and irrational behavior are to be understood and their roots exposed for analysis and recognition of their power by the conscious register of thought. Again, this does not mean that through psychoanalysis the analysand will come to master their unconscious and thereby their conscious actions, rather it promises to bring the analysand to the point where they can understand that the unconscious will make demands that influence their thoughts and actions and to recognize how those demands impact their lives in ways both good and bad so that adjustments can be made.

Early on Freud's larger focus on the libidinal drive and the unconscious impact of the id on behavior can be explained on the one hand by his medical background and psychobiological approach and on the other hand by his desire to enter bourgeois life. What modern bourgeois society in Austria opened for Freud was an avenue that had previously been foreclosed socially because of his Jewish heritage, and personally because of his father's limited success as a merchant. Seeing that opening and witnessing the new technological and economic means of achieving social mobility, coupled with a deep-seated drive to make his mark in science, Freud's early embrace of and desire to join bourgeois life is understandable; in his circumstances it was a pathway to greater freedom. However, he came to recognize its Janus-faced nature by understanding what was gained by this revolution and what was lost when he wrote: "If there had been no railway to conquer distances, my child would never have left his native town and I should need no telephone to hear his voice; if traveling across the ocean by ship had not been introduced, my friend would not have embarked on his sea-voyage and I should not need a cable to relieve my anxiety about him" ([1930] 2010, p. 61). In this manner, Freud expresses the ecstasy and the anxiety of technical vertigo under the logic of capital as it disrupts life by providing new avenues of pleasure and new freedoms which are co-constructed with new avenues of pain and new structures of domination and unfreedom. While his method prior to this stage of his intellectual development was critical from a psychodynamic perspective, it was not yet fully critical from a sociodynamic perspective, which lent psychoanalysis to helping individuals but at the cost of reinforcing the structural logic of modern societies that were in large part responsible for the incompatibility between libidinal desires and civilizational checks on behavior in the bourgeois order. Toward the end of his life, however, he came to place a greater import on the super-ego and the death drive, and therefore had to incorporate more of the sociological dimension, especially as modern society violently erupted and the hidden dark features boiled up from the depths to encompass the human experience in World War I.

Unable to remain uncritical of the interconnectedness of the psyche and the social, and therefore of the psychosocial, Freud began to adopt a viewpoint of society that mirrored the classics of modern social thought in a manner that recognized to a greater degree the impact of the totalizing logic of religion and of capital on the psychic makeup of the individual. This is evident in a shift found in his diagnostic remarks in *The Future of an Illusion* ([1927] 1961), a book that critiques the religious logic in society, when he addresses the social order and how it touches on the psychic order:

> Human civilization ... includes on the one hand all the knowledge and capacity that men have acquired in order to control the forces of nature and extract its wealth for the satisfaction of human needs, and, on the other hand, all the regulations necessary in order to adjust the relations of men to one another and especially the distribution of the available wealth. The two trends of civilization are not independent of each other: firstly, because the mutual relations of men are profoundly influenced by the amount of instinctual satisfaction which the existing wealth makes possible; secondly, because an individual man can himself come to function as wealth in relation to another one, in so far as the other person makes use of his capacity for work, or chooses him as a sexual object; and thirdly, moreover, because every individual is virtually an enemy of civilization, though civilization is supposed to be an object of universal human interest. It is remarkable that, little as men are able to exist in isolation, they should nevertheless feel as a heavy burden the sacrifices which civilization expects of them in order to make a communal life possible. Thus civilization has to be defended against the individual, and its regulations, institutions and commands are directed to that task. They aim not only at effecting a certain distribution of wealth but at maintaining that distribution; indeed, they have to protect everything that contributes to the conquest of nature and the production of wealth against men's hostile impulses. Human creations are easily destroyed,

and science and technology, which have built them up, can also be used for their annihilation. (pp. 6-7)

Here Freud recognizes the dual-sided nature of the psychosocial and the need for each to be kept in check against the intrusions of the other. If the social and the psyche are not in balance, then the pains of alienation, anomie, and a life guided by instrumental rationalization become too painful to bear. Given the totalizing logics that shape modern society, the cost of this pain is fully born by the individual and while society has so far continued to function while refusing to check itself against the reproduction of these social illnesses, the cracks in social order are visible across the globe and most noticeable in the most advanced modern societies of the Western world. Freud pushes his diagnosis further along this line in *Civilization and Its Discontents* ([1930] 2010):

> [Religion] is so patently infantile, so foreign to reality, that to anyone with a friendly attitude to humanity it is painful to think that the great majority of mortals will never be able to rise above this view of life (p. 39). Life, as we find it, is too hard for us, it brings too many pains, disappointments and impossible tasks. In order to bear it we cannot dispense with palliative measures ... powerful deflections ... substitutive satisfactions ... and intoxicating substances (p. 41). [E]very man must find out for himself in what particular fashion he can be saved (p. 54). The development of civilization imposes restrictions on it, and justice demands that no one shall escape those restrictions ... The urge for freedom, therefore, is directed against particular forms and demands of civilization or against civilization altogether (p. 72). If the loss is not compensated for economically, one can be certain that serious disorders will ensure (p. 75).

Freud here recognizes that given the terms set forth by modern society, it is economic compensation that, at least in a limited fashion and according to the terms of the totalizing logic of capital, alleviate some of the pains of living in this kind of society. As in his own case, it was not merely the rewards of scientific discovery that drove him, because those rewards are often only experienced in modern society in an individualistic fashion as pride; they are not a reward in and of themselves in this configuration of the social, rather it is the economic rewards of bourgeois life that result from the making of such discoveries that signal the level of success one has achieved and which enable the freedom to enjoy the ecstasies that life offers while ignoring the anxieties to a greater degree. Once those anxieties become immanent features of modern life, as in the case of World War I, even the economic rewards are insufficient to alleviate their presence. Recognizing this, Freud suggests a new project for those who would follow in his footsteps of revolutionizing the scientific enterprise in light of the changing socio-historical circumstances with the following question: "If the development of civilization has such a far-reaching similarity to the development of the individual and if it employs the same methods, may we not be justified in reaching the diagnosis that, under the influence of cultural urges, some civilizations, or some epochs of civilization—possibly the whole of mankind—have become 'neurotic'?" (p. 147). If we answer this in the affirmative, then it would require the analysis and diagnosis of the cultural super-ego of modern society and the undertaking would require the synthesis of Marx, Weber, Durkheim and Freud's perspectives to see exactly how and from what sources modern society has constructed its super-ego, how the structure of society influences and co-constructs the libidinal and death drives at the level of the social, what impact this has on us as individuals as we internalize these structural forces, and finally, what impact the *social unconscious* has on our planetary existence.

The call for a critical socioanalysis is precisely what is needed if this project is to take on the totality of human life in its social, cultural, economic, political, and technical dimensions. It is distinct from psychoanalysis in that its focus is not on the libidinal drive of the individual as a biological instinct, but rather more closely follows Freud's later shift in focus toward the death drive which recognizes how the social is internalized in a psychosocial dimension that also comes to regulate and enforce certain libidinal tendencies through the construction of a cultural super-ego. While both the libidinal and the death drive are present in the internal structures of the mind, they are also visibly mirrored in the social structures of modern society. The object of critical socioanalysis is not, however, simply to aid people in embracing desire to match the ways that bourgeois society channels acceptable releases of libidinal pleasure. Rather the object is to confront the ecstasy and the anxiety produced by modern society as it pursues the logic of capital through technological means, whose ecstasies are overemphasized in their production while the anxieties are dismissed by the wealth extraction principle. At the level of the social, capital knows nothing of the multifaceted nature of desire. It follows a pleasure principle that resides at the early anthropological—we might substitute the word mechanthropological here to illustrate that this is an evolution of technology and technique that follows a machinic logic—stage of its development, a stage when pleasure is overwhelmed by the needs of survival; even its reproduction is oriented to the present, not the future, and to its own survival, not that of the many species that call Earth home whose lives are subject to the paternalistic dictates of capital. The libidinal drive of capital is not a human drive, it is not one that has learned of pleasure for the sake of pleasure, of delaying gratification to increase pleasure, and it has not reached, and may never reach, the stage of development in which its survival is so assured that it can explore the pleasures of leisure. Simultaneously it has ignored raising to a conscious level its tendencies toward aggression and the death drive of modern society is deeply embedded in the social unconscious of the totalizing logic of capital. Whereas the death drive under the totalizing logic of religion was based on the subject only (as judgement for the afterlife is presupposed at an individual, not a social or institutional, level), the totalizing logic of capital is materially engrained in the whole of our planet, which it consumes as it carries out its mass coding of reality.

With this greater understanding of Freud's critical method, it serves us to examine in brief his understanding of anxiety, first, because it provides an example of how Freud's thought transformed over time in light of the socio-historical material reality due to his commitment to the critical method, and second, because he was the only one of the classics to study anxiety directly as a concept of central concern. The shift from Freud's biological to sociological approach in his psychoanalytic system is evident in the transformation of how he understood anxiety over the course of his career. This work began prior to his full development of psychoanalysis as a method. Starting in 1894 Freud wrote a draft paper called "How Anxiety Originates" (1954, pp. 88–94) in which he offers his most biologically oriented view. In this paper he does not venture far from one of the avenues of anxiety covered by Kierkegaard fifty years earlier ([1844] 2014), insofar as it relates to the linkage of anxiety with sexuality and the act of coitus.

Kierkegaard was writing in 1844 under the domination of the totalizing logic of religion, so he treats sex from the standpoint of procreation in line with the religious logic. In doing so, he links anxiety to the male in the sexual act, for "[i]n the moment of conception, spirit [in the Hegelian sense] is furthest away and for that reason the anxiety is at its greatest" (p. 88). On the one hand, this means that for Kierkegaard, anxiety arises at the moment when we are furthest from the place where we might consider our authentic self to be—that is, the idea of the civilized and modern human is most absent in the moment when we embrace our biological imperatives and transform momentarily into the bestial state of nature. On the other hand, using Freudian language, what Kierkegaard is getting at is that the moment of sexual release is the moment when the ego (while recognizing that this is not the seat of the 'authentic self,' but merely where the conception of it resides as the conscious idea of an 'ego-ideal') has suspended its function to the libidinal drive of the id. The subject is at that moment farthest away from himself as an individual, consumed by the animal nature of the act, and is, if only momentarily, lost in the process. The French use of the phrase, *la petite mort* (the little death), to describe the orgasm, reflects this moment of anxiety in which pleasure and pain are suspended in the deciding moment of whether the sexual act has come to fruition (be it orgasm or successful fertilization) or has ended in frustration (lack of orgasm or failed fertilization).[6] In the act of procreation this suspense is not merely toward the pleasure of the sexual orgasm, but hangs in suspense over the successful/unsuccessful fertilization of the egg and the inception of the zygote. "In this moment of anxiety, the new individual comes into being" (Kierkegaard, [1844] 2014, p. 88). If the process is successful, then the question of the male's role in the formation of a new life is now passed onto the female, leading Kierkegaard to conclude that "[i]n the moment of birth, anxiety culminates a second time in the woman, and that instant the new individual enters the world" (p. 88). Therefore, in this model geared toward the biological function of coitus, Kierkegaard places anxiety at the moment of conception for the father and the moment of birth for the mother, as he accords each moment to their function in the reproduction of life (ignoring the woman's anxiety of carrying the fetus successfully to term). Furthermore, in each moment (granting that the burden on the female of the sex in the process lasts considerably longer than the male's "moment") the question of the identity of the individual involved in the sex act is suspended as the question of becoming a parent hangs in the balance at both points.

Freud's, more modern, biological approach is rather focused on the question of sex as pleasure and his observation that "inevitably *coitus interruptus* practiced on a woman led to anxiety neurosis" (1954, p. 88). For Kierkegaard this would mean that the circumnavigating of the man's anxiety by foreclosing the possibility of conception suspends the question of assuming the parental identity as a result, leading, therefore, to the circumstance in which the woman could not achieve the release of anxiety in birth (he does not consider the release of the orgasm from the female perspective). Therefore, anxiety, lacking the means to be discharged, would turn into a "*hysterical* symptom." Freud, however, quickly rejects this notion because he discovers that this sexual anxiety occurs both in women who can and women who cannot experience sexual pleasure, leading him to conclude that anxiety must lie beyond "psychical events … in physical events" (p. 89). The similarity between the two psychical situations of the women lies in the lack of achieving orgasm, whether possible

---

6 See Bataille's (1985, pp. 235–239) essay "The Practice of Joy before Death," on this link of sex and death.

or not, because there is "a physical accumulation of excitation—*an accumulation of physical sexual tension*" (p. 90) and this "tension, not being psychically "bound" [as it would become in the achievement of orgasm], is transformed into—anxiety" (p. 91). In this way, Freud surpasses the religious idea of sex as only for procreation and that the only anxiety in the sexual encounter is one related to biological reproduction; now it is linked to the sex act as a libidinal desire centered on the notion of pleasurable release. The tension of not getting that release becomes anxiety, because as Freud defines it here: "Anxiety is the sensation of an accumulation of another endogenous stimulus—the stimulus towards breathing—which cannot be worked over psychically in any way; anxiety may therefore be capable of being used in relation to accumulated physical tension in general" (p. 93). Just as breathing cannot ultimately become fully subject to the whims of the psyche, neither for Freud can the moment of sexual release which he sees here as leaning toward a purely physical accounting.

By 1907, after Freud made the move toward the development of psychoanalysis, he makes a shift toward a more psychological explanation of anxiety. Here he links anxiety to repression, claiming that "during the process of repression itself anxiety is generated, which gains control over the future in the form of expectant anxiety" (Freud, S., [1907] 1959, p. 124). This move gives Freud's theory wider explanatory power than the limitation to the sex act itself, since according to the psychoanalytic view, sex is not the only release mechanism for the libido. What this means on the one hand is that the accumulation of libidinal tension can occur outside of the sex act in the general field of pleasure, and on the other hand, it emerges as a side-effect of repression. This new theory is applied in 1909 to the case study of a small boy, known as 'Little Hans', who suffered from a number of phobias (Freud, S., 2003). The boy becomes afraid of leaving the house around the same time that he has discovered the pleasures associated with touching his genital region, which Freud links to an increase in the affection he receives from his mother and which is complicated by her demand that he 'not touch himself.' With the phobia and the anxiety arising at the same time, Freud believes that "Hans's anxiety, which corresponds to a repressed erotic yearning, is initially without an object" (p. 19). Without having a known object upon which the anxiety can release itself, the libidinal tension grows, but even when an object is substituted for the one (an unknown object, which Freud hypothesizes is the mother as a part of the Oedipal complex) that has sparked the libidinal longing, "the anxiety remains ... and can no longer be transformed back entirely into libido; something holds the libido back in the state of repression" (p. 20). In other words, it is not merely the anxiety that must be dealt with, but it is the underlying repression, which left untreated simply recreates the anxiety anew. The authoritative force that encourages repression of the drive to pleasure is not raised to conscious thought, so it evades understanding and causes a situation of pain even when there are momentary substitutions that allow brief interludes of pleasurable release. What is learned from this case study is that "once a state of anxiety has been created, anxiety devours all other feelings; as repression takes its course and those once-conscious ideas to which strong feelings become attached move more and more into the unconscious mind, all the associated emotions may be transformed into anxiety" (p. 27). This is why Freud comes to place such a focus on anxiety in his career over the other emotions and affects that the patient may experience. This is not to the detriment of other emotional experiences, but because all emotions are potential carriers of anxiety and anxiety has the power to consume their impact and dominate the emotive experience of the mind.

The following year, in 1910, Freud continues to develop his theory that dreams are based on wish-fulfillment and he incorporates the concept of anxiety into this theory. He must at least mention anxiety here, because the anxiety-dream challenges the idea that dreams are wish-fulfillment, since they are experienced as torment rather than pleasure. Freud touches on, but does not expand on the theory here, merely stating that: "Anxiety is one of the ego's reactions in repudiation of repressed wishes that have become powerful; and its occurrence in dreams as well is very easily explicable when the formation of dreams has been carried out with too much of an eye toward the fulfillment of these repressed wishes" (Freud, S., [1910] 1961, p. 38). The logic remains the same: unable to consciously attach the anxiety to the specific object of its generation, the repressed wishes bubble up from the unconscious in ways that do not satisfy the anxiety with release, but rather amplify it because their true nature has been repressed, and so, the satisfaction, denied the object of its release, attempts to expel the anxiety on objects that do nothing but reaggravate the repressed wish. Again, the problem that Freud focuses on is that repression not only leads to anxiety, but it is what prevents anxiety from finding an object for its release. In his introductory lectures given between 1915–17, he summarizes these two positions on anxiety and distinguishes between fear and anxiety (Freud, S., [1915–17] 2012). "[A]nxiety," writes Freud, "is used in connection with a condition regardless of any objective, while fear is essentially directed toward an object" (p. 234). This confirms the notion that if anxiety is to have an object for its release, the object must be found in the repressed unconscious material, out of the unknown knowns, where it is linked primarily to the notion of pleasure (the pleasure is not guaranteed, but it suggested by the nature of the object as a possibility); whereas fear is a conscious process that knows its object and the object signals danger.

The book Freud finishes right after these lectures is *Beyond the Pleasure Principle* ([1920] 1990), which comes out in 1920 after the end of World War I and signals his move toward the sociological dimension. The War plays a major role in Freud's shift toward a greater examination of the death drive which, given the material reality of the War that he witnesses firsthand, necessitates a greater accounting of sociological material in his theories. The problem that the War presents for psychoanalysis is the issue of accounting for aggressive tendencies and, ultimately, the death drive as it manifested socially and played out across the globe, but especially in Western modern societies. Since the id makes up the major part of the unconscious and is guided by the pleasure principle, Freud cannot locate the death drive in the unconscious mind and must attribute it to the super-ego and the meditative function of the ego. Whereas fear is fixated on death as an object, anxiety is more closely linked to the repressed cathexis of death as it is fragmented in aggressive behaviors that ultimately flirt with the concept of death but deny it a place in the conscious mind as the motivating object. For instance, there is a cultural fixation on gladiatorial style entertainment in the United States, particularly in the sports of boxing and mixed martial arts, whose purpose is not to reach death, but to flirt with it by getting as close as possible to the object without proclaiming it as its focus. Death is repressed by the participants and the spectators, who view the matches and gain pleasure by the amplification of anxiety in the bout which denies that death is the object of beating an opponent senseless, therefore, the object of death is repressed by an anxiety (a reversal of Freud's earlier position) that refuses to transform into fear because the masses who consume this form of entertainment do not want to raise the specter of death to the conscious mind in the course of the event; they are fixated on experiencing the anxiety of the death drive while repressing the fear of death at the same time. What is repressed into the unconscious is precisely that fear of death, and what is held conscious is the anxiety. This is the logic that leads Freud to claim that the relationship of the ego "to the super-ego is perhaps the most interesting," because "[t]he ego is the actual seat of anxiety" ([1923] 1990, p. 59). The ego gives something up to the super-ego in the formation of anxiety "because it feels itself hated and persecuted by the super-ego, instead of loved" (p. 61). If to the ego "living means the same as being loved" (p. 61) then the ego is making an offering to the super-ego without giving in to its ultimate demands; the ego therefore accepts the generation of anxiety instead of succumbing to fear in exchange for the protection of the super-ego which it translates as love mediated by aggression. What this means is that "anxiety is reinforced in severe cases by the generation of anxiety between the ego and the super-ego" (p. 62), which signals a shift, not only from repression causing anxiety to anxiety causing repression, but also from placing the id and unconscious forces at the heart of anxiety to placing it in the relationship between the ego and the super-ego, which means that anxiety has a complex relation to the unconscious, the preconscious, and the conscious mechanisms of thought.

These new positions on anxiety are then expanded into a wider theory of anxiety in 1926 with Freud's publication of *Inhibitions, Symptoms and Anxiety* ([1926] 1959). In this book Freud spells out his full reversal of his early position, claiming now that "[a]nxiety is not newly created in repression; it is reproduced as an affective state in accordance with an already existing mnemic image" (p. 12). In other words, the seat of anxiety is established before the emergence of the super-ego and the process of repression; anxiety, for Freud, now precedes repression as something that emerges in the "primal repressions" which are the repressions of nature, not those of social authority produced by the emergence of the super-ego. Anxiety appears in this text to be far more closely aligned with Freud's definition of fear—"Anxiety is a reaction to a situation of danger" (p. 57). Anxiety "is obliviated by the ego's doing something to avoid that situation or to withdraw from it ... [S]ymptoms are created so as to avoid a *danger-situation* whose presence has been signaled by the generation of anxiety" (p. 57). What this means for our purposes is that when the danger situations created by the totalizing logics of modern society are made manifest, in the attempt to stave off the anxiety produced by them, there is a new generation of pathological symptoms that arise (taking the form of alienation, anomie, and instrumental rationality). This is a particularly troubling conclusion, because it suggests that the root causes of the anxiety are prone to be ignored as the symptoms of, in this case, alienation, anomie, and the spirit of capital, overwhelm us. The root cause of the symptoms is neglected in favor of treating the symptoms superficially in the system of capital, for instance, through economic rewards. Rather than answering our questions as to how to treat this anxiety, then, the understanding of anxiety only produces more questions given the "*gravity concrete and specific socio-historical conditions and circumstances exert on endeavors*" (Dahms, 2017a, p. 48) taken by subjects in modern society to improve our lives or, at the least, to try to understand what problems prevent us from making improvements.

In Freud's final direct treatment on the subject of anxiety, in 1933, he summarizes the problem of anxiety in all its dimensions across the structures of the mind:

> Thus the ego, driven by the id, confined by the super-ego, repulsed by reality, struggles to master its economic task of bringing about harmony among the forces and influences working upon it; and we can understand

how it is that so often we cannot suppress a cry: 'Life is not easy!' If the ego is obliged to admit its weakness, it breaks out in anxiety—realistic anxiety regarding the external world, moral anxiety regarding the super-ego and neurotic anxiety regarding the strength of the passions in the id.

FREUD, S., [1933] 1989, pp. 97–98

And he makes clear the two major transformations of his theory of anxiety as it appears in its final form: "first, that anxiety makes repression and not, as we used to think, the other way round, and [secondly] that the instinctual situation which is feared goes back ultimately to an external situation of danger" (p. 111). What this means, is that the more anxiety we have, the more we will be compelled by the structure of our mind to repress confrontations with the sources of our anxiety, and that anxiety crosses the psychosocial divide, precisely because it is a response to something external to us, something out in the social reality which provokes the 'instinctual situation' of anxiety and psychical repression. The agenda of socioanalysis, therefore, must be concerned with uncovering how the social structures of modern society feed into the production of these anxieties as outlined here by Freud. However, it is not simply enough to take Freud's theories and apply them to current circumstances, as made evident by the transformation of Freud's thought over time. As new social conditions produced new forms of anxiety that then rose to the surface and complicated his theories, they necessitated a self-critical approach that changed the theories to explain the new phenomena. After spending a career examining these forces, Freud writes that he "cannot promise that [the question of anxiety] will have been settled to our satisfaction, but it is to be hoped that we have made a little bit of progress" (pp. 115–116) at least to the point where he could claim that "if we look at it purely psychologically, we must recognize that the ego does not feel happy in being thus sacrificed to the needs of society" (p. 138) which means that the psychological problems are sociological problems and vice versa. The social demands for aggression and pleasure must be dealt with if we are to find a rational basis in modern society upon which we can strike a balance between our psychological anxieties and ecstasies in a life that satisfies the needs of the individual as well as those of civilization.

Freud pushed the concept of anxiety as far as he could in his lifetime and opened the door for a psycho-sociological investigation of this material concept from the standpoint of a synthesis of psychoanalytic and sociological perspectives. Given the socio-historical circumstances of his own life he experienced the bourgeois ascendance and the explosion of the dark side of modernity in World War I, and the rise of the Nazi power prior to the start of World War II which necessitated his immigration to London where he died in exile.

He was only able to escape because he had entered bourgeois life and enjoyed the comforts it had to offer (such as, smoking cigars, which ultimately caused the cancer that killed him), but he was not an apologist for the mode of social organization that allowed these dark forces to rise up in threat of the social order precisely because they were produced by the same forces that granted him the benefits of that class lifestyle and made them worthy of critique. Rather, Freud's example, like those of Marx, Durkheim, and Weber, is a lesson in the benefit of employing the critical method. The Freudian contribution furnishes the socioanalytic toolkit with talk therapy as a means of interrogating the psychosocial divide and for subjecting the anxieties and the ecstasies of life to a rigorous and systematic analysis so as to better understand how rational thought gets warped by irrational impulses and threatens the good life for individuals and societies, especially as those societies encourage the repression of conscious thought on the causes of our social ills. With this final piece of the puzzle gained from the classics, our next task is to understand precisely how the individual and the social continued their modern transformation in the 20th century, and how the critical method evolved to keep pace with the changed material circumstances in which we now find ourselves.

Conclusion to Part 1

A final two questions must be raised here in conclusion of Part 1: (1) if the psychoanalytic enterprise made these discoveries about anxiety and recognized that the social dimension increasingly played a larger role in their development, why should we make a shift from either sociology or psychoanalysis to a new field of critical socioanalysis? And, (2) should the treatment of anxiety not be left to the medical profession which has already staked its claim on this turf? At the current point in this study, we must content ourselves with partial answers by way of explaining why sociology did not move in this direction in the first place and what Freud's own view was on the continued development of psychoanalysis.

Since Marx wrote before Freud, psychoanalysis had not yet been developed and so it was not an avenue open to him for synthesis with his perspective. Furthermore, Marx was not a sociologist and he was right to be skeptical of the positivistic sociology advanced by Comte and Spencer that dominated any discussion of the science in his lifetime. At the time Marx was writing the individual was still in its infancy. People were still testing the boundaries of the possible, in terms of their development as individuals, that had opened briefly in the enlightenment but were, by Marx's time, rapidly closing again with the spread of capital, without realizing precisely what avenues for the development of authentic individuality were being shut down as a result of that spread. So, the bourgeois individual was not fully fashioned yet by bourgeois society, just as bourgeois society was not yet fully fashioned by these new bourgeois individuals in the insidious feedback loop they were building, as exposed by Marx. Given those sociohistorical circumstances, Marx was only able to examine how that mode of social relations emerged historically and what effect its logic would have on those whose lives it transformed, both in terms of social conditions and psychological affect. We may draw the conclusion that the historical material reality of Marx's time did not yet provide a fertile ground for either a practice of psychoanalysis (which only emerged as a part of bourgeois development with Freud), and therefore it was especially not fertile for the development of socioanalysis which is a recombinant method built upon both psychoanalytic and sociological knowledge which were lacking in Marx's time.

Although there is some historical overlap between Freud's work and that of Durkheim and Weber, their attention was on building a critical sociology that could stand as its own discipline with its own object of study. This meant drawing disciplinary boundaries between sociology and its scientific neighbors, notably psychology and economics, so that it would be distinguished in method and object from the work going on in those areas. However, their research also demonstrates that sociology was not uninterested in psychological and economic questions, rather it broadened the horizon of the questions that those disciplines could ask by recognizing the power that the social had on shaping both our internal structures and those of modern society as a whole. In doing so they both demonstrated the need for the sociological perspective in the study of modern social ailments found in all domains of modern life, be they cultural, political, economic, social, psychological, etc. At their time the most visible forces guiding the production of society and the individual were found in the passing of the torch from the totalizing logic of religion to that of capital, which explains their focus on affairs both religious and economic in the shaping of modern peoples and modern societies. The question as to whether the organization of modern societies could be affected at the level of the social or at the level of the psyche was still unanswered, although Weber clearly believed that it could not happen at the social level without the appearance of unsurmountable material barriers which would cause enough psychological shock to necessitate a rational reexamination of the organizational principles of modern society.

Freud was less preoccupied with drawing boundary lines on psychoanalysis than he was in developing it as a critical scientific method that could theoretically account for the hidden structures of the human mind and the persistence of irrational behaviors given the elevation of rational thought and reason in the enlightenment as the necessary guiding principles of, and method for, human organization. Freud did, however, recognize the dangers of letting those who had only a superficial understanding of psychoanalysis claim to engage in the psychoanalytic method. In order to practice psychoanalysis, one must "be well versed in its technique" and this technique "is best learned from those who have already mastered it" (Freud, S., 1912, pp. 205–206). There are two reasons for this. First, "a psychoanalytic intervention assumes from the beginning a longer contact with the patient" (p. 205). It cannot be rushed and does not fit into the 'fast-food' model of doctor/patient interactions that have gotten so popular in the age of corporate insurance control over medical decisions, wherein the goal is to make a diagnosis as quickly as possible so as to make use of psychopharmacological tools to treat the symptoms and get the doctor onto the next patient as soon as possible to maximize billable hours and patient flow. Psychoanalysis does not make use of those tools—and it ignores the modern capitalist logic of speed—rather, the diagnosis and treatment are undertaken at the same time within the psychoanalytic intervention, and this requires patience and time, and must be learned by those who have already come to understand that this is the case through trial and error, the road which was paved by Freud. Second, failure to learn this from those who mastered the technique, would require a continuous reinvention of the method, which would demand "sacrifices of time, effort and success" (p. 205). This is why it is important to work our way through Freud, in addition to the sociologically minded classics of the critical method, if we are to practice critical socioanalysis; for although his intention is different with its singular focus on the psyche, it is his work that laid the foundation for talk therapy which is borrowed by the critical socioanalytic enterprise.

However, Freud's argument above is directed at other medical professionals, those who already believe themselves experts and specialists of psychological knowledge, but who, remembering above, Freud claims have nothing special in that knowledge to bring to the understanding of hysteria, and who, in following the medical model as it is constructed by capital are prone to take shortcuts that fit with its framework of practice, not the analytic model. Socioanalysis, unlike many current forms of psychoanalysis that have deviated from Freud's teachings, does not claim to be a medical practice. It does not claim to offer any sort of 'cure' for the social ills that manifest themselves as anxieties in the individuals it treats. Rather, it is more akin to a process of critical reflection, in which the socio-analysand learns how to do the work that Mills claimed was central to the development of a sociological imagination: linking the personal to the public, the biography to history, and ultimately our thought patterns, attitudes, and emotions, to the social structures that constrain our lives for good and bad. The goal of socioanalysis, is akin to the goal of psychoanalysis, in that it wants to transfer the desire of the critical method to others so as to agitate the places where the totalizing logics of modern society have calcified in the mind of the individual. If it is successful in this process, what it promises is not a cure for anxiety, but a means of understanding what the anxiety is signaling to the person and how this relates to the social structures that guide their life, so that, if cracks ever appear in that social structure, their anxiety can find its object and trigger the actions that are currently unavailable to them for its release. Freud warned against the full on medicalization of psychoanalysis precisely because he thought that the case may be "that in this instance the patients are not like other patients, that the laymen are not really laymen, and that the doctors have not exactly the qualities which one has a right to expect of doctors and on which their claims should be based" (Freud, S., [1926] 1978, p. xxviii). Freud's challenge against doctors claiming mental health, and in the case of socioanalysis social health, as their domain appears ludicrous in a situation in which "[n]othing takes place between [the analyst and the analysand] except that they talk to each other. The analyst makes no use of instrument ... nor does he prescribe any medicines ... The analyst agrees upon a regular fixed hour with the patient, gets him to talk, listens to him, talks to him in his turn, and gets him to listen" (p. 6). That is the whole of the analytic session. What greater understanding of the complexities and varieties of life can a medical professional offer in this setting, since their training orients them toward a view of the patient as an object, not a subject with a unique psychosocial reality? "[A]nalysis," like society in the Durkheimian sense, "is a procedure *sui generis*, something novel and special, which can only be understood with the help of *new* insights" (p. 9). These new insights are both produced in the analytical session and by the unique life-histories brought into the session by the analyst and the analysand. Gaining the widest possible understanding of the human experience across modern societies, and of the various language games that proliferate in those societies, is the first necessary precondition to becoming a socioanalyst. Medical knowledge, as such, is therefore not a necessary or needed requirement of socioanalysis given its method and its stated goals.

Two things are required, however, for being a socioanalyst that parallel Freud's requirements for being a psychoanalyst. First, "everyone who wants to practice analysis on other people shall first himself submit to an analysis. It is only in the course of this 'self-analysis', when they actually experience as affecting their own person—or rather their own mind—the processes asserted by analysis, that they acquire the convictions by which they are later guided as analysts" (p. 20). This requires a full commitment to understanding how we are ourselves alienated, guided by the spirit of capital, and subject to anomic forces, and how these have shaped our personal biographies. The process is often a painful one as it demands that the person who undergoes it is committed to uncovering every element of themselves that is shaped by the social, and ultimately in recognizing the painful truth, that our lives are not our own, that we are not as free as we often imagine ourselves to be, and that opportunities and choices available to us are at times foreclosed precisely because of our personal position in society regardless of our will, ambition, and abilities. This also demands a reckoning with how, in our own lives, we have acted as a tool of the social structure to limit the freedoms, choices, and individuality of others. Second, since socioanalysis is geared toward a critical understanding of the effects that social structures have on shaping our thoughts, the analyst must first and foremost have learned the critical method of sociology, the critical theories of society, and how to recognize the latent force of the social unconscious and the effects of the cultural super-ego; especially as the totalizing logics of society use us to reproduce conditions of alienation, anomie, and the spirit of capital in ways that threaten the future of modern society and the well-being of ourselves and others. It is to this later task that the current text is oriented for would-be socioanalysts.

To answer the questions posed in this conclusion in a more satisfactory manner, we will have to continue on in our next section where we examine the changed historical situation of the 20th century, after the lives of the classics of modern social thought had ended, and the development of the critical method into a program known as 'Critical Theory' by the members of the Frankfurt

School in the late 1930s. It is during that stage in history when the forces of capital and the production of new technologies come to play an even greater role in the production of our anxieties and our ecstasies, and in the process further confuse the two as alienation, anomie, and the spirit of capital, transform alongside material reality producing new means for domination and repression both in and of the mind and society in general.

PART 2

Technology, Modern Wars, and the Rise of Consumer Culture: The Frankfurt School Revisits the Critical Method

∴

Introduction to Part 2

The concepts and symptoms outlined by the classics of the critical method had to be transformed to keep pace with the dynamic nature of modern society as it continued its material development economically, politically, culturally, technologically, and to a lesser extent, socially in the 20th century. As evidenced in Freud's own evolution on the topic, the way that the affect of anxiety came to manifest itself in the 20th century was also transformed by the rapid pace of change effecting all spheres of society. As such, thought on anxiety as well as its diagnostic criteria needed to be continuously updated to incorporate new dimensions. An increasingly heavy reliance on technology in modern society, fueled at a torrential pace by the logic of capital, was behind many of these changes. Anxieties in Marx's time centered on the difficulties of beginning and adjusting to life under the capitalist mode of production. By Durkheim, Weber, and Freud's time, they largely emanated from an increase in psychological repression and social anomie caused by the internalization, or lack thereof, of ethical behaviors that aligned with the spirit of capital rather than the spirit of humanity. However, by the mid-20th century anxiety became infused in the production and technological transmission of a mass culture that channeled the spirit of capital to new heights and stratified individuals along new lines of inclusion and exclusion.

Although Marx's focus was on the structure of capital and the effects of political economy on modern life, he recognized that one of the primary ways that capital effected the social and our modes of thought, was technology. He wrote, "Technology reveals the active relation of man to nature, the direct process of the production of his life, and thereby it also lays bare the process of the production of social relations of his life, and of the mental conceptions that flow from those relations" ([1867] 1990, p. 493). Technology plays a crucial role in how we think, how we live our lives, how our emotive experience of that life is translated, and how anxiety changes its shape and function. Marx came to this conclusion by witnessing the transformational effects of the spread of technology in his own life.

Railroad development accelerated after his death, but Marx lived to see its expansion across Europe and the United States. He also witnessed the invention of the telegraph in 1837—which connected countries, cities, and towns across the old and new worlds—the photograph in 1839—which allowed images of space to be frozen in time—and the telephone which was patented in 1876 just a few years before his death—which came to directly connect homes and businesses to each other (in our contemporary milieu with the cellular phone, it connects individual bodies to each other) (Casson, [1910] 1922).

Industry developed automation machines—both those powered by humans and animals and those powered by artificial means—transforming the process of labor. These industrial technologies especially drew his critical gaze as they were largely responsible for ripping people from the fields and relocating them to the factories, altering human relations in the process. The life altering effects were further amplified by the development of techniques for managing and disciplining those laborers to function in accordance with the capitalist mode of production and to get the most out of that labor by pushing the limits of efficiency and productivity through the control of time and motion.

Despite the vast improvements in productive efficiency as novel techniques structured human labor in more rational ways and machines contracted space in time by shrinking the temporal distances between sites of extraction, production, and consumption, these technical forces were major new sources of anxiety. They caused interruptions in daily life that had previously not existed and now exerted a degree of control over peoples' actions as they instilled new behavioral patterns whose obedience or violation was subject to new sanctions. Weber, for instance, wrote of "the first 'shattering' event of my life: the train derailment ... the sight of a locomotive ... lying like a drunkard in a ditch" (cited in Radkau, [2005] 2009, p. 12). Having wittnessed this as a child, the memory traumatically effected him as an adult. His biographer Radkau also reports that the "telephone rang constantly in [Weber's mother] Helene's house ... a still novel disturbance of the peace which ... used to drive Max ... mad" (p. 16). Often the anxieties caused by these technological intrusions manifested as repeated outbreaks of anomie because the technical innovations often directly eroded the skillsets of human labor, making many redundant and superfluous to the production and reproduction process.

The problems of scale, the limitations of human ingenuity, and the applications of technology to the development of yet more technology, further added to the increased feelings of uncertainty and anxiety about the future. By the late 1880s photographs were put in motion with the invention of cameras that could capture motion and by the 1890s the Lumière brothers were leading the way in developing film technology and bringing motion picture shows to the masses (Serban, 2016). By 1893, Nikola Tesla "made the first public demonstration of the radio" and soon entertainment and news could be piped directly into peoples homes in real time (Federal Communications Commission, 2003-2004, p. 1). Weber experiencing these technologies as disturbers of peace and quiet was symptomatic of many who first encountered them. Increasingly, however, it became impossible not to invite these new machinic devices into their personal spaces; leaving the factories and offices people returned home to an environment made up of yet more machines. Additionally, industrial technologies created the conditions for artificial disasters. New avenues for tragic accidents—caused by loss of human control, failures of their predictability, their unprecedented speed, and the cultivation of dependency on their successful functionality—increased as businesses were forced to adapt to these technologies if they hoped to stay competitive in exchange markets. News of such disasters even reached the ears of those who claimed to maintain the old lifestyles in rural settings as they tuned the radio dial to hear of the world's happenings, which were too tempting a curiosity to ignore.

The accelerating pace of these material changes appeared to many as either genuine progress or as a series of obstacles waiting to be overcome, rather than as a structural feature of the constitutional logic of a modern society dominated by the totalizing logic of capital. Even for Freud it was not until World War I that he began to recognize to a greater degree that the pathologies he encountered in his patients' lives could not be fully explained without accounting for the increase in social pathologies occurring across the modernizing West. Freud recognized, when commenting on the impact of technology in his own life, that the introduction of technology necessitated further technologies as they transformed the social landscape and created new barriers to individual freedoms at the same time that they broke down the barriers of traditional life and eased some of the challenges associated with that mode of social organization. However, at the same time sociology as it continued to mature as a science tended in its mainstream variant in the direction of the positivist approach to the study of society, rather than in keeping with the critical approach of the classics. The effect was that those who engaged in sociology as a mainstream practice did not recognize the dialectic of the social engine that technology and capital represented, and how, as Marx phrased it, the bourgeoisie continuously revolutionized both the means of production and as a consequence the whole of our social relations.

It was precisely because critical forms of sociology accounted for the fact that they were necessarily always behind the curve of modern social transformations that the institutions of modern society could not recognize the use-value of the practice. To justify and rationalize a use-value for the capitalist system was to align the practice with the eclipsing of the social by the totalizing logic of capital. While practitioners of the critical method sought to avoid this fate, mainstream sociologists made use of their instrumental rationality and had few qualms about institutionalizing a form of the science that aligned with the furtherance of that logic. As Daniel Bell ([1960] 2000) claimed, modern society did not develop as an intellectual society, but rather as a business society. The result was that academia changed in kind as it sought relevance in a society that did not value contributions that could not immediately translate into use value as defined by the business class. Even as mainstream sociology adapted to the demands of the status quo in the 20th century it still struggled to gain relevance as a modern science because it could no more keep pace with modernity than critical sociology. Rather than address this fundamental problem, the mainstream versions of sociology succumbed to societal pressure to produce reports that aligned with the constitutional logic of modern society. To do this, its practitioners had to describe reality as if it was static and, therefore, they operated under the implicit assumption that either change was controlled and unidirectional or change would wait until it could be rationally orchestrated once their descriptive work was complete.

By becoming a science that was more dedicated to the mere description of reality, rather than the critique of it, much of sociology came to embody an idealistic approach to history; especially as it related to the power of capital and technology in shaping modern society. Toward the mid-point of the 20th century those German scholars, who came to be known as the Frankfurt School, challenged this dominant paradigmatic approach in the discipline. They saw that the conditions in modern society were in fact not advancing socially in the progressive manner that its apologists proclaimed, but rather each advance contained the seeds of its own demise. As the dark side of modernity expanded its reach in ever more manifest and latent ways that shunted the growth of the social, that darkness erupted in the most horrendous atrocities of war, economic crises, the continued tragedies of nature, and the new tragedies arising from the 'accidents' of artificial interventions.[1] Recognizing the need to keep the critical tradition alive for future generations, they set out in the spirit of the classics to update the diagnoses of modern society to reflect the changed material circumstances. On the one hand, this required a synthesis of the classical perspectives and an analysis of the ways that society had come to accommodate these changes through overt mechanisms of discipline. On the other hand, they had to pay increased attention to the more insidious latent forces of control that played on the desires of humanity to maintain compliance and complacency. Specifically, this latter development demanded a thorough understanding of how mass culture successfully sold a conformist ideology and bought obedience in exchange for consumerist comforts without acknowledging the individual and social costs.

While they did not take the classics as gospel, recognizing that those theories were appropriate to specific times and spaces, they paid careful attention to the way that the critical method was developed as a tool for tracking

---

1 On the 'accident' see Virilio & Lotringer, 2002 and Virilio, [2005] 2007.

the totalizing logics that guided the structuring of society and would continue to do so as the modernization project forged ahead. Since the social was not developing in the same manner as the other major spheres of influence in modern society, they could not invest their hopes in the social as a force for positive change in the same way that the mainstream variants of sociology often proclaimed. Rather it was their task to find out how these other spheres had come to stifle the growth of the social and why its progress was stagnating in modern societies. Before diving into the scenic landscape of the Frankfurt School and their contribution to the critical method, a little more stage setting is appropriate in this introduction to explain why this is an important component in the development of critical socioanalysis.

As the original developers of the critical method the classics of modern social thought were concerned with modern society insofar as the material reality of everyday life in those societies was qualitatively different than that of the premodern human. It was their goal to explain the differences of this mode of life by comparing them to what was lost, what was gained, and what the opening of this historical possibility for transformation could mean for the future of humanity. Sociology was to be the rational and systematic study of modern society because the modern individual witnessed first-hand how the application of rational thought to the problems of nature could reshape reality. They experienced new benefits that arose from these artificial interventions and inventions and came to recognize that the old ways of thinking about reality that had served the human species for millennia were no longer adequate for navigating the changed circumstances they faced in everyday life; but with those new benefits new problems arose. These problems were examined in their depth by the classical theorists who developed the critical method, as outlined above, and for critical sociology and psychoanalysis the problems led to several visible symptoms which took the form of alienation, anomie, instrumental rationality as the spirit of capital, and increased psychological repression, all of which can be traced to the anxiety that is produced by the organizational structure of this modern mode of life. However, since modern society is founded on a dynamic mode of social organization, as the totalizing logic of capital spread out across the Western world and eventually encompassed the whole globe in the 20th century, the problems that were tied to the structures of modern society, their sources and symptoms, were also subject to that same dynamism, which means that they too changed their shape, appeared in new forms, produced new symptoms, and were reproduced as structural features across all areas of modern development. This is why the critical method was developed, so as to give social scientists the tools to study a dynamic reality.

The political economist Harry Braverman ([1974] 1998) framed part of this problem in relation to the increased technological embeddedness prevalent in modern societies:

> The evolution of machinery represents an expansion of human capacities, an increase of human control over environment through the ability to elicit from instruments of production an increasing range and exactitude of response. But it is in the nature of machinery, and a corollary of technical development, that the control over the machine need no longer be vested in its immediate operator ... What was merely *technical possibility* has become, since the Industrial Revolution, an *inevitability* that devastates with the force of natural calamity, although there is nothing more 'natural' about it than any other form of the organization of labor. (p. 133)

Surely, if humanity could rationally tackle the problems of nature and alter them with science and the development of technical solutions, then there must be benefit in a science whose object was the source of these transformations which would focus on the problems pulled in the wake of the creation of an artificial, socially shaped and constructed, reality that came to replace the "natural" order. Alternatively, the more humanity came to learn about the external world they inhabited, the more unbearable the realization was that they knew so little about themselves and what effect that technological embeddedness had on their lives as they were guided by the totalizing logic of capital. The problem that these sciences faced was just how easy it was to confuse the artificial order with the natural order and the tendency to think that the same methods for subduing the natural world would work on the "second nature" of the artificially constructed modern world. As Braverman illustrated, the problem is that these technologies are increasingly harder to control, frequently assume the managerial role as they take on a life of their own, come to control the operators rather than the other way around, and are fictional constructs made material by the species' drive for domination over nature, and therefore, over humanity itself. What this means, then, is that the structural processes exposed by the classics became ever more embedded, that is, hidden and cemented, in our material existence, making our actions ever more regulated and dominated, regardless of thought or the ways that the power structures—dominated as they are by their own technologies—presented them as, and often convinced themselves that they were, advances of "freedom." The problem was exacerbated by the capitalist control of technological development because it was in their favor to promote the attitude that Kant had already exposed in 1784 as symptomatic of the anti-enlightenment: "I need not think, if only I can pay—others will readily undertake the irksome work for me" (2007, p. 29).

If the critical method is first and foremost a method to encourage thought to recognize the contradictions between itself and reality, and critical thought on the effects of humanity's artificial interventions is precisely what is needed as a precondition to understanding and potentially altering the conditions and organization of modern life, then it runs counter to the trends of modern society which is structured precisely in ways that discourage critical thought. This is especially complicated when the demand is that we think self-critically to root out the most intimate contradictions to challenge the assumptions we have about our identity and engage with the uncomfortable truth that our modern existence is more fragile, more damaging, and more controlling of our actions than it claims to be on the surface level which emphasizes the freedoms of consumerist choice while ignoring the structural unfreedom of modern life itself. In other words, the very logic of capital guiding modern society is one based on its own survival and its own needs, and since it feeds vampirically on the human elements that fuel its growth, the self-interest found in that logic is one which structurally attempts to negate any modes of thought that would threaten its existence. However, whether this is best accomplished through authoritarian means or through a mode of fetishized exchange in "free" societies, was yet to be determined and the experiment was to be performed in the 20th century.

Sociology and psychoanalysis were historically founded as sciences that had the potential to illuminate a method for developing "critical reflexivity with regard to the constitutional logic of modern society that the latter is prone to discouraging" (Dahms, 2018, p. 160) and the effects that logic has on thoughts and actions of people in that society. If the classical theorists could be said to offer a hope for the future, then their hope was placed in the enlightenment tools of reason and rationality (while also recognizing the limitations of those tools) and the possibility, if not the probability, that humanity could eventually learn how to wield them in a manner that would allow them to act socially on the problems facing the world. The precondition for this would be recognizing how the structure of modern society lent itself to the production of certain pathological trends in the social and reproduced in the individual, which amplified the dark side of modernity when those societies uncritically accepted and integrated all the modern world had to offer without evaluating the associated planetary, social and individual costs and potential for deepening inequalities, destruction, and conditions of unfreedom (Alexander, 2013).

Since the underlying goal of the critical method could be expressed as the advance of freedom, the authoritarian means of control could not be the solution, but neither was the Western democratic method of self-governance—entwined as it was in the political economy of capital—proven as a sure means for paving a path forward to meet the goals of genuine freedom for individuals supported by the power of the social. By Freud and Weber's time the focus on developing industrial technologies expanded to include institutional techniques based on the instrumentalization of psychology. The goal of these techniques was to code the mind to conform to a predetermined set of behaviors in an increasingly mechanistic fashion. These techniques were no less prevalent in countries operating under the banner of democracy than they were in authoritarian regimes. Rather than encouraging thought and critical reflexivity, modern society doubled down on the machinic logic and the structuring of the individual's mind against critical thought, calling into question the faith in reason and rationality that had appeared so promising in the enlightenment and continued, albeit in diminished fashion, to shine in the early period of the modernization process.

As the totalizing logic of religion collapsed and the grand narratives of beginning along with it, the totalizing logic of capital began to reinforce the future-oriented disposition of the moderns. The future was dangled in front of the masses as a carrot in exchange for obedience and the continued suffering of their present symptoms of repression, alienation, and anomie, with the spirit of capital as the primary driver of their instrumental rationality. The ascendance of the bourgeoisie, their overturning of the aristocratic order and transformation of political economy—including the means and modes of production—were held as proof that the carrot was real, but other than dangling the carrot to gain the self-policing obedience of the workers there was no inclination to structure the logic of capital in a way that would facilitate the sharing of power with the laboring masses. The problem was that the future was uncertain, no less than the past was unstable. Traditional life was presumed to be rigidly fixed in a static past, but it was only accessible through cultural artifacts which were in turn subject to interpretation and revision, ideological bias, and incomplete and missing data. As the literary critic and historian Hayden White would treat it, "history didn't exist as an object but as a concept, something that could only be accessed through, and reconstructed as, narrative" (Elias, 2018). Although it was a material fact, history could not rest, even after it became relegated to the past, supposedly beyond our ability to intervene in that material order. Poked and prodded, the corpse of history was repeatedly dragged from its grave and hoisted upon the autopsy table by critical social scientists who rejected the modernist proclamation of "no past," as an attempt to ignore the lessons that could be gleaned from it, and who were eager to understand why modern society assumed the form it did, why they were born on this path of modern society under the dictates of the bourgeoisie, and why, despite evidence of temporal malleability, it was so difficult to steer the course of history using the navigational charts provided by reason and rationality (assuming, of course, that the charts were even accurate). They were interested in history precisely to the extent that they were interested in the future. If the future appeared to be foreclosed by the 20th century, then anxiety became the ever-present affect and motivated the turn to critically examining the prehistory of this modern constellation.

Critically oriented social scientists could only do this by examining the cultural artifacts that continued to speak the narrative of history long after the storytellers who lived it had died. Not only did these autopsies reveal some of history's secrets—such as the tools power wields, how it wields them, and how those spheres of power are maintained—but the narratives were far more alive than the corpse of the past appeared on the surface as the bourgeoisie sought to control the narrative of history to their continued advantage. The cultural artifacts reflected the socio-historical conditioning of those who recorded and lived it; meaning that the manifest content of the narratives could not be taken at face-value. Rather it was the latent content that needed to be revealed by reading the narratives in their historical context as a form of immanent critique. Those who followed in the critical tradition, often turned to Hegel, who "provides a critical analysis of reification (as objectification) aimed at demystifying the *human* construction of history. It is an *immanent critique* because *its critical standards are ones given in the historical process*" (Antonio, 1981, p. 332). They then supplemented Hegel with Marx, looking at history as a dynamic process of transformation which contains within that dynamic "the basis for an immanent critique that turns the treasured values of bourgeois ideology against the unfreedom, inequality and misery of developing capitalism" (p. 334). For Adorno, the task is "[n]ot: to confront capitalist society with a different one, but: to ask if society conforms to its own rules, if society functions according to laws which it claims as its own" (2018b, p. 5). In other words, if the bourgeois order claims that the future is always open because there is an open system for the advance of any individual through the sale of one's labor power, why does the system still, in its advanced stage, tend toward a social reality composed of the "haves" and the "have nots"? By performing an immanent critique in the way proposed by Adorno, history could be used for the critique of the present, not as a tool for usurping and maintaining power by a new class of intellectual elites at the cost of keeping the masses submerged beneath a veil of ideological ignorance, but as a tool for enlightening the masses to the hidden features of the social structure that prevents them from achieving a life of social emancipation and authentic individualism, but only if and when the material conditions are such that the masses become willing to undertake the irksome business of engaging in social and self-critical thought.

While Marx was a necessary addition for understanding the logic of capital, his work was not sufficient for understanding the full range of its totalizing effects on the transformations of modern society that proceeded along with the development of that logic: the attacks on reason, the psychological desires of the masses which were subject to cultural exploitation, and the persistent acquiescence of the exploited and the rejected to the various systems of domination that proliferated across modern societies in countless forms. To understand the totality of modern society and to keep tracking the impact that it had on the minds of modern individuals, Marx's work had to be synthesized with that of Weber and Freud (and to an extent, Durkheim). This was precisely the program implemented by the members of the Frankfurt School—most notably by the director of the Institute for Social Research, Max Horkheimer, and the social philosophers Theodor W. Adorno and Herbert Marcuse—examined below. Socioanalysis was not the result of their synthesis, instead psychoanalysis was largely integrated through its theoretical insights, not its method, and was used to supplement the structural theories of society with the psychological dimension provided by Freud's insights to give a more complete view of modern society in terms of the totality of its effects. Only by this route could history be used as a material basis for the construction of a critique of society that continued to track the evolution of the social and the individual, and the effects that modern society had on the minds and thought processes of its subjects.

Sociology and psychoanalysis did not, however, simply offer a new method for the study of history in order to explain the past. Rather, as indicated by Marx's method of immanent critique, they were at their core developed as future-oriented sciences. Although necessarily grounded in history, their ultimate orientation is toward the future because the point of having a science of the social, a science of modern life, is ultimately to be able to guide that modern life in ways that are favorable. The philosopher Alfred North Whitehead spoke of the need for this orientation, claiming that "[i]t is the business of the future to be dangerous; and it is among the merits of science that it equips the future for its duties" ([1925] 1948, p. 208). The critical method, which embodies this belief, must, therefore, never take its eyes off the ever-changing material conditions in which it operates, because at each moment history changes, those changes have effects that ripple out in the future yet to come. Each new feature that modernity installs in society both rewrites and reinforces the code of the totalizing logics that push the development of these new features. In order to focus on the future, then, the method of these sciences was comparative and historical, but also grounded in the concrete material reality of the elusive present, self-critical, perspectival and scalar, so as to be able to track the various paths of possibility that opened and closed as each new feature of society revealed paths not yet discovered and made others obsolete, whether explored or not.

The significant difficulty of having a science oriented to the future is that the future is often seen as a repository of ideological thought. The trap of a positivist sociology, which the mainstream version of the science increasingly came to follow after the time of the classics, is that it often either imagines the future as an indefinite continuation of the present or sees "a progressive future defined in terms of the extension of liberal democracy and social welfare, and the power of technoscience to meet human needs and wants" (Tutton, 2017, pp. 478-479). Richard Tutton, borrowing from the Science, Technology, and Society literature, proposes instead that we think of the future, neither as an imaginary concept in the sense that it would form an ideal, nor as a static continuation of the present, but as a material reality in which "matter and meaning are entangled with each other" (p. 486). Since thought cannot directly access matter as the thing-in-itself, but only the semblance of matter, as Adorno had it, "semblance and the truth of thought entwine" ([1966] 2007, p. 7). This entwinement, this entanglement, is the nexus of understanding how technology points beyond itself at the same time that it warps thought in a mechanthropomorphic direction that aligns with the administered world of modern society. This perspective encompasses the mode of thinking proper to the critical tradition by recognizing that the future must be thought alongside the logic of capital and its technological progeny because the concepts cannot be thought as separate from the material reality capital has wrought. This approach makes sense for two reasons. On the one hand, because one of the primary stated goals of technologists is to shape the future (whether the capital motive is fueling the development or not), and on the other hand, because in the 20th century the focus of technological development shifted from an industrial basis to a largely consumerist one. With regards to this latter development, consumerist technology became entangled between conceptions of the material world and the meanings individuals were able to draw from their everyday lives, as it came to shape not only the social structures of modern society but also the mental structures of modern individuals.

Maintaining an orientation toward the future—one that was not ideologically motivated and followed the critical method—became an increasingly challenging enterprise in the early 20th century because these transformations not only effected the world out there, but also the mental structures of those who wanted to study these effects. As the structure shaped thought to reflect its priorities, history became a battle of competing narratives, and as the pace of technological change accelerated alongside the spread and maturation of the logic of capital, it complicated the ability to incorporate temporality in any critique of the present. Feeding off each other, the entanglement of technology and capital caused dire consequences for the social, which also imperiled notions of futurity based on the social as the economy and the culture of modern societies began to eclipse the social and control its directionality, rather than the other way around. Before examining the ways that the Frankfurt School drew on the classics to synthesize their perspectives into a program known as Critical Theory, which they used to study these changes and trace the eclipsing of the social, I will examine the scenic landscape of modern society at this stage in the history of modern society and the impact of technology on this process. Doing so will provide an outline of the problems that the members of the Frankfurt School had to confront as they set out to update the critical method in a way that could respond to the material situation of the modern society they inherited. Furthermore, it will aid in our understanding of how anxiety became ever more embedded in the structure of modern society and therefore further illuminate the objects of socioanalytic inquiry and how the method must respond in kind.

CHAPTER 3

Scenic Landscape 2

1   The Totalizing Logic of Capital Comes of Age: The Path to Technological Embeddedness and Mass Society

From the start of the modern age to the period in the 20th century when the members of the Frankfurt School dedicated themselves to updating and synthesizing the critical method of the classics—turning it into the program they referred to as Critical Theory—there were three major stages or forms of technological development that led to its deep embeddedness in the everyday life of modern societies and began to massify societies. This massification implied a new configuration of society in which the quantity of avenues to develop the individual and the social grew but only along avenues which supported the totalizing logics. The effect was that the essence of the individual and the social began to disappear, thereby homogenizing them to the extent that the qualitative content of their concepts was largely lost.

The first form of technological development emerged in the 18th and 19th centuries. It was during that period when the logic of capital began to shape the production of the technical crafts primarily around the development of industrial technology and artificial sources of power, which embedded technology in the labor process. Examples such as the steam engine and the telegraph had drastic impacts on social organization by effectively shrinking space through time and by increasing beyond the state of nature the rate of the productive and communicative forces. However, these technologies were costly, so they were generally limited in scope to applications in the economic sphere or to large state or private sponsored projects for public use. This led to the mobilization of the masses and a concentration of population in urban zones whose structure was organized around the use and availability of these technologies and the jobs that developed around them (Gibbs & Martin, 1958; 1962).

The second form emerged in the late 19th and early 20th century when technical rationality was applied to industrial and managerial techniques. These techniques produced systems of behavioral control whose function was to rigorously modify human behavior in ways that mimicked the restricted motions and functions of the industrial machines, thus embedding a technical logic in the psychology of laborers (Doray, 1988; Pruijt, 1997). This form of mental and bodily control was implemented in the workplace as a hierarchal system of bureaucratic management, and since its success depended on the recoding of individual behaviors, as this coding became internalized its effects extended beyond the working day into private life and ways of organizing the self and its immediate environment. These technical forms of organization restructured the spheres of public and private life in a mechanistic fashion. Technology was no longer merely an external appendage of the human which extended its potential, once it assumed the form of institutional techniques it was internalized in ways that furthered the mechanthropomorphism of the species in modern societies. The disruptions and restructuring of everyday life that occurred as a result of industrial technologies and institutional techniques were easily observed and felt because they were externally imposed by means of disciplinary forces and coercive tactics, as the laboring class who had to rely on the sale of their labor-power for survival in societies organized around the logic of capital had few, if any, alternatives. Getting and maintaining a job in those societies meant compliance with these structural demands imposed by the agents of capital.

The third major form, appeared in the 20th century with an increased attention on consumer technologies which were initiated with developments such as electricity, the combustion engine, the telephone, radio, and film. However, this form only found wide-spread adoption once the population had sufficiently adapted to the social conditions ushered in by the previous two technological revolutions. Furthermore, the masses had to have obtained a level of income that went beyond meeting basic survival necessities to afford the adoption of these technologies. In this stage technology became embedded in the everyday activities of adults and children, regardless of their role in the labor-force, as it assumed a central place in the culture of modern societies and became a structuring agent of the social fabric.

A common feature of these three forms of technological development is that there was a lag between the time of the technical innovations, including the requisite scientific knowledge and engineering designs that paved their way, and their widespread adoption. One reason for the usage lag relates to the characteristics of the adopting country, both in terms of the level of economic development and the historical point at which the country began to follow the modernization model of development (Comin, Hobijn, & Rovito, 2008). As such, there is a reciprocal relationship between technological adoption, gross domestic product (GDP), and income per capita, with "the empirical estimates suggest[ing] that 70% of the differences in cross-country income per capita can be explained by differences in technology adoption" (Comin & Mestieri, 2010, p. 31). Since Western countries, led by America, England, Germany, and France, were primarily responsible for the original development, production, and diffusion of these technologies, they were the first to benefit from the increased GDP and income per capita enabled by their adoption, which allowed them the ability to implement newer technologies faster than those countries who began the modernization process in a later historical period. Furthermore, since these Western countries nearly exclusively controlled the production of these technologies, they set the implementation costs which allowed them to dictate to a large degree the rate of implementation in the rest of the world so as to maintain a competitive advantage on the world market. However, these lags also exist within countries as development was also stratified internally and tended to concentrate in urban zones, making adoption by rural populations lag that of urban populations.

Unlike the industrial and institutional modes of technological development, consumerist technology was far more insidious in terms of restructuring the social order because its implementation was presented as a reward to laborers in exchange for their compliance with the administered life. Like other technologies, these too required a higher level of income necessary for their adoption among the masses, but unlike those other technologies, consumerist technology does not necessarily translate into increased income at the individual level, although it does stimulate the productive/consumptive relationship which translates to increased GDP at the country level. Meaning that the benefits of adopting technology are unequally distributed across populations with the disproportional benefits funneling to the capitalist class. As such, adoption of consumerist technologies into everyday life appeared on the surface to be optional, as they were primarily sold on the basis of personal desires and secondarily as ways to 'buy-back time' that had to be spent selling labor-power to purchase the devices in the first place. This implicitly meant that individuals' time was not their own, it was the capitalists, and it was something to be exchanged for in their market of consumer goods (Baudrillard, [1970] 1998). This made the latent effects—such as increased anxiety—of adopting consumerist technologies into everyday life far less easy to identify as the goods came draped in the cloak of 'free-choice' and the illusion of temporal freedom.

The classics of modern social thought had primarily witnessed the industrialization project through the transformation of the economic system which relied on the continuous revolution of the means of production through a process of recombinant innovation and generational advance. For Marx's time, the focus in modern societies was almost exclusively on developing industrial technologies; by Durkheim, Weber, and Freud's time, there was a new focus on institutional techniques as a form of technological control for office jobs that emerged in support of industrial activities and for assembly-line jobs that were the result of the restructuring of the factory system as an evolution in the division of labor. These techniques impacted the lives of the rising middle classes who had come to share in more of the benefits of the capitalist system, and as such were the first class targeted by consumer technology, in exchange for their working as the primary social representatives of bourgeois consciousness (Weber, 1978).[1] The dream of these first two forms of modern technology were two-fold, and each aligned with the logic of capital. On the one hand, they prioritized efficiency and speed to maximize production output as part of the project of transforming nature, managing labor, and accelerating the growth of capital, and on the other hand, they sought to surpass nature, primarily by the authoritarian application of rational thought to human action and by discovering artificial sources of power that could augment and replace that of humans, animals, and environment (such as water or wind powered mills) to be stored for use, or produced, in spaces and times when and where these sources were not available or were viewed as hindering the growth of capital. These represent two sides of the automation process that were present in the first modern machines. First, the efficiency of movement with machined (or human) parts engineered (or managed) in configurations that minimized wasted movement, thus saving time. And, second, the ability to continuously make those movements without the natural need for rest to replenish energy stores or the unpredictability and limitations of individual attributes, environment, geography and topography, or by the use of the alternating shift system

---

1 A good example of the workings of this class is found in the science fiction novel (Morgan, 2002) and television series *Altered Carbon*. The character Oumou Prescott serves as a lawyer to the elite class. Her entire life and the meaning she derives from it is filtered through her job. She considers herself to be one of the elite, but no matter what sacrifices she makes on their behalf she always retains a second-class status in their minds. She could not locate the object of her anxiety, which related to her social status, because it was necessary to repress it to carry out her duties. When her wealthy employers needed a scapegoat, she was quickly sacrificed and discarded, revealing that so long as she worked for them her self-identity as one of them was nothing but an illusion they supported in exchange for her obedience. Without locating that anxiety, she failed to act before it was too late. Marx labeled this kind of person as a member of the petite-bourgeoisie. In an essay on the science fiction author, H. G. Wells, the Marxist writer Christopher Caudwell defines this class: "[The] *petit bourgeois* … of all the products of capitalism none is more unlovely than this class. Whoever does not escape from it is certainly damned. It is necessarily a class whose existence is based on a lie. Functionally it is exploited, but because it is allowed to share in some of the crumbs of exploitation that fall from the rich bourgeois table, it identifies itself with the bourgeois system on which, whether as a bank manager, small shopkeeper or upper household servant, it seems to depend. It has only one value in life, that of bettering itself, of getting a step nearer the good bourgeois things so far above it. It has only one horror, that of falling from respectability into the proletarian abyss which, because it is so near, seems so much more dangerous. It is rootless, individualist, lonely, and perpetually facing, with its hackles up, an antagonistic world" (p. 76–77).

which broke humans out of the circadian rhythms of nature and disciplined them in accordance with clock time.

According to Marx, it was not the discovery and development of artificial power sources that launched the industrial revolution, but rather "[i]t was the invention of machines that made a revolution in the form of steam engines necessary" (Marx, [1867] 1990, p. 497). He labels "the self-acting mule" as the invention that "opened up a new epoch in the automatic system" (p. 563).[2] Automation machines increased productivity to such an extent that the number of human laborers could be reduced in industries that adapted machinery to the production tasks. They also aided standardization because they eliminated variation in the production of goods that resulted from human laborers possessing different skill sets and degrees of mental and psychical acuity. The successful growth of the capitalist system depended on these developments and its mantra became 'adapt and revolutionize, or die.' Where technology successfully improved productivity the technical logic was pushed to the extreme out of capitalist necessity, to the point where its primary objective was not the immediate ends to which the technology was applied, but rather for the technology to surpass itself so as to gain, if for a brief moment, a competitive advantage in the market.

Modern technology, therefore, is built on the principle that it never reaches the point of culmination, the point at which we can proclaim that it has arrived once its stated ends are met. The logic of the modern, which is always

---

2 Marx identifies the Scottish engineer Peter Fairbairn as the discoverer of "several very important applications of machinery to the construction of machines as a result of strikes in his own factory" (footnote on p. 563). Fairbairn was important to the process of applying machinery to the production of machinery, thereby finetuning the technological appendage by applying it to itself. For example, shortly before his death in 1861, Fairbairn patented an invention for "an improvement in rollers for preparing hemp and flax," which *Scientific American* described as surprising given the cost of obtaining a patent, but noted that "[w]hile the *great* fortunes are made from great inventions, like the sewing machine, the reaper, the electric telegraph, &c., those which are most certain to pay moderate sums of a few hundred or a few thousand dollars, are modifications in the details of mechanism, made by practical mechanics who see the objections to the machinery in use, and who happen to think of a way of overcoming them" (p. 71), indicating a whole new avenue for the making of profit in intellectual property through the process of technical refinement and generational advance of existing technologies. However, in terms of the self-acting mule and the machines involved in the textile industry that Marx saw as a fundamental cause of the revolution of the steam engine, the inventions of the English engineer, Richard Roberts, were far more consequential than Fairbairn's in the inauguration of production engineering and the design of mechanical tools that were necessary "to achieve standardization in ... manufacturing" and thus, precision in the manufacture of machines and machine produced goods that required constant motive power.

future-oriented, is contained in the technological arts and sciences, so that the concept of each device and system points to its next version, its next generational leap; thus, like capital, technology (in its informational guise) represents a totalizing logic. The move toward artificial motive power illustrates this point and served two purposes in the system of capital. First, in recognizing that technology allowed industrialists to downsize their labor force without sacrificing production output, it appeared logical that these expenses could be further reduced by automating more of the tasks that remained. In other words, the costs of technological development and implementation had to be less than the costs of human and animal labor-power, and once the early forms demonstrated that this was possible, the continuation of this process was a necessary furtherance of the logic of capital. Second, as these machines scaled in size to meet industrial demands, it called "for a more massive mechanism to drive it; and this mechanism, in order to overcome its own inertia, requires a mightier power than that of man" or animal (p, 497). Not only was overcoming the inertia of these automation machines a technical problem to be solved with additional technological solutions, but technology came to embody the very notion of overcoming itself in a dialectic of concept and materiality.

While those early automation machines revolutionized the labor process, the steam engine was a technology with societal consequences that reached beyond its industrial roots. What this meant was that the dialectic effected more than just the immediate materiality of the technological implements. Possibility itself was embodied in their material form which gave them the potential to reshape thought and restructure social organization. Although the steam engine ultimately had revolutionary consequences, it did not come into existence fully formed, rather it required the unfolding of the dialectic for its potential to be actualized. Understanding the material and historical process of technological development is necessary in order to counter the idealistic ways that thinking about technology is encouraged by its conceptual form; which includes the notion that it is in the nature of technology to overcome its own weaknesses. That is, by understanding the history of how technology is developed we can illuminate the gaps between the positivist takes on technological determinism of the future and the critical views that illuminate the latent effects these devices and systems have on material reality including how they produce anxiety.

Like a child taking its first steps, technology is the accumulation of learned knowledge in material form and the refinement of *technical* knowledge gained by experience, and recorded as information, from interactions with that material form. In the case of the steam engine, the proof of concept dates to the work of Hero of Alexandria in the first century AD, but it was not until the Renaissance that the practical applications of using steam power generated enough interest to develop it as a technological device with economic and labor-saving consequences and not merely as a cultural curiosity. The industrial development of the steam engine began with solving a practical problem caused by the turn toward extractive industries which were rushing to provide the raw materials needed to produce industrial and commercial goods in the early development of capital. Mining was a particularly dangerous form of labor and the mines were prone to flooding, which worsened working conditions and often prevented the work from being done. Early development of the steam engine, notably by the Frenchman Denis Papin around 1690 and the Englishman Thomas Savery from 1695–1702, was focused on using the technology to create a hydraulic steam-powered pump that could drain the mines (Nuvolari, 2004). While their inventions did, in a limited fashion, serve the purpose they were designed for, they were not suitable for widespread adaptation. The Savery engine "was highly uneconomical," due to its inefficient design which consumed large quantities of fuel, "the metallurgical techniques" were not sufficiently advanced to safely maintain the high pressure the device required, and "in practice" the device could not pump the water from very deep in the mines (p. 14).

The historian of science, Donald Cardwell, credits the English inventor Thomas Newcomen's 1712 invention as the "first successful steam engine in the world" (1994, p. 121), insofar as it could pump water at greater depths than its predecessors and it could safely maintain higher internal pressure. But it was not until the Scottish inventor, James Watt, set himself the task of making the engine fuel efficient that its wide spread industrial use became feasible, and this was only possible once there were advances in metallurgy that allowed for precision machining of parts to achieve the requisite pressure needed to compel the engine's motion, a feat which was accomplished by Newcomen. In 1776, the year of American Independence and Hume's death, Watt's steam engine came to replace the Newcomb pump in most commercial mines and following its success Watt continued to make improvements on it until he received a patent for the device in 1781.

The shift from a mindset of nature to one of an artificial world constructed by modernity is apparent in how these new technologies were described in relation to the world of nature they sought to leave behind. Watt coined the term horsepower to compare the work done by his steam engine to that which was previously done by horses. Because of the variables in 'horse' power output, the calculation was a best estimate achieved in the following manner:

> According to Dickenson (1967), in the early 1780s Boulton and Watt were manufacturing rotary steam engines that replaced horse gins. Quite

> naturally, payment for the engine was an annual premium based on the number of horses needed to do the equivalent amount of work. In discussions with millwrights, Watt learned that during a day's work a horse would walk an average of two and a half times per minute around a 24-ft diameter mill wheel. Dickenson (p. 145) says Watt assumed a horse exerted a tractive effort of 180 pound force (lbf), yielding a power estimate of 33,929 ft-lbf min-^{1} (power = force x distance/time). In Watt's blotting and calculation book this number was rounded to 33,000 ft-lbf min-^{1}, equivalent to the more familiar definition for HP of 550 ft-lbf sec-^{1}. (The US Bureau of Standards gives a different account of Watt's calculation that says he considered engine friction.) By either calculation, Watt's measure of power output is clearly based on a rate that horses could maintain for a full day, not a peak performance.
>
> STEVENSON & WASSERSUG, 1993

Watt formulated the price of his invention in terms of animal labor-power to justify its expense to the industrialists in terms they could comprehend. Furthermore, he marketed it in terms that could appeal to both capitalists and laborers. On the one hand, the manifest objective of the steam-engine was that it could maintain peak performance giving it an advantage over human and animal labor, and on the other hand, it could contribute to the protection of workers from the unsafe conditions of flooded mines. The bottom line was that it allowed miners to work longer hours without interruption.

One effect of embedding technologies like this into the labor process was that the anxiety of nature was increasingly transformed into the anxiety of artifice. Anxiety over natural hazards was partially assuaged by the steam-pumps which made working conditions safer, but machines that could replace animal-power could replace human-power, and a job whose dangerous conditions made for a wretched existence was only marginally improved at the cost of threatening the job upon which the workers relied on for their daily survival. Marx, citing Gaskell (1833), illuminated these latent effects: "the steam engine was from the very first an antagonist of human power, an antagonist that enabled the capitalist to tread under foot the growing claims of the workmen, who threatened the newly born factory system with a crisis" (Marx, [1867] 1990, p. 475). In other words, Marx examined technology from a dialectical perspective by drawing in the effects that this had, not only on capital, but on humans, recognizing that in "revolutionizing the instruments of production" technology developed under the logic of capital also revolutionized "the relations of production, and with them the whole relations of society" (Marx, [1848] 1988). The purpose of the machine was, from this perspective, not to alleviate the anxiety of the workers but primarily to alleviate that of the capitalists who would lose profit when there were work stoppages. Introducing more technology into the labor process thereby transferred the capitalists' anxiety over lost profits to the workers over their loss of jobs.

The ways this revolutionized humanity, beyond the material reality of mine workers, was made evident in all the ways that this technology came to be deployed. Watt quickly found avenues for selling his engine beyond the mines, such as to the mills, and in the process revolutionized the means of production across capitalist enterprises. By 1807 Robert Fulton adapted the technology to power ships and by 1837 they were adapted to ocean liners cutting the travel time across the Atlantic Ocean from New York to England from approximately 30 days to 15 (Fry, 1896; Hydrographer of the Navy, 1973). Finally, in the 1820s it revolutionized transportation across land in the form of the steam locomotive.[3]

Two economically viable areas for further development were opened by the success of the steam engine, but they were not solved directly by steam engine technology. In this way technical solutions led to the discovery of new technical problems; problems that would not have risen to consciousness unless the material circumstances had changed. These changes were ushered in by the application of the steam engine. By means of its use people began to recognize that it opened future avenues of possibility. The first was the application of artificial modes of power to an expanded industrial and commercial arena. The second was related to portability, that is, to the size of these generators of power so that they could be easily moved and thus have applicability beyond large public works and industry. These problems were solved along two main avenues of technological development. The first required taking the collective knowledge of electricity and refining it to a level that allowed for its capture and manipulation. Western knowledge on electricity dates to the German philosopher Leibniz's observations on electrical phenomena in 1671, the English astronomer Stephen Grey's discovery that metal conducts electricity in 1729, the American polymath Benjamin Franklin's experiments and design of the lightning rod in the 1750s, and in 1799 the Italian chemist and physicist

---

[3] "[I]n 1800, it took a whole day to barely get outside of [New York] city; two weeks to reach Georgia or Ohio; and in five weeks you could just about get to Illinois and Louisiana. About 30 years later, in 1830, train travel in the U.S. was almost twice as fast … Rather than taking two weeks, going to Georgia or Ohio from New York took one week, and in two you could get to the state borders of Louisiana, Arkansas and Illinois … By 1857 … [y]ou could now do in a day or two what used to take a couple weeks. With a week's travel you could get to the eastern border of Texas, and in about four weeks you could get to California. [By] 1930 … [i]t now only takes two days to get across half the United States by train, and three to four days to get to the other coast from New York City" (Richard, 2012).

Alessandro Volta's invention of the battery (Kryzhanovsky, 1989). These contributions were not practical, but rather formed a kind of basic research in that they sought to build a body of knowledge that could potentially have practical applications in the future by virtue of their general advance of human knowledge. The modern age of electrical development began in earnest in the 1820s, when the German physicist Georg Ohm "bridg[ed] the gap ... between static electricity of previous ages and the new era of electricity that includes current flow" (Morgan, R. B., 1991). In 1831, the English scientist Michael Faraday then discovered "the exact relationship between the current, magnetism, and motion" which allowed him to formally demonstrate "the laws of electromagnetic induction" (Carlson, 2013, pp. 35-36). With the scientific knowledge now in hand, electricity could be put to more practical applications and it served as the basis for the invention of the telegraph, with the American Samuel Morse gaining the fame with his proof of concept in 1837.

Social consequences arising from the revolutionary advances in electricity came once the system of capital took notice of the potential that these technologies had for transforming the market and the consumerist landscape. In the 1870-80s the charge was led by Thomas Edison, whose technological brilliance was only surpassed by his willingness to play the capitalist game as ruthlessly as necessary. Setting the stage in 1869, Edison invented an "improved stock ticker" that would "establish him as perhaps the premier electrical inventor of his day," because the invention had significant value for the capitalist class which made them take notice of him (Friedel, Isreal, & Finn, 2010, p. 1). As the historian of technology Thomas P. Hughes (1979) argued, what set Edison apart from other inventors and engineers of the day was his holistic approach as a systems builder that was guided in equal measure by Edison the inventor and Edison the entrepreneur. "Edison invented systems, including an electric light system" which included a "generating station and distribution network" and its success can be traced to his willingness to "reach out beyond his special competence to research, develop, finance, and manage his inventions" (Hughes, 1983, p. 18). This required partnering with others who shared his vision and ability to see how a vast sum of seemingly disconnected parts could come together to form a *sui generis* whole. Edison, the inventor of systems, found what he was looking for in Samuel Insull, who "managed systems," and S. Z. Mitchell, who "financed their expansion" (Hughes, 1979, p. 124).

The other titan of electricity was the Serbian-American inventor, Nikola Tesla, who briefly worked for Edison's company in the early 1880s but felt disrespected by the management and set off on his own. Edison and Tesla approached technological development in different ways, with Edison "preferring to develop his ideas by physical means" and Tesla "who called himself a 'theoretical inventor' since he preferred to edit and shape inventions in his mind" (Carlson, 2013, p. 10). After leaving Edison's company, Tesla received funding from a series of American investors and successfully developed an AC generating motor. This led to his securing backing from the Westinghouse Electric & Manufacturing Company in support of his AC method of electrification, which was in competition with Edison's DC method. In 1890, "the failure of a major London brokerage house, Baring Brothers, set off a financial panic and prompted Westinghouse's creditors to call in their loans" (Carlson, 2013, p. 130). The effect was that Westinghouse claimed they could no longer pay Tesla royalties on his patents, but he let them continue to use them in exchange for the marketing recognition of his inventions. This financial crisis also depressed the value of stock in Edison's company, so Edison went on the offensive to protect his interests. He began a propaganda campaign against Tesla's AC model and Westinghouse, but he couldn't compete with it on the cost, nor could his DC method solve the problem of converting higher and lower voltages necessary to accommodate the full range of industrial and consumer technologies entering the market that was solved by the AC model.

After successfully integrating the logic of capital in his earlier ventures, Edison made a costly mistake by trusting in his abilities as an individual over the power wielded by the logic of capital, and in 1892, with his shareholders frustrated by his inability to turn the value of their shares around, he was ousted from his company. The financial titan J. P. Morgan swooped in and bought out Edison's company and the other of the big three electrical companies, Thompson-Houston, and merged them to form General Electric. Recognizing that AC was the future, Morgan also secured a patent-sharing agreement with Westinghouse—who because of their financial troubles convinced Tesla in 1897 to settle for a lump sum payment for use of his patents, permanently forgoing royalties and effectively denying him the millions he would have earned once General Electric led the charge of building electric plants across America (Cheney, 1981). With Tesla losing out on the fortunes made off his work, and Edison ousted for refusing to adapt to the market, the future of electrical development was out of the hands of its inventors and controlled by finance capitalists who had wrested it from their control.

General Electric set its sights on the electrification of modern societies and with J. P. Morgan financing the venture their success was all but assured. The rapid and visible transformation of society occurred so quickly and with such a radical impact that government saw a need to intervene and regulate the electric industry. By 1914 "state regulation of utilities became commonplace" with 45 states "establishing government oversight of electric utilities" (The National Museum of American History, 2014). This fused the nexus of political economy and technology dictating their reciprocal relationship and the necessity of future technologists to pay allegiance to business and government interests if they were to successfully integrate their technologies across modern societies.

The second revolutionary device was designed to solve the problem of portability raised by the steam engine. This came in the form of the combustion engine which, while developed alongside the steam engine, faced technical challenges that were not solved until innovations in the steam engine took off (Cummins, 1976). In 1860, the Belgian engineer Jean Joseph Étienne Lenoir was the first to successfully patent an "internal combustion engine fueled by coal gas" and in 1862 the French civil engineer, Alphonse Beau de Rochas, "patented but did not build a four-stroke engine" (Ratiu, 2003, p. 146). Throughout the remainder of the 1800s, several European and American engineers built on and improved these early designs. In 1876 the German engineer, Nikolaus Otto, developed the four-stroke engine that solved the portability problem, but not the efficiency problem. In 1885, another German engineer, Gottlieb Daimler, solved the efficiency problem with his design and, based on this technology, in 1886 "the first commercially successful automobile was built by Karl Benz in Germany" (Laurent, 1998, p. 140) which he patented the following year.

Once the automobile became commercially viable as a means of transport in modern societies, it made the horse all but obsolete in modern economies. But, the notion of horse-power has remained the primary measure of engine power as an anachronism that harkens back to a life embedded in nature, despite the nearly ubiquitous dismissal of horses from modern labor and the lack of frame-of-reference for modern subjects who know nothing of a horse's power. These innovations in transportation not only further distanced humanity from nature by embedding artificial technologies in everyday life, they connected geographically distant cities and countries which expanded the range of commerce and transformed familial and communal relations by opening avenues for people to live further apart. Which in turn necessitated an increased reliance on technology to stay in touch and maintain those now distant relationships. What was gained by these technologies was a more robust foundation for the development of capital and greater access to the spaces of the globe; what was lost was the intimacy of close contact and a larger portion of the power previously held by the human laborer which was now held by the owners of the means of production.

With the introduction of electricity and the automobile, as the most revolutionary artificial modes of power for mass consumption, technological development began to point toward the consumerist phase. Before it could get there, it had to pass through the phase of institutional and managerial techniques to solve the problems of mass production and cost reduction necessary to meet the demands of the masses at a price-point that would allow them to participate in the consumer market. The two men who led the way in developing the institutional and managerial techniques that would underpin the system which made the rise of consumerist technology possible, were the mechanical engineer and management consultant, Fredrick Winslow Taylor, and the industrial and automotive magnate, Henry Ford.

In 1895, Taylor made his first foray into the field of "scientific management" when he published a paper that argued for a "piece-rate system" of worker pay. He based his analyses "on a standard time and output to be determined 'scientifically' through detailed job analyses and time and motion studies of the work involved" (Chandler Jr., 1977, p. 275). This method of determining the variable nature of labor-power was a significant leap over Watt's calculations of horsepower. Best estimates were replaced with careful statistical analyses which fueled the capitalist desire for the social sciences to move in a positivist direction, thereby creating a tyranny of the probabilistic center wherein those who fell on the left side of the curve were either subject to increased disciplinary tactics or eliminated altogether from the mainstream labor-force and had to find whatever undesirable work was leftover. Taylor argued that the way to overcome the antagonism between workers and employers was to introduce a differential and variable pay system directly tied to production output that rewarded the most productive workers and penalized the least productive workers. Thus, he suggested decoupling the rate of pay from "the kind of work each man performs" and linking it to "the accuracy and energy with which he fills his position" (Tayler, 1895, p. 356). If the task required a group of individuals and one could not keep up with the others, then "the drone will surely be obliged by his companions to do his best the next time or else get out" and "the low rate [of pay] should be made so small as to be unattractive even to an inferior man" (p. 356). The target of his argument is clear. It is a furtherance of the totalizing logic of capital justified by the totalizing logic of information, applied to the behavioral lives of individuals, who if they cannot comply are left to rot in the streets as victims of their inability to perform at the levels dictated by the capitalists in the most machinic manner possible.

In 1911, Taylor published an extended collection of his views in a book titled *The Principles of Scientific Management*, which was written to advance three goals:

> (1) to illustrate "the great loss which the whole country is suffering through inefficiency in almost all our daily acts;" (2) to argue "that the remedy for this inefficiency lies in systematic management;" and (3) "[t]o prove that the best management is a true science, resting on clearly defined laws,

rules, and principles, as a foundation. And further to show that the fundamental principles of scientific management are applicable to all kinds of human activities, from our simplest individual acts to the work of our great corporations, which call for the most elaborate cooperation." ([1911] 1919, p. 7)

By the application of systematic rational thought to the behaviors and actions of individuals, Taylor held the belief that we could maximize human efficiency. The key point here is that Taylor's goal was to extend this logic across all domains of human action, whether in the workplace or in the private lives of individuals.

Since industrial technologies had demonstrated the ability to increase efficiency to maximize production output, but since it was not possible to eliminate the human element entirely in the labor process, the next best thing was to manage human behaviors in ways that made their motions more machinic; that is, more rationally orchestrated. This would serve two purposes: first, it would increase the efficiency of the human element by maximizing their productive potential, and second, by structuring human actions in machinic ways, it would provide a blueprint for technologists in the future as to how and where their machines could plug into the labor process to replace more of the variable and difficult to predict human element. Even though Western countries proclaimed themselves the guardians of freedom and proponents of individualism, Taylor's model that they embraced was opposed to these freedoms:

> The idea, then, of taking one man after another and training him under a competent teacher into new working habits until he continually and habitually works in accordance with scientific laws, which have been developed by someone else, is directly antagonistic to the old idea that each workman can best regulate his own way of doing the work ... [T]he man suited to [this kind of labor] is too stupid properly to train himself. (p. 63)

Neither did Taylor shy away from Marx's point that technology was a direct antagonist of human labor, rather he embraced it, writing that "sympathy" for those who lose their job as a result of the implementation of these techniques "is entirely wasted" (p. 64). Taking a functionalist stance, Taylor presumes that it was a kindness for them to lose these jobs "because it was the first step toward finding them work for which they were peculiarly fitted" (p. 64), ignoring the desperation through which most laborers are forced to choose a mode of employment.

The effect of implementing this model of technical management was a drastic increase in worker anxiety and alienation, since if they could not keep up they were left without a job and a way to sustain themselves and their families. None other than V. I. Lenin, the head of the Soviet Union, called out the hypocrisy of "free" countries whose capitalists were allowed by their "free" citizens to enforce this mode of authoritarianism over the labor process. He had this to say about Taylor's system of management:

> The result is that, within the same nine or ten hours as before, they squeeze out of the worker three times more labor, mercilessly drain him of all his strength, and are three times faster in sucking out every drop of the wage slave's nervous and physical energy. And if he dies young? Well, there are many others waiting at the gate!
> LENIN, 1974

Even those who did keep up with the physical demands suffered the psychological effects of these "rationally" enhanced disciplinary tactics, and while there was ostensibly a higher pay rate that the worker could receive, this was negated at the social level by decreased levels of employment and at the individual level by the emotional disturbance it caused.

There are two main differences in approach that separate the "scientific management" style of Taylor from that of Henry Ford. First, Taylor was a consultant who had to sell his ideas to industrialists, not to laborers, so he wrote from the standpoint of the capitalist class with a voice that shows a clear antipathy for the plight of the working class. Ford was an industrialist, so while he had the ability to enforce his ideas in the top-down authoritarian fashion espoused by Taylor, he also had to deal with the practical side of employment which at a minimum required taking into account employee satisfaction for purposes of retention, as training new employees was time consuming and costly. Second, Taylor was primarily concerned with improving production processes, not improving working conditions or worker pay. Ford, on the other hand, recognized that increasing production was a worthless endeavor if the potential consumers of those goods were unable to purchase them. Ford's approach, therefore, paid equal attention to the production of consumers as a class as it did to the production process of material commodities.

In the 1890s, Ford, like Tesla, got his start working as an engineer at Edison's company. In 1903, Ford incorporated The Ford Motor Company and began producing the Ford Model A. This and other early automobiles suffered from high production costs and questionable reliability, but less than three months after selling the first Model A, the company had "turned a profit of $37,000," already more than the "$28,000 cash investment" that got the company off the ground (Ford Motor Company, 2018). The high costs of these early automobiles, their less than reliable technology, and the fact that "[i]n 1908, there were only about 18,000 miles of paved roads in the US" meant that there were multiple barriers to diffusing this technology. Ford reasoned that it was not social and political revolutions, but rather revolutionary advances in industry that were the way to solving social problems, including that of poverty (thus, lack of a consumer class), and this could be accomplished by dealing with two issues. The barrier to the distribution of wealth was, for Ford, that "the waste is so great [in industry] that there is not a sufficient share for everyone engaged," and "the product is usually sold at so high a price as to restrict its fullest consumption" (Ford & Crowther, 1922, p. 185). This is not to suggest that Ford was an altruist who was concerned with the general plights of humanity or a desire for a more equitable society. His rabid anti-Semitism and belief that it was unnatural to think and treat all humans as equal negates any arguments made in that light and places him firmly as a devout representative of bourgeois consciousness. Rather, as an innovative agent of the logic of capital, Ford thought that the employer not only has to create products, but also "has to create customers, and ... his own workers are among his best customers" (Ford, 1926 [1988], p. 154).

Ford solved the problem of unreliable technology with a radical redesign of the automobile in the form of the Model T, which took advantage of advances made in the combustion engine, and then proceeded to address the problem of high adoption costs by innovating on two fronts. First in 1913, Ford installed "the integrated moving assembly line" in his factories, which took advantage of the new electrical networks and sped up the construction of the Model T chassis "from 12.5 to 1.5 hours" lowering the development costs and the market price of the vehicle. In this manner, Ford's project resembled Taylor's in that it "imposed discipline from above through new methods of technical control designed by managers and engineers" (Antonio & Bonanno, 2012, p. 582) in the form of the assembly line model of production which took the division of labor and automated it. Second, in 1914, Ford doubled the pay of his workers to $5 per day and "reduced the work day from nine to eight hours" which "allowed Ford to run 3 shifts a day instead of 2" (Ford Motor Company, 2018). This high-wage doctrine (Taylor & Selgin, 1999) of Ford's was based on the notion that if consumers had high enough wages they would be able to consume more. Included in this increased consumption was the purchase of the products he manufactured, thus boosting the economy by soaking up excess consumer goods that were previously only affordable to the bourgeoisie class. The effects, in terms of employment, were that from 1913 to 1915 the turnover rate decreased from a whopping 370% per year to 16% per year (Slichter, 1919), training times for 79%

of employees took less than 1 week, and in terms of sales, these innovations led to a profit of $541,744 per week, more than double that of his nearest competitor, General Motors (Raff, 1988).

The third problem was solved by the government in 1916, who, despite Ford's view that government was a negative enterprise only good for removing the barriers to business, recognized the potential of the automobile for the American economy. This led Congress to allocate millions to build the urban road system, thus wielding social power to create the necessary conditions to fully realize the potential of Ford's consumerist project. However, it was not until 1956 when "President Eisenhower signed into law the federal aid highway act" that "[t]he modern era of roads" began laying the millions of miles of roads that now pave America, making the automobile practical for rural consumption as well (The National Museum of American History, 2018).

The effects of Ford's system were revolutionary insofar as they changed the social relations of the working class to consumer goods, transformed the labor market, forced other manufactures to change their production models if they were to have any hope of competing against Ford, and changed the landscape of modern America.

> Once Ford started designing new machine tools and new assembly lines, and once General Motors had created its Research Division [to keep pace with Ford], all the larger companies employed highly trained engineers and skilled craftsmen to search through the technical literature and conduct laboratory and field experiments designed to solve specific motive problems. Thus after 1920, most innovations in the mechanics or design of American automobiles were the result, essentially, of managed development, rather than invention.
>
> COWAN, 1997, p. 232

Just as had occurred in the electrical industry when J. P. Morgan took the helm, and in the automotive industry when Ford implemented his production model, technology became intrinsically linked to the logic of capital. No longer were the names of inventors held up as romanticized icons in the development of technology, rather the financial titans, the CEOs, and those who managed corporations took control of the development and dispersal of technology. It was they who came to dominate the modern imagination as the pavers of the road to the future, but it was done on the basis of technology developed in the service of capital and beneath that on the backs of laborers who were often ignored as the forgotten base of the commodity market. Whereas prior to this point science and technical crafts were generally viewed as distinct enterprises, at this juncture, business and government became more involved in the structuring of science along positivist lines to make it more pragmatically oriented to the development of technology and technical solutions for capitalist ends.

The totalizing logic of capital gained massive strength and power in shaping world affairs by virtue of this nexus of economics, politics, and technology. By 1916, president Woodrow Wilson had adopted the Fordist mentality and signaled a shift in foreign policy in a speech he delivered to the Salesmanship Congress in Detroit, Michigan, where he called for "the peaceful conquest of the world" (Wilson, 1916). This would be accomplished by embodying the spirit of American exceptionalism fully punctuated by the spirit of capital, which he riled up in his speech, telling those gathered, "you are Americans and are meant to carry liberty and justice and the principles of humanity wherever you go, go out and sell goods that will make the world more comfortable and more happy, and convert them to the principles of America." As Lenin critiqued Taylorism, the Italian Marxist, Antonio Gramsci, recognized the global implications of this Fordist approach being sold and imposed on the world through U.S. policies. He came to label the method of Ford's innovations "Fordism," and critiqued it on the grounds that it

> requires a discrimination, a qualification, in its workers, which other industries do not yet call for, a new type of qualification, a form of consumption of labour power and a quantity of power consumed in average hours which are the same numerically but which are more wearying and exhausting than elsewhere and which, in the given conditions of society as it is, the wages are not sufficient to recompense and make up for.
> GRAMSCI, 1971, pp. 311–312

While the high pay did translate into more workers staying at the Ford Motor Company, as Gramsci critically pointed out Fordism was still based on the logic of capital and, therefore, depended on the continued exploitation of workers and the maximization of the use-value of their labor-power. Despite their increased pay, Ford received an even greater rate of profit as a result of making the jobs less free and more mechanized, meaning that the workers could not set the pace of the labor but were compelled by the possible loss of employment to work at an ever-increasing rate to satisfy production demands. Furthermore, this increased pay was funneled back into the hands of the capitalists as consumer technologies changed the standards of living, making survival more expensive and complicated than the necessities of premodern life. As Antonio and Bonanno (2000) argue, Ford "fostered a social psychological climate that harnessed workers to their jobs and contributed to the rise of an emergent Fordist regime of capital" (p. 35) which attached itself to the American model of politics.

Anxiety, therefore, came to be used in a new way as a mode of control over workers. Whereas previously employment anxiety related to the possible loss of employment for failing to perform, Ford's regime, which pulled workers into a new class of consumers as a result of their higher pay, meant that loss of employment did not merely mean finding a new job, but likely finding a new job that would pay less, thereby knocking consumerists back down the social ladder. This anxiety was explored by the novelist Aldus Huxley ([1932] 2005) in his science fiction dystopia *Brave New World*. Huxley saw Fordism as a such a revolutionary force in capital that he based the new government in the book, called the World State, on the principles of Fordism. The word 'Lord' was replaced by the word 'Ford' as Huxley imagined a new modern secularism that would complete the transference of the power held by the totalizing logic of religion to that of capital. The dates in the book are listed as A.F., or 'Anno Ford,' replacing the Gregorian calendar use of A.D. or 'Anno Domini' (year of our lord), with year one on the A.F. calendar beginning when the first Model T was produced. The dystopian nightmare explored in Huxley's novel was caused by the antagonistic social psychology of Fordism and it required the use of indoctrination techniques and psychosomatic drugs to compel obedience through the manufacture of artificial pleasure. Control was not maintained through fear, as in Orwell's science fiction dystopia, *1984* (1950), based on the overt disciplinary tactics of authoritarianism, but rather it was based on an anxiety arising from the implicit control of a populace through the supposed "gifts" of consumerist life given to those who were obedient to the ruling class. As the landscape of mid-1900s America took shape, it was Huxley's warnings that appeared to ring true to the greater degree in modern Western societies.

Beginning in the 1950s the market for and development of mood-enhancing and stabilizing drugs took off (Hillhouse & Porter, 2015), but the true drug of choice that overtook the working class was consumerist technologies and cultural commodities which began their ascent in the 1920s. For example, the automobile was marketed and sold using techniques that attached its symbolic value to the status of modern man, putting social pressure on the need to own a car to gain social standing. Supplementing the masculinized technologies, sporting events became a prime cultural product sold to these modern men as socially approved emotional outlets to regulate the psychological burdens of labor related anxieties. With the success of Major League Baseball in the late 1800s and early 1900s, the National Hockey League was formed in 1917, the National Football League in 1920, and the National Basketball Association in 1946. "Employers found team sports to be useful in controlling worker attitudes and behavior, and politicians and social workers believed organized play a useful tool to help "Americanize" immigrant children. Religious leaders believed sports properly played would create better Christians, while newspapers used college and professional sports as a means of reaching wider audiences" (Davies, 2012, p. 60). Men were the first target of these cultural transformations because they formed the majority of the paid working class, but quickly it was realized that women represented a significant untapped market with a unique consumer potential.

Since many women (especially white women) were not in the paid labor force, by lack of opportunity, they were seen as having the time to spend the income their husbands earned, so they could fill the other side of the equation by becoming the largest consumer class. In the 1920s corporations overwhelmingly began to target women with consumerist technologies, playing on their status fears in the same way that they had successfully done to men. "In 1917 only one-quarter (24.3 percent) of the dwellings in the United States had been electrified, but by 1920 this had doubled (47.4 percent—for rural nonfarm and urban dwellings), and by 1930 it had risen to four-fifths percent" (Cowan, 1976, p. 4). This enabled a whole host of technical 'solutions' for the tasks that were traditionally handled by women in the early 20th century home. Some of them, such as the electric iron, did help reduce the time needed to perform the domestic labor, but others, such as the early washing machine, were rudimentary and required constant supervision. These domestic technologies served a larger function as a status symbol of the rising, largely white, and suburban, middle class than as a problem solver for women's labor. Much along the lines of Lenin and Gramsci's critiques of scientific management, empirical studies on these consumerist technologies aimed at women and household labor, showed that "[t]echnology may play a role in changing the time women allocate to housework but it certainly does not decrease the hours spent in housework" (DeFleur, 1982, p. 411). Rather, with the introduction of technology in the domestic sphere, women then had the added anxiety of status living, and the increased standards of living meant that everyday life had greater psychosocial costs. The result was a new form of alienation, which Betty Friedan ([1963] 2001) referred to as 'the problem that has no name,' because although these women had more material goods in their lives than previous generations, had children and the means to provide for them, and owned suburban homes in predominantly White and Protestant neighborhoods, they still felt something missing from their lives which somehow made their lives feel inauthentic and suffocating. Alienation was not simply a byproduct of laboring under the capitalist mode of production, but so too this stage of technological development demonstrated that it also was a result of the capitalist mode of consumption. To maintain the new standards set by modern consumerist societies, women became increasingly drawn into the paid labor force which often meant a decrease in the family size, as men overwhelmingly did not assume an increased share (if any) in the household division of labor. Instead the burden was placed on women's time as they overwhelmingly continued taking care of household duties as well as those of their limited paid labor opportunities thus doubling their sense of alienation (Bose, 1979) and making them a prime target for the prescription mood altering drugs designed in the 1950s.

Following this path, technology grew from an industrial base to managerial techniques to consumerist forms, and became embedded across the modern landscape effecting men, women, and their children, by altering social relations, changing the size and shape of the family unit, continuously revolutionizing the means of production and thus the mode of employment, and ultimately, adding to the social sources of anxiety. As more people willingly bought in to this configuration, or were compelled and coerced into it, societies began to exhibit mass tendencies which evaded rational orchestration. The only thing standing in the way of a fully consumerist society was the incessant crises of capital punctuated by war, which at this stage also took the form of mass war, but as the totalizing logic of capital had demonstrated even war can be good for furthering its interests. While Wilson proclaimed a 'conquest through peace' with the spreading of this new American ideology, in the years following his speech, America found war to be just as efficient a vehicle for bringing its mindset to modern Europe (de Grazia, 2005).

2 The Darker Side of Modernity: World War

Despite the increase of day-to-day anxieties in 20th century Western life, the prevailing mood in America and much of Europe at the turn of the 19th century was one anchored to a belief in a better tomorrow and faith that science and technology would be the vehicles to get us there. Emboldened by revolutionary transformations of material life, increased opportunities for economic growth, and a rise in the standard of living, many believed that the society of peace was just around the corner. It was believed that the peaceful society would be an industrial one which would pick the peoples of the world up by putting them to work on the project of modernization. Since technology had solved some actual problems, positivist thought inspired by those solutions led more people to believe that technology and technical thinking could solve most problems, and many came to see the barrier to the good life as a technical problem not a psychosocial and political one. The critical insights of the classics of modern social thought, which demonstrated a recognition that these problems were features of the structuring logic of capital in modern society, largely failed to imprint those warnings on the cultural superego of the West. Instead the sources of anxiety, woven as they were into the fabric of the social, were repressed by the distracted subjects of capital. Socially, economically, and politically, the divisions persisted between the haves and the have nots. The rising middle class were in actuality closer to the bottom tiers of society than the bourgeois overlords, but they had bought into the bourgeois project, trading the possibility of creating a world based on genuine social and individual freedom for all, for an administered life that provided them with greater access to a wide-range of consumer commodities.

Although materially the challenges and rewards of life in modern societies were comparable between America and Western Europe, since the end of the American Civil War in 1865, the United States was the more politically stable of the two. On the European continent the borders ebbed and flowed as political alliances rose and fell with the passing years and the outbreak of numerous wars. While the artificially imposed invisible borders were fluid and shifted accordingly, ethnic and culturally linked communities were divided by imaginary lines on maps that came with real consequences. Tensions over religious, ethnic, and cultural identities, coupled with persistent economic inequalities, political unrest, and a rapid pace of technological change led to an unstable Europe.

The United States, on the other hand, did not face the same degree of fluidity in borders as European countries, and in structuring itself around embedded notions of white identity they weathered internal struggles by ideologically imprinting their problems onto the identities of minority and marginalized populations. Mexico was embarrassed militarily and economically in skirmishes and land grabs by the United States in the 19th century harming cooperation between the countries. But by 1911, under the rule of Porfirio Diaz, Mexico had constructed nearly 17,000 miles of rail lines (Donly, 1920) which directly accounted for "between one fifth and one quarter of the total income per capita growth" (Herranz-Loncán, 2011, p. 28). In getting a taste of the economic consequences of modernization offered by the American way, connecting the countries' economies and peoples via railroad, and by threat of a superior military might who had demonstrated a willingness to use it to get their way, in the 20th century Mexico increasingly came to acquiesce to American authority on the North American continent and stabilized political relations.

Meanwhile, tensions between Canada and Britain in the late 1800s led many authors (Monro, 1879; Smith G., 1891; Moffett, 1907) to speculate that the future of Canada did not lie in maintaining a relationship with Britain, but rather in "promot[ing] a morally superior North American civilization supported by close economic integration" (Bow & Chapnick, 2016, p. 294). As with Mexico, however, the relationship between the United States and Canada was a paradoxical one. On the one hand, anti-Americanism featured strongly in the Canadian election of 1911 due to a rise in "English-Canadian optimism concerning Canada's future, an upsurge of British imperialism in Canada, and American outspokenness and ingenuousness regarding Canada" (Baker, 1970, p. 448). On the other hand, "Canadians persistently embraced ever-increasing levels of economic and cultural integration with the United States" leading to a situation where they became "dependent on" the U.S., in ways that the British could not rival, consumed as they were with more urgent and pressing European concerns on the continent (Nossal, 2005, p. 12). The relationship was finally cemented by a convergence of mutual interests between Canada and the US due to the circumstances surrounding World War I, leading to a relatively stable, peaceful, and economically prosperous North America.

In the years leading up to World War I, Britain was still the world's economic superpower. They secured this status from centuries of commitment to a colonialist project centered in ethnonationalism and white supremacy, and as the first empire to start down the path of industrialism. But their answers to the question of what to do with the accumulated capital they gained from these policies was part of the problem that led to the War. Building on Marx's definition of capital as value in motion, David Harvey explains that "the circulation of capital is ... a spiral in constant expansion" (2018, p. 4); if it faces a barrier to its growth then the value coagulated in capital begins to rot, meaning that capital must continue to expand itself if it is to maintain the value congealed in it. Capital cannot tolerate barriers to its growth and must find ways to either break them down or transcend them. In the 18th and 19th centuries, the modernization project in the West was able to soak up the increase in capital, enabled by the productive impact of the division of labor, by heavily investing in industrial technologies. Furthermore, vast quantities of capital, acquired through taxation of corporate profits, was used to build the national infrastructures required in support of the capitalist mode of production. However, in the late 19th and early 20th centuries, the acceleration of production by institutional techniques and advances in the assembly line process led to enormous reserves of capital surplus in the hands of a small class of elites, which could not easily be put in motion unless there was a realignment of the global order, the emergence of new industries, or a major boost to the development of a consumerist class who could aid in the circulation of capital.

Ford's vision of the consumerist society started to develop in America in the early 1900s, but it was slow going convincing capitalists who were loath to spread their wealth with the working class that there was a counterfactual conditional argument that held that if workers had more wealth at their disposal it would actually increase the wealth of the capitalist class by placing capital in motion, increasing demand and thus production, thereby allowing the spiral to expand. Georges Bataille called the excess of accumulated capital *The Accursed Share* ([1949] 1991), because "if a system can no longer grow or if the excess cannot be completely absorbed in its growth, it must necessarily be lost without profit; it must be spent willingly or not, gloriously or catastrophically" (p. 21). Without committing to the production of a well-paid consumerist class quickly enough, the surest way to soak up the excess was through the development of military technologies as an offshoot of the three forms of technological embeddedness explored above. Then by using those technologies in actual warfare they could be tested and refined on the battlefield, and the material destruction of infrastructure and the culling of the working class whose bodies would be sacrificed in the battles would create a new series of growth opportunities as a byproduct of the mass destruction. Short of spreading the wealth around with the masses, the alternative for the elites was war and this is the path they embarked upon.

The sociologist Richard F. Hamilton and the historian Holger H. Herwig (2003),

> define a world war as one involving five or more major powers and having military operations on two or more continents. Wars of such extent are costly ventures. The principle "actors" therefore have to be rich nations and ones with substantial intercontinental outreach. Rich, of course, is a relative term. The masses in a given nation might have been poor, but that nation, relative to others, could be rich, sufficiently so as to allow it to sustain large armies and navies in distant struggles for extended periods. (p. 2)

Although the prima facie trigger for World War I was the Serbian-nationalist inspired assassination of the Austrian Archduke Franz Ferdinand in 1914, even that event can be traced to the larger logic of capital, which Hamilton and Herwig's definition makes clear is the primary necessity of countries who choose to engage in a sustained war effort on multiple fronts. The Serbian group, Black Hand, which orchestrated the assassination, had deep ties with the Serbian state. There is some debate as to whether the actors were individually motivated by anxiety over foreign policy or desires for a nation-state of unified Serbs (Sahara, 2016), but the antagonism between the Serbs and Austria-Hungary that led to the founding of the group had economic roots. In 1906–1911 "Austria-Hungary closed its borders to Serbian pork, the most important export of the Serbian economy" which caused a "kind of economic nationalism" and intensified the making of plans for "the unification of all 'Serb-lands' under Serbian leadership" (Roudometof, 2001, p. 170). In 1908, Austria-Hungary annexed Bosnia-Herzegovina putting a kink in the plan for a unified Serbia and the Black Hand was formed in response as a group dedicated to the nationalist project at all costs.[4] In 1909, under pressure from the more powerful Austria-Hungary, the Serbian government "was forced to disavow" the Black Hand further radicalizing the group and making their leader, Dragutin "Apis" Dimitrijević, come to see "the archduke as a threat to greater Serbian nationalism" (p. 171).

With the successful assassination of the archduke in 1914, the nationalist concerns of the Black Hand quickly evaporated. They lit a powder keg in international politics and when it exploded with a force hitherto unseen in history they were consumed in the shockwaves and left as naught but a footnote to the war. What transformed the assassination from an isolated state level conflict to total war had everything "to do with the investment of constant capital in equipment, industry, and the war economy" (Deleuze & Guattari, [1987] 2007, p. 421). This metaphorical pyre was largely, but not exclusively, built by the British in an attempt to manage the systemic economic crises of capitalism of which they were the main protagonist. By the late 1800's these crises were directly linked to the surplus capital that accumulated around Europe by means of advances in productive industrial technologies. An armament race between European powers and price inflation helped soak up some of the surplus value, but as Arrighi ([1994] 2010) summed it, "the cure proved worse than the disease" (p. 277). After all, what is the point of developing the technologies of war, if not to put them to the test on the battlefield? Two major sides formed alliances in the subsequent war, on the one side were the Allies, with the combined forces of Russia, France, and the United Kingdom, and, on the other, the Central Powers of Austria-Hungary and Germany.

Because the United States had largely closed the door on the question of national identity in its Civil War, did not experience existential threats on its borders, and was pursuing the Fordist doctrine of building a consumerist society, they chose a non-interventionist stance and refused to enter the war. Despite having friendly economic relations with Europe, "[a]t its beginning, [World War 1] was considered to be a sign of European backwardness as compared to American modernity: war was a feudal relic, an expression of

---

4 In Serbian, the group was officially called Уједињење или смрт, which translates to Unification or Death.

European senility and decadence which America intended to avoid" (Joas, 1999, p. 460). But this did not stop America from selling "more than $2 billion worth of goods ... to the Allies" (Zinn, 1980, p. 353), which supported the war effort and boosted their import/export economy at the same time (Jefferson, 1917). These sales served as a much-needed boost to the American economy, which J. P. Morgan characterized as in a depression at the time. With limited investment opportunities at home in America, Morgan financed both the British and the French war efforts despite America's non-interventionist approach (Horn, 2000). Furthermore, non-intervention did not stop America from spending large sums of tax dollars on their own military to innovate new technologies and techniques for use in the theater of war, just as their European counterparts were doing.

"Development along three lines significantly, if not decisively, affected the course of the First World War, though holding out still greater promise of future use: motorized transport, armored fighting vehicles, and fighting aircraft" (Hacker, 2005, p. 258). The first major developments along these lines were in the navies, because their technologies had long since been battle tested and it was a simple matter of upgrading the wooden ships of yesteryear to modern armored ships made of iron. Broadside cannons were replaced with rotating turrets, and by the early 20th century naval technologies included the development of battleships and battle cruisers by the Allied forces, and extensive development of submarines by Germany of the Central Powers. England had ruled the seas since the 18th century, but America and Germany were catching up as they modernized their navies. "Germany [came to be] viewed as the principle challenger" of American might and was used to justify the development of naval technology out of a fear of "the ominous portents of American naval inferiority to the Kaiser's fleets" (Smith, D. M., 1965, p. 11).

On the battlefield traditional warfare changed with the widespread use of machine guns. Soldiers no longer met in open battle but dug themselves into the ground establishing a new form of trench warfare which highlighted the defensive rather than offensive tactics needed to cope with these new technologies. The combustion engine and the automobile served as the basis for the armored tank, which was then developed as a means of breaking through the trenches which had the additional cost of having destroyed the landscape making it impossible for traditional vehicles to navigate the battlefields (Terrell, 2016).

In 1903, with the successful proof of concept by the Wright brothers, aviation and specifically manned flight became a new area of technological development (Meyer, 2013). The potential military applications were immediately apparent (Spaight, 1914), but the technology did not have time to mature before it was deployed so it had limited success (Johnson, 2001). In 1914 with the outbreak of World War I, major financial support began to flow into this industry, with France leading the way. In France, for example, the number of airplanes produced rose from 57 in 1909 to 796 in 1914, and the pace of their construction greatly accelerated during the war (Chadeau, 1987). The first aircrafts that were deployed in the war were used for reconnaissance; the next generation enabled air combat. With this new form of combat, navies also began to design and build aircraft carriers to extend the limited range of these early aircrafts. These aerial technologies were not limited to the Allied powers, and Germany developed its own flying machines keeping pace with the advances made by the Allies. The war instituted a technological race with each side gaining a momentary advantage when a new technology entered the battle, but the advantage would evaporate as the other side quickly released their next generation technology in response. Modern war was clearly a different enterprise than those fought before the 20th century, as the anxiety war produced now featured the dynamic introduction of technologies in the laboratory of the battlefield.

With a budding stockpile of military technology, America was poised for war but still uncommitted to entering the fray. In 1909, several years before the war, Captain Paul B. Malone wrote a paper at the War College that "described Germany as surpassing the United States in many areas of economic competition," ultimately warning that "while war may never result between the United States and Germany yet the student of history must recognize the existence of causes [economic][5] which tend to produce it" (Smith, D. M., 1965, p. 11). Then in 1915, after the war began, the newly appointed Secretary of State, Robert Lansing, jumped on this line of reasoning and argued that "Germany must not be permitted to win this war" because then "the United States would be confronted with a hostile naval power threatening its interests" (p. 19). While the American public originally demonstrated an aversion to entering the war, those in positions of power were busy drafting arguments to justify America's entry as they were eager to put the new military technologies to use.

The classics of modern social thought considered in Part 1 did not write much on the topic of World War I,[6] but the American sociologists, W.E.B.

---

5 Brackets included in original.
6 Marx had been dead for over 30 years, but he wrote extensively on the wars in his time from a historical perspective, concluding from his empirical analyses that the bourgeoisie would use any means, including that of deploying military might, to maintain their power. His work provides the best critical foundation of the classics for understanding the circumstances leading up to World War I (for a more extended discussion of Marx on war, see, Gilbert, 1978). Durkheim's publications on the War are far more polemical than sociological in their

Du Bois and Thorstein Veblen both wrote critiques of the situation leading to the War. Du Bois (1915) argued that the war had racialized origins as a result of the colonialist project. He wrote that "Africa is a prime cause of this terrible overturning of civilization" because the wealth of the West "comes primarily from the darker nations of the world. The present war is, then, the result of jealousies engendered by the recent rise of armed national associations of labor and capital whose aim is the exploitation of the wealth of the world mainly outside the European circle of nations." Veblen, on the other hand, took a narrower and less critical—from the perspective of the social totality—view of the war. This echoed the same belief as Durkheim (Durkheim & Karsenti, [1915] 2017), namely, that German backwardness made them pursue war and posed a constant threat to peace. In a book whose object of critique was the German intellectual class who supported the nationalist agenda, Veblen argued that "[t]he German ideal of statesmanship is, accordingly, to make all the resources of the nation converge on military strength" ([1915] 2003, p. 64). Veblen viewed Germany "as straying from the normal path of modernity" (Joas, 1999, p. 461); the true path for him, represented by England and America, had supposedly left war behind in the feudal age and replaced it with industry and economic discipline.

The argument taking shape, both politically and intellectually, was that Germany posed a possible existential threat to the modern project of America and the rest of the West. Veblen continued his line of reasoning in a book dedicated to the theme of peace which he wrote while the war raged in Europe:

> Germany is still a dynastic State. That is to say, its national establishment is, in effect, self-appointed and irresponsible autocracy which holds the nation in usufruct, working through an appropriate bureaucratic organization, and the people is imbued with that spirit of abnegation and devotion that is involved in their enthusiastically supporting a government of that character. Now, it is in the nature of the dynastic State to seek dominion, that being the whole of its nature. And a dynastic establishment which enjoys the unqualified usufruct of such resources as are

---

content, and Weber's views were largely transmitted through corespondence with friends and family; both reflected nationalist stances in favor of the positions taken in the war by their home countries (see, Cotesta, 2017). Freud, at first, also took a nationalistic view of the war, but within six months of the opening hostilities he grew disillusioned with it and wrote an essay expressing his views against it . Beyond that essay, Freud's views on war were best articulated in his correspondence with the physicist, Albert Einstein, who, on behalf of the League of Nations, asked Freud if there might be a psychological solution to war .

> placed at its disposal by the feudalistic loyalty of the German people runs no chance of keeping the peace, except on terms of unconditional surrender of all those whom it may concern. No solemn engagement and no pious resolution has any weight in the balance against a cultural fatality of this magnitude. (1917, p. 103)

Veblen did not advocate for an outright war with Germany, but his critique cogently reminds the reader that the historical record points to these problems being solved through force, rather than diplomatic negotiations for peace. Nothing less than the system of capital was at stake in this war, and Veblen saw only two alternatives available to America and the West. The first was that the interests of capital could yield to the social project of building a society of peace, which would demonstrate beyond a doubt that the American model of modernity was a genuinely new system that had turned its back on the traditional historical recourse to force in situations of conflict. The second was that the Western nations could "conserve their pecuniary scheme of law and order at the cost of returning to a war footing and letting their owners preserve the rights of ownership by force of arms" (p. 336). Although Veblen correctly linked the structural causes of the war to the logic of capital, he "believed that Western ideas, like Western technology, were essentially in conflict with the reactionary dynastic state and, so, would eventually undermine the latter" (Loader & Tilman, 1995, p. 342). This points to a failure on Veblen's part to recognize how technology had been so totally coopted by the logic of capital, forming a nexus with political economy in the late 19th and early 20th century, that if one was to make the argument that its use was socially constructed, then to do so without accounting for how political economy had subsumed the social and the structuring function would be an abandonment of critical insight and a move toward a positivist take on reality. The positive ideological pull of technology and the belief that it inherently points beyond itself largely clouded the view of material reality in science as in politics and public opinion, none of which clearly saw technology as a furtherance of the logic of capital which had eclipsed the social.

At the height of World War I, when the Russian revolution commenced in 1917, overturning the Tsarist system and installing the communist Lenin as the leader of the Russian state, the arguments for America entering the war now appeared on two fronts, following the claim that the war was "a battle between democracy and autocracy" (Joas, 1999, p. 460). By playing the war off as a direct attack against American values by the Germans and the Russians, President Wilson hoped that the public would willingly support the war, but the death tolls were already running into the millions by that point and "six weeks after

[America's] declaration of war only 73,000 volunteered" (Zinn, 1980, p. 355) for military duty. Falling far short of the 1,000,000 that Wilson believed were needed for a decisive victory, Congress passed the military draft and instituted the selective service to compel young American men by force to join the war effort. American bodies and American technology entered the war achieving global status and impact. Modernity had not led humanity down a new path to sustained peace, but it had led the world to a new kind of war. Veblen was right that technology would be used to undermine the dynastic state, but he was wrong to think that it would not come in the form of militaristic interventions, gift-wrapped by the titans of capital.

The most succinct and important attempt to sociologically understand the war, not the causes, but the actual mechanisms by which World War I differed from the long history of human warfare, was "On the Sociology of World War," written by the Jewish-German sociologist Emil Lederer in 1915. Here he captures the way that technology transformed warfare to devastating impact:

> The combination of advanced war technology, expanded man power, and intensified massification of forces, stems from the nature of the military apparatus. Every military apparatus has as its aim the defeat of an enemy in war; there is, and can be, no military complex that does not have this aim. But as soon as this aim is fixed and held constant, the technology employed in the service of this end acquires an immanent necessity of its own. A search for increased destructive capability and quantitative superiority is intrinsic to military life. Relative to this end, the military complex becomes a dynamic formation with its own immanent logic. Its capabilities never need to be absolutely but only relatively more effective than the enemy's, and therefore there arises—long before the advanced capitalist economy—an early form of competition. Every advance in military technology requires ever greater masses of men, both for the managing of the apparatus of attack and for the repulsion of the increased violence of the enemy. Machine power and manpower interact with one another reciprocally, because increased manpower also in turn demands more and more destructive technology. ([1915] 2006, pp. 247-248)
>
> We stand today perhaps before a paradoxical moment in history. Organized life as it has arisen in all states now measures itself in this war ... In its assimilating of all life's forces to machines, the war spells gigantic intensification and transmuting of problems much discussed in these last years in terms of dangers of objectification, depersonalization, and mechanization. But perhaps at its end, the war will summon all who believe in living in a society to make a renewed stand against abstract

organization. Perhaps once people behold the essence of war, its ideologies will unveil themselves to us. (pp. 266-267)

The war, however, raged for another year and a half after the American forces entered and engaged the Central Powers, ultimately helping secure the victory for the Allies. The financial costs of the war were estimated at $80,680,000,000 in 1913 dollars (Fisk, 1924), or $2,089,481,608,081 in 2020 dollars after accounting for inflation.[7] When the smoke cleared in the final months of 1918, nearly 70,000,000 people had either voluntarily or by means of compulsion been drawn into the fighting. Between military and civilian populations an estimated 17.6 million people lost their lives, but even with that shocking essence of war so brutally exposed, the ideologies and abstract organization of modern society overwhelmingly remained unexamined.

The effects of the war also carried over into post-war life as soldiers and civilians attempted to return to their everyday lives. "Before 1914, mental illness was generally thought of in [the biological] terms of heredity and degeneration but by 1918 [and the winding down of the war], many clinicians had acknowledged that the environment could have an important role" (Jones & Wessely, 2014, p. 1712) on mental health. In the United Kingdom, for example, 6.3% (84,681) of the soldiers returning from the war suffered from a variety of neurological and mental disorders (Jones & Wessely, 2014), including the new diagnosis of 'shellshock' which would not become a formally recognized diagnosis by the psychiatric profession until the 1980 publication of the DSM-III which then labeled it as Post-Traumatic Stress Disorder (PTSD) (Loughran, 2012).

In April 1919, three months before the ink dried on the Treaty of Versailles—officially ending the conflict—the French poet and essayist Paul Valéry wrote of the resultant "Crisis of the Mind" facing intellectuals at this new juncture in the quest for modern life. The grand contradiction of modernity that rose to the forefront of critical modern thought was that "so many horrors could not have been possible without so many virtues" (Valéry, 1919), and yet the death tolls and psychological trauma only seemed to prove for many that Nietzsche had been right all along and God was dead. Modern society required a new critical reckoning with the modern mind and the void of spiritual significance in the materially focused life that arose with the decline of the totalizing logic of religion. The totalizing logic of religion was officially subsumed by the totalizing logic of capital in modern societies, and the power of capital was driving the disenchantment process which was further amplified by the emergent logic

---

7   Calculated using the tool provided by www.officialdata.org.

of information in its totalizing form. Lederer was right in his prediction that this military technology, like industrial, institutional, and consumerist versions developed in service of the logic of capital, pointed beyond itself to the next more efficient versions. This made technology into something enchanting at the same time it served a disenchanting purpose, liberating as it likewise subjugated. However, rather than pointing to a social utopia these technologies increasingly pointed to future wars and permanently heightened states of anxiety over the anticipation of where and when those wars would erupt (which was only a transfer of the anxiety in peace time related to employment and consumer status). Lederer's hope that people would rally against the imposed organization of the capitalist mode of production and recognize that there was another, better, way forward, remained mere hope. The numbers of the dead and wounded, the mass destruction, and the interruptions of family and social life, were too massive for most to comprehend. Before the anxieties of what humanity had wrought could be reckoned with by the cultural super-ego, before the needed critical accounting could take place, mass repression set in. Soon enough the logic of capital demonstrated anew its totalizing power as the survivors resumed their lives as wage-laborers.

It was widely recognized that the root causes of the war stemmed from the economic issues in Europe as nations fought for prominence in the modern world order, despite the narrative justifications that centered on the moral failings of the defeated nations. The reckoning that Valéry hoped would take place in the collective consciousness was warped by the social structuring of the minds of the would-be reckoners. On some level modern societies, however, required an accounting of what had happened to justify the continuation of their mode of organization after the war. The sociologist George S. Painter (1922), for example, reconsidered the question of progress and whether the organizational principles of modern society were actual improvements over the previous forms it had deemed as inferior. His analysis concluded that,

> the horrors of war are always liable to warp our judgement. It is of course certain that the fundamental instincts and passions of men have not been changed. Neither is it possible to fancy what such a transformation might mean. Our enlarged knowledge may be used to any end whatever, good or bad. Man's progress in intelligence has only multiplied the diabolical agencies of war. And it is well said that the atrocities of one war become the established agencies of the next. Mechanical, chemical, and all scientific knowledge have been turned into the services of war with a shrewdness hardly equal in relation to the peaceful walks of life ... Conferences may outlaw and proscribe the submarine and poison gases,

as now proposed, but when war actually breaks out it is certain that all nations will resort to any means by which they may save themselves ... Selfishness rules supreme ... War is a reversion to barbarism, and it is destined to become ever more terrible because of our greater mastery of the mighty forces of nature which are enlisted in such struggles ... The Great War had not ceased until profiteering and industrial war were found raging with greater fury than ever before ... Many worship the almighty dollar more than they worship Almighty God. And out of the struggle for life there evolves an aristocracy of wealth that is one of the most offensive kind. In haughty selfishness it gloats in exclusiveness. And it often knows no moral restraints in the execution of its greedy aims ... Progress means a rationalizing of life, a subduing of the instinctive impulses to the higher rational nature, and the enthroning of the good will ... It profits a man nothing if he gains the world but loses his soul. (pp. 278-279)

Unable to comprehend the full role of the social, as it was a force that had been coopted and consumed by the totalizing logic of capital, Painter, a sociologist, concluded that it was war that had warped judgment, not the organizational mode of modern life. Then, in a historically deficient manner, he fell back on binary thought by mourning the replacement of religious logic with capitalist logic, as if that prior guiding logic did not lead to the same destructive outcomes. And, finally, ignoring any insights from the classics, he concluded that "[e]ducation can do much, but in the last analysis it is an individual matter" (p. 279). This individualistic mindset was firmly planted within the discourse of the logic of capital. America and its allies used the individualistic argument to justify their turning a blind eye toward the ways that they had shaped the social landscape: a landscape that produces individuals whose minds are not structured to develop a consciousness that is inherently critical of the organization of modern life. As that would be the necessary precondition to reorganize society in a peaceful manner, one which would account for the psychosocial needs of all and thus would be an existential threat to bourgeois society.

On the American home front this meant ignoring that "the year 1919 was marked by the most widespread strikes in U.S. history, headlined by the Seattle General Strike, the Boston police strike, and the Great Steel Strike" as well as "more than twenty major race riots," the start of "the Red Scare" and "an ensuing set of mass deportations of so-called 'alien radicals'" (Jensen & Nichols, 2017, p. 241). For the elites in power controlling the anxiety of the white masses was far too successful a stratagem for maintaining their positions, regardless of the occasional global and/or economic costs. They simply displaced these costs back on the masses, or they were orchestrated in such a way that the costs were borne by those whose identities were structurally and systematically excluded from genuine social integration by the elite classes who continued to reap the rewards plundered from the sweat of those whose labor built modern society. Ultimately, blame, if there was any to be assigned for the failings on the side of the victors, was placed on individuals and groups who were subjected to the "othering" process and treated as outsiders, while blame on the losers was placed squarely on their supposedly national and ethnic failings.

The Allied forces that won the war could, therefore, only find genuine sociological fault on the losing side and refused to account for their own mode of social organization in precipitating the conditions that proved fertile ground for conflict. In the mainstream social sciences and in the public discourse there was no attempt to reconcile the racist legacies of their empires with the colonialist project. Even America, which bragged about being a melting pot and believed itself to represent a new form of politics based on individual freedom, was full of contradictions that it refused to face. The Italian social scientist, Vilfredo Pareto (2014) called this out with a biting tongue when he wrote: "'Democracy' in the United States of America has, as a principle, that all men are equal; that is why in that [civilized][8] country Negros and Italians are *lynched*, and Chinese immigration is forbidden, whereas war would be declared on China if Americans were excluded from that country" (p. 54). Furthermore, there was no attempt to critically engage with the system of capital to learn how it reproduced inequalities, led to crises, and was propped up on the continued exploitation of the masses; that was, after all, the very purpose of the system and it was controlled by the people who benefitted most from it. And there was little conscious acknowledgement of how an economy structured around the continued development of more advanced military technology left modern societies on a sure footing toward new and more terrible wars, as the massive investments those projects required only seemed justified when they had practical use against enemies real and imagined. By parading as the moral victors, the Allied forces reproduced the conditions that led to the War in a way that allowed them to continue to feel morally superior while crippling the future of the "othered" masses who lived as second-class citizens in their own countries or in the foreign lands exploited for their resources and cheap labor who bore the brunt of the capitalist quest for mass wealth extraction.

American intervention in the war also began to affect the balance of global power, shifting it toward the comparatively new nation, making them feel

---

8 Brackets in original.

justified in the sociopolitical course they pursued. Despite the massive costs of the war by putting capital in motion:

> [t]he war was ... a watershed for the U.S. economy and the nation's banks. The United States was a debtor nation when the war began in 1914. After the war, with many parts of Europe in ruins and desperately in need of reconstruction loans, the United States supplied much the capital and became a net creditor nation. In the process, New York emerged as the world's leading capital market [taking the throne away from London].
>
> J. P. MORGAN CHASE AND CO., 2018, p. 8

Although America shared in the human and psychosocial costs of the war, they avoided fighting battles on their soil and with their industries intact America was poised to assume a greater role in shaping global affairs to their advantage; primarily by leveraging their economic capabilities over the project of European reconstruction.

The reproduction of the social conditions that led to war began anew with the reconstruction process as outlined in the peace treaty that ended the conflict, which had optimistically been dubbed, the War to End All War. The economist John Maynard Keynes was appointed as a representative of the British treasury to attend the Versailles peace conference. Although he was a defender of capitalism he recognized that without democratic intervention and government regulation of economic affairs the structural tendencies of this mode of social organization would continuously divide people in ways that had dramatic societal, and therefore psychological, consequences. As a social scientist who had studied the socio-economic conditions that precipitated the war, he was opposed to making a treaty that was punitive to the German peoples. If peace was the goal, then the cultural, national, and economic tensions that led to it would only be amplified if the Allied demands for the conditions of peace exacerbated the conditions that caused the war in the first place. But Keynes approach was overruled, and rational thought was superseded by nationalistic pride and notions of revenge as the other British representatives swayed President Wilson to their way of thinking. Blame for the war was planted squarely on the losing countries. Like the European Allied nations, Germany was also facing a devastated landscape, infrastructure, and industrial base, but Article 231 of the treaty insisted that they accept responsibility "for causing all the loss and damage to which the Allied and Associated Governments and their nationals have been subjected as a consequence of the war" (Paris Peace Conference XIII, 1921, pp. 137-138). In Article 232, despite knowing that "the resources of Germany are not adequate ... to make complete reparation for all such loss and damage[, t]he Allied and Associated Governments, however, require, and Germany undertakes, that she will make compensation for all damage done ... with interest" (p. 138). To ensure that Germany would pay, the treaty granted the commission the power to record data on German commercial activity and set the levels of their national taxation rates. America lent money to France and England to rebuild, and when they could not pay those debts, America lent money to Germany to pay the reparations to France and England, who then sent the money back to America.

The consequence of this decision rattled those who had tuned their minds toward a more critical analysis of modern society and the structural, as well as historical, causes of this monumental war. Although defeated, Keynes (1920) was not silent about his criticisms of the treaty, writing that

> the spokesmen of the French and British peoples have run the risk of completing the ruin which Germany began, by a Peace which, if it is carried into effect, must impair further, when it might have restored, the delicate, complicated organization, already shaken and broken by war ... perhaps it is only in England (and America) that it is possible to be so unconscious ... the earth heaves and no one but is aware ... of the fearful convulsions of a dying civilization. (pp. 3-4)[9]

While critical of the economic conditions that led to the war, and willing to attack President Wilson "as an old man ... who neither expects nor hopes that we are at the threshold of a new age" (p. 36), Keynes did not link the resulting crises to the core structural logic of capital and its totalizing effects. Instead, Keynes believed that the problem with capitalism was that it was being held in check by the protestant ethic (or, as he called it "those instincts of puritanism"

---

9 These sentiments were also shared by Vilfredo Pareto, who wrote: "In our own days the rivalries of great business interests played no inappreciable part in causing and in prolonging the World War; and there is reason to fear that these interests are today preparing the way for new conflicts ... Germany's prostration, Russia's chaos, the menacing revival of Islam, and other mortal ills weigh heavily upon the world ... Europe has fallen into inexplicable contradictions. We know positively that no country can meet heavy periodical payments to another country unless it can export its merchandise. Yet we expect Germany to pay enormous sums to neighboring countries, although we prevent her from exporting the products of her labor, fearing lest she flood our markets (p. 447). So our visions of prosperity after the war have proved fallacious; and the mirage of a universal political concord vanishes the moment we seek to grasp it. Dark clouds lower on the Eastern horizon. Germany is not penitent, nor will she relinquish her projects of revenge ... Common political interests will sooner or later make Germany and Russia allies. An armed invasion of Western Europe by these countries is not immediately to be feared, but it remains a future peril" (p. 449).

(p. 20)), which demanded work but forbid the consumptive side of the equation on moral grounds. On the one hand, "it was precisely the *inequality* of the distribution of wealth which made possible those vast accumulations of fixed wealth and of capital improvements which distinguished that age from all others" (p. 19). But, on the other hand, by not placing enough of the capital in motion to secure the continuous development of the productive forces, "[t]he duty of "saving" became nine-tenths virtue and the growth of the cake the object of true religion" (p. 20). Since the system was built on the unequal distribution of wealth, and since there is a fixed material limit on the consumptive capabilities of any one person (especially with a puritanical stance against the consumption of luxury goods), the class that had the most capital was the least likely or able to consume it. They were only able to amass the capital so long as the laboring classes maintain a belief that there is not enough for all to enjoy the fruits of labor, so amassing capital becomes the end game of all. However, according to Keynes,

> The war has disclosed the possibility of consumption to all and the vanity of abstinence to many. Thus the bluff is discovered; the laboring classes may be no longer willing to forego so largely, and the capitalist classes, no longer confident in the future, may seek to enjoy more fully their liberties of consumption so long as they last, and thus precipitate the hour of their confiscation. (p. 22)

In other words, Keynes bought into a version of the Fordist project which would require the development of a consumer class and, if not the complete redistribution of wealth then, at least a greater share of the wealth going into the hands of the laboring classes as a means of pacifying the masses to prevent them from overthrowing the capitalist class by force.

Although America found an economic boost by becoming a creditor nation to the European rebuilding effort, the effects on the national economy were limited because the government could not maintain wartime spending levels on American production which had boosted the economy in those years. Prior to the war, as pointed out by J. P. Morgan above, the U.S. was facing a recession. Selling goods to the Allied war effort and then entering the war in 1917, the US faced an economic boon during the war years and "unemployment declined from 7.9 percent to 1.4 percent in this period" (Lozada, 2005). But this was not enough to cover the costs of war, and tax rates were raised primarily on the wealthiest individuals and on corporate excess-profits which "accounted for about two-thirds of all federal tax revenues during World War I" (Brownlee, 2004, pp. 64-65). In the immediate aftermath of the war, however, without sustaining the increased governmental spending on the war effort or the increased need for the mobilization of the productive forces, the U.S. fell back into recession and then into a brief depression until 1922. "Factory employment dropped 30 percent from March 1920 to July 1921. Unemployment rose above 4 million. Two years after the end of the war, more workers were out of jobs than ever before in the United States" and the cause was seen as "run-away inflation" (Woytinsky, 1945, p. 20). In an attempt to get the Fordist project back on its feet, electrified factories and assembly lines were introduced in industries that wanted to copy the success of the automotive industry. This provided a new production boon in 1922, but socially there was unrest in America caused by mass economic anxiety. The problem was that "[m]ass unemployment ... had left bitterness and frustration, particularly among ex-servicemen, who sincerely believed that they had fought to make the world safe for democracy and found themselves without jobs after they came home" (p. 20). With a mass of trained military men, many still suffering from the psychological effects of the war, who were bitter about returning home to the anomic conditions that awaited them in everyday life, the U.S. had a potentially hostile political climate on the home front.

Rather than thinking long-term about how to shape a better American society that could meet the continued needs of all its citizens, business interests fought back against the high tax rates imposed during the war. A long line of Republican politicians took power and shaped policy to meet the demands of the capitalist classes. President Warren G. Harding, Wilson's replacement, appointed one of the richest men in America, Andrew Mellon, to the position of Secretary of the Treasury. Having vast reserves of capital, the elite class that Mellon represented did not share the same anxieties as the laboring class; their anxieties related to how they would grow their piece of the economic pie and retain control over the means of production, while the laboring class was again faced with the anxiety over how they would find a job to meet their bare requirements for survival. The elites could weather economic downturn in a way that the masses could not by drawing on their financial reserves, but they could not tolerate social restraints placed on their capitalist ambitions. In his published views on taxation, Mellon (1924) called upon Congress "to remove the inequalities in [the tax] structure which directly injure our prosperity and cause strains upon our economic fabric" (p. 14). Since the vast majority of working Americans did not pay a federal tax at that time, it is clear that he was referring to those of his ilk as being the bearers of this "inequality." This supply-side economic policy was based on an empirically erroneous belief in a trickle-down effect: that by allowing the wealthy to keep more of their money they would invest more of it in ways that would stimulate the economy.

Mellon successfully convinced Congress to cut taxes in 1921, 1924, and 1926. Those years came to be known as the *Roaring Twenties* as the immediate economic impact had the desired short-term effect of boosting the top line numbers of the US economy. However, rather than reinvesting in the productive and consumptive forces, the elites found financial speculation on Wall Street to be a quicker and easier route to increasing their capital, meaning that the average American did not see much economic gain from these policies. Without building a solid material basis for the American economy, the gains of the 20s were short lived and by 1929 the stock market crashed as it never had before. The economy was in shambles and the United States found itself in the midst of a Great Depression with effects that rippled across the globe.

When Franklin Delano Roosevelt became president in 1933, he embraced the economic policies advocated by Keynes and began a series of social reforms known as the New Deal, with the goal of righting the economic ship and finally creating a society based on a consumer class. Among the laws enacted in Roosevelt's policy shift, were: "the Emergency Banking Act, the Economy Act, the Federal Emergency Relief Act, the Agricultural Adjustment Act, the Emergency Farm Mortgage Act, the Tennessee Valley Authority Act, the Truth-in-Securities Act, the Home Owner's Loan Act, the Glass-Steagall Banking Act, the Farm Credit Act, and the Railroad Construction Act" (Dobin, 1993, p. 10). With these actions the pendulum made a full swing from an economic strategy that emphasized the self-regulating market to a system based on social protection through increased government regulation of economic affairs. This is what the economist Karl Polanyi ([1944] 2001) would come to call the "double-movement" between the desires of business interests for deregulation and the simultaneous fears that arise when the economy starts to become disembedded from the social mechanisms of the political system. With protections and interventions coming from Roosevelt's administration up and down the social ladder, across all industries and consumerist markets, he enjoyed mass popular support and won four consecutive terms as president. However, while many Americans were emerging from the Great Depression to a marginally better standard of living, fears stemming from the Great Depression did little to pull America toward the consumerist society mentality. The day of reckoning warned by Keynes and Pareto was finally coming to a head in Europe where the specters of fascism and communism were rising to challenge the democratic capitalist model of the West.

In the post-war years, while America doubled down on the capitalist mode of organization and continued to affirm a commitment to their version of the democratic project, several European nations were experimenting with alternative forms of modern politics. In contradictory fashion, the American messaging about the development of military technologies and involvement in war cautioned restraint, while their actions rang a different tone. As early as 1909, in Italy the Futurist movement led by the poet Filippo Tommaso Marinetti counseled throwing restraint to the wind by embracing a radical form of modernism that preached technological violence and war as the only vehicles for paving a new path forward. In Marinetti's manifesto he celebrated "the use of energy and recklessness as common, daily practice" and glorified "aggressive action … speed … man behind a steering wheel … violent assault upon the forces of the unknown … war—the sole cleanser of the world—militarism, patriotism … and scorn for women" (Marinetti, 2006, pp. 13–14). This hyper-masculinist celebratory orgy of the darkest forces of modernity was founded on an impatience with the slow pace of social change in democratic life as it dragged the burdens of the past along with it (burdens that even Marx acknowledged weigh "like a nightmare on the brains of the living"). It was a contradictory vision at once conservative and progressive. Conservative in the desire to maintain the traditionally aggressive, conflict-based, patriarchal mode of life that was amplified by the creation of masculinist technologies of war. Progressive in the sense that it wanted to overturn the slow pace of traditional life (and in a contradictory fashion many features of bourgeois life as well) and replace it with whatever the modern world could offer by way of new possibilities; no matter the social, psychological, or biological costs. This attitude paved the way for the rise of a fascist politic in Central Europe that would emerge in the interwar years as a rival to the democratic vision of the West and the communist vision of the East. Its roots were in the aesthetic movements of the avant-garde, such as Marinetti's Futurists, which the German social critic Walter Benjamin would describe as a form of "self-alienation" that experiences "its own annihilation as a supreme aesthetic pleasure" (Benjamin, 2008, p. 42). Specifically, this model of thought rose to prominence in three European countries in the 1920s and 1930s.[10]

---

10 A similar transformation occurred in Japan when Shōwa nationalism became the guiding political ideology in 1926. This shift in Japanese politics followed a similar economic pattern as the countries in Europe but it lies outside the scope of this project. My focus here is more narrowly limited to consideration of societies that emerged from within or regularly engaged with the Western model of modernism, which only took root in Japan during the post-war years. For more on the Japanese move toward fascism in the lead up to World War II, see Tanin & Īogan, 1934; Shillony, [1981] 2001; Fletcher III, 1982; and Hofmann, 2015.

First, in the immediate aftermath of World War I, Italy was facing economic catastrophe. Between 1919–1920, there were nearly 4,000 strikes across industry and agriculture with millions of participants expressing their economic anxieties which they blamed on the bourgeois order. Riding the anti-capitalist wave, the pro-labor Socialist party became the "largest single political force in Italy" (De Grande, 1982, p. 24). In 1922, Benito Mussolini led a reactionary faction to power in Italy and provided state-backing to a new brand of political ideology, personally "morphing ... from [a] socialist agitator to the leader of a new revitalization movement called Fascism" (Griffin, 2007, p. 204). As a youth Mussolini experimented with several political philosophies, including the work of Marx and Pareto, and his early thought was marked by an antinationalist stance. By 1909, though, his views shifted as his goals took on a more pragmatic dimension centered on the mobilization of Italy toward a new form of modern life. Mussolini determined that the surest way to accomplish this goal "was a form of national socialism that was at once elitist, voluntaristic, moralizing, mass-mobilizing, and antiparliamentarian" (Gregor, 1979, p. 99). His approach to modernism built on the cultural groundwork laid by the Futurists and was in stark contrast to the Fordist/Keynesian model in the United States.

For Mussolini, capitalism was the vehicle for obtaining an industrial base, but the consumerist version in America was a distortion of the social power that capitalism unleashed and representative of bourgeois decadence. If the anxieties of the masses could not be solved by the nexus of capitalism and democracy, and if socialism was rejected by those who would gain power at all costs thereby distorting the project, then whatever outcome fascism produced was a secondary concern, because at least by embracing an unrestrained modernism it would change the world. Mussolini's fascism presumed to abandon any notion of anxiety as a guiding force altogether. Of course, anxiety gave rise to his fascism, but in practice it was at the same time an attempt to repress anxiety at the highest levels of social organization by providing a closed narrative for society to follow; something that had been lost when capital overtook the logic of religion.

The perceived flaw of the democratic model, which Mussolini dismantled in Italy in 1925, was that it was ultimately directionless because it allowed too much freedom for individualist pursuits at the expense of the grand projects of social mobilization. And social mobilization was necessary for any attempt to build a different future than could be imagined from within the material reality constructed by the Western model. Although Mussolini's politics were anti-democratic, they were more amenable to capitalism than the communist alternative. When push comes to shove, the West has always chosen the economic logic of capital over the political logic of democracy, as demonstrated when "both presidents Hoover and Roosevelt expressed their approval of Mussolini's regime" (Tooze, 2016).[11] It is not altogether clear what world he imagined his fascist politics would build, as he never wrote a manifesto or programmatic outline of his ideas, the only clear point was that it would use the social to cause change and disrupt the form of modern life that many were rejecting for its rampant alienation, anomie, and blind following of the capitalist spirit, even if it meant creating a world with more alienation, anomie, and repressive tendencies.

Next, in Germany unemployment varied after World War I and was largely kept at bay by the rebuilding efforts, but by 1932 the unemployment rate had grown to a catastrophic 43.8 percent (Walter & Zeller, 1957). Playing on the utter chaos this economic situation caused for the Germans, in 1933 Adolf Hitler became the Chancellor of Germany with the support of his National Socialist (Nazi) Party. As a youth, Hitler participated in a failed coup which led to his imprisonment and the writing of his manifesto, *Mein Kampf* (1941). Therein he laid out his vision of a Germany once again in control of its own destiny and free from the punitive measures of the Treaty of Versailles. The evidence that the treaty had taken its toll on the German peoples was visible in all areas of German society, but perhaps nowhere was this more clear than in the mass unemployment which carried an enormously negative toll on the psychological and social state of German citizens as predicted by Keynes and Pareto. Like Mussolini, Hitler (1941) studied social democracy and Marxism as a youth, but came to view the projects as having ulterior motives hidden beneath the words which he interpreted as a global Jewish conspiracy and marked his turn "[f]rom a feeble cosmopolite ... into a fanatical anti-Semite" (p. 83). His political project was to reunify the German empire, and it involved motivating a psychological revolution in a way that the political revolution of communism had failed to mobilize the productive capabilities of a people unified by a singular political project. Although Hitler played up the economic anxieties and inequalities externally imposed on Germany by the West after World War I, the central thesis of his manifesto was decidedly based on racial unity which could only be achieved socially by creating a common enemy in the minds of the German people. Since there was still a lot of international attention and control over Germany as a result of the treaty, at first Hitler did not turn his ire against the forces in the West who drafted the terms of the treaty, rather the enemy that Hitler identified was the German Jewish population. They served

---

11   On the American views of Mussolini's politics, see also Diggins, 1972 and Migone, [1980] 2015.

as an easy target for his plan because there was already a long and deep-seated history of anti-Semitism in the region, the Jews had no real foreign allies to call on for support, and Hitler claimed a Christian religious backing of his project of racial cleansing (p. 84). The result was that there was not much condemnation of his views by the supposedly "Christian" nations of the West whose own relationship with Jewish populations was only marginally less vocal with their anti-Semitism.

Despite claims that Hitler's brand of fascism was anti-capitalist, anti-modern, and not a result of the structural logic of capital—because of the overtly hostile, racially charged, and anti-Semitic language in his manifesto (Pellicani, 2012)—it is wrong to conclude that his project was not structurally linked to the logic of capital. First, the economic conditions that proved fertile ground for his rise to power were a direct result of international protectionist policies meant to prop up their capitalist markets while blocking German exports. Second, Hitler (1941) not only praised Henry Ford as a "great man" (p. 930) but Nazi engineers, German auto manufacturers, and industrial and political bureaucrats turned to the Fordist model of industrial rationalization in the construction of their political economic program. Where Hitler differed was not in the industrialization and modernization aspects of capital but, like Mussolini, in the creation of a consumer class which in emphasizing the freedoms of the individual ran counter to the mobilization of the social which he deemed necessary to achieve his ambitious ends of an Aryan society (König, 2004). Hitler's vision required mass industrial capabilities to build his war-machine and obtaining these had proved difficult in the communist model which he had rejected on the grounds that it originated in Jewish thought. Ford's model, however, had proven to be the most efficient vehicle to obtaining advanced industrial capabilities, so Hitler willingly adopted it.

Furthermore, the official American response to Hitler's rise to power was not an outright rejection of his politics as illegitimate in the modern capitalist world order. As a result, they did not bear the brunt of his polemics, nor did they intervene with his rise to power as they were more concerned with the anti-capitalism of the Soviets. Following the contradictory nature of American sentiment, although there were public protests against the Nazi's for their anti-Semitic stances in America, there was no mass demand for political action to back up those words as the American public was still war-weary from World War I and again viewed Hitler's Germany as a European problem. Roosevelt came into power only a few weeks after Hitler and was busy dealing with domestic issues stemming from the American Great Depression, so "the keynote of his response to Hitler beginning in 1933 was appeasement" (Marks III, 1985, p. 970). The plight of the Jews was at best dismissed as an exaggeration, and at worst considered an acceptable sacrifice if it meant postponing war on the European continent and the continuation of business as usual in the West. Emboldened by their non-interference with his fascist politics, Hitler consolidated power in the 1930s and put Germany on a new footing in the international world order.

Finally, there was Spain. Spain was a neutral country in World War I and boosted its economic reserves through "[m]assive exports to both sides of the conflict" but after the war "exports … dropped by 39%, while imports grew by 33% between 1919 and 1922" (Giménez & Montero, 2015, p. 201). This led to a divide between the laboring classes and the industrialists with each side fighting for political power to protect their interests when national industry slowed to a crawl. There is not enough reliable data from this period in Spain, but the available data does suggest that by 1930 the fights between the unions and the industrialists over the preceding decade caused a rise in unemployment, as strikes and lockouts exacerbated the situation, and "in many industries work was limited to two, three, or four days a week" (Garner & Benclowicz, 2018, p. 8) making the conditions of the labor class precarious at best. Per capita growth from 1930–1935 went into the negative (-0.97) causing a depression and ushering in the conditions that led to the outbreak of a Civil War lasting from 1936–1939. That war placed a coalition of leftist oriented political factions, led by the Republicans who supported labor rights, against the right-leaning nationalists who had wide Catholic support. Although the "Republicans in Madrid were in power through legitimate election" (McVeigh, 2009, p. 261) the tensions of the Great Depression in America had split support of the war and Roosevelt chose a neutral stance, which effectively gave a tacit support to the anti-democratic nationalists. France and the United Kingdom also assumed positions of non-intervention, which left the forces in Spain to turn elsewhere for support. The Republican's received support from the Soviet Union who sent "the latest-model planes and tanks, accompanied by hundreds of Soviet military advisors and specialized personnel" and the Comintern sent "approximately 42,000 foreign volunteers" (Payne, 2008, p. 7) to aid in the fight. The nationalists, led by General Francisco Franco, secured the support of the Italian Fascists and the German Nazis. In the bitter war that followed, the Nationalists won. This effectively crushed the forces of labor and allowed Franco to become a military dictator. He then chose to adopt a blend of the Italian and German political styles with fascist politics at the core, finishing out the triad of fascist regimes that came to power in Europe.

Beyond the threat of fascism and the democratic model of the West, the other political contender in Europe at the time was communism. After the Russian Revolution in 1917, without instituting the capitalist model, Lenin set the Soviet Union on the path to industrialization through a centralized and planned economy. Having read Marx's theory of class exploitation and come to recognize the alienating condition of life under the system of capital, Lenin believed that by having this knowledge one was compelled to act on it and should therefore make every attempt to skip the stage of capital development to spare the peasantry from the painful process of becoming a proletarian class. To justify this, Lenin elevated political consciousness over the economic order, saying that the economic "*framework is too narrow*" (1969, p. 78). To accomplish his politicization project, he determined that the theory must come from outside of the proletariat because they are trapped within the economic order and are unable to develop political consciousness internal to it. Lenin, therefore, posited a vanguard—that is, an intellectual class acting on behalf of proletarian interests—who thinking the totality "must "go among all classes of the population" as theoreticians, as propagandists, as agitators, and as organizers" (p. 81). Without having the proletariat class at his disposal, Lenin assumed that his vanguard theory could be developed in the Soviet Union and exported to the Western laboring classes. If successful in totalizing the minds of the Western proletariat, it could then pull them under its umbrella as part of a global revolution. Lenin went so far as to say that "an incipient movement in a young country can be successful only if it makes use of the experiences of other countries" (p. 26) thereby resolving the Soviets from having to become proletarian themselves, while still allowing them to play a leading role in bringing about the end of history with communism.

Starting in 1919, however, with the post-World War I defeat of the proletarian revolutions in Germany and Hungary, Lenin's theory appeared either as a failure or indefinitely delayed. Rather than admitting defeat and abandoning the Soviet project, Lenin adopted the latter mentality and saw his as an incomplete project that was to remain always partially there in becoming total at least until historical circumstances changed, which was deemed better than doing nothing at all to advance the cause of communism. However, in doing so, by the time of his death in 1924, he had left the door wide open for Stalin's shift away from thinking the totality to thinking its perversion of totalitarianism. This theoretical leap toward a ruthless pragmatism turned a humanist ideology into a statist one. Socialism under one-state was developed theoretically by Nikolai Bukharin in 1925 and was adopted by Stalin as the official Soviet policy in 1926. It served as a final attempt to justify the continuation of the communist solution as a partial totality, ignoring the contradictions of such a stance that were already philosophically demonstrated to be wanting by Marx and Lenin (although Lenin was wavering on this by the end of his life). In other words, the perversion of Marxist ideology created a material condition that was open to further perversions. The material historical account then, rather than bringing liberation to the oppressed masses that spared them from modern anxieties, shows that the Stalinist project amplified its own brand of modern anxieties and reproduced conditions of exploitation and oppression, leading to the deaths of millions of innocent victims swept up in Stalin's hunger to maintain power (Wheatcroft, 1999).

Prior to the revolution in 1917, the country was primarily populated with an agrarian peasantry, but after the war, Lenin made efforts to restructure the economy. By 1928 it was "third among oil producing nations" with an output that "was nearly 25% above pre-war production" (Soviet Union Information Bureau, 1929, p. 83). As with the case of Spain, the Soviet Union did not publish unemployment rates, as their political ideology was based on the notion that it provided full employment, but the US based Soviet Union Information Bureau estimated that unemployment hovered between 1 and 1.35 million between 1924 and 1928 (1929, p. 190). The realignment of the Soviet economy did not accelerate the pace of growth, as gains made in industry were canceled out by losses in agriculture. As more peasants flocked to urban zones to join the industrial economy, the unskilled labor force grew too large to be absorbed in the newly founded industries. Without first industrializing the agricultural base to increase its efficiency, the loss of agricultural workers hit the animal industries and grain production hard. The result was "a great famine in Russia in 1932 and 1933, with up to several million victims" (Smirnov, 2015, p. 138). Under Stalin's leadership, the Soviet Union assumed a far more pragmatic approach to the goal of maintaining power and balancing the global hegemony of the West, rather than attempting to build the good life in a modern society for Soviet citizens. This program, therefore, required similar tactics as fascist governments, which included the silencing of protest and the authoritarian exercise of control over most facets of everyday life.

(1) With increased global anxiety over the seemingly ever-present crises of the capitalist mode of production and the insistence on sticking to capitalism no matter the psychosocial costs, (2) with decreased standards of freedom spreading across the European continent and questionable standards in the West, (3) with Hitler's ambitions for a racially based society and the West's persistent ambivalence on its own racist practices: global tensions in the 1930s reached an all-time high eclipsing those that had led to World War I. In 1939 the hostilities broke out when Hitler invaded Poland and began his genocidal project against the European Jewry. There has been so much written on World War II analyzing and diagnosing its causes from nearly every perspective imaginable and the history of the war retains an enormous place in Western cultural consciousness due to massive propaganda efforts made to humanize the

Allied forces and paint them as the saviors of the world from the evils of alternative modes of modern sociopolitical organization. Because of the plethora of available literature and cultural artifacts related to the event, I will limit my comments to a few key and pertinent features that distinguished this event from World War I, briefly emphasizing the structural impact of the technological, political, and economic dimensions and the effects these had on the general societal anxiety.

It has almost become a cliché to argue that World War II was simply a continuation of the hostilities of World War I. When adopting this viewpoint, the major difference was that in the Second World War everything was amplified and intensified, taking on larger and more earth-shattering proportions. The first amplification was in the number of countries who directly participated in the war. As in the previous War, America at first took a non-interventionist stance while implicitly supporting the Allies by selling them supplies in support of the war effort (Department of State, 2018a). At the outbreak of the war the Allied forces were made up of Poland, France, and the United Kingdom. Germany joined forces with Italy and Japan forming the Axis powers. Spain again claimed a neutral stance and, as Franco was more concerned with Spanish nationalism than with Hitler's crusade, he offered tacit support to the Axis powers but fell short of signing onto the war because of the pressure Spain was under to keep their economy afloat with imports from Allied nations (Bowen, 2006). After Germany invaded the Soviet Union in mid-1941, the Soviets joined the Allied powers in an uneasy alliance where both sides viewed the German fascist threat as a greater existential danger than that which existed in the tug-of-war between Western capitalism and Eastern communism (Department of State, 2018b). After Japan attacked Pearl Harbor at the end of 1941, the United States officially joined the Allies. Similarly, due to persistent conflict with Japan in the 1930s, China officially joined the Allied war effort at the end of 1941 (Mitter, 2013). By the time the War ended, nearly 50 countries from around the globe had joined the Allied forces, 9 had joined the Axis powers, and fighting encompassed the European and Asian continents, dipping into Africa and Australia, stretching across the Pacific and its islands, and impacting every corner of the globe.

The second amplification and the greatest destructive impact on World War II came from engineers and scientists who had 20 years to perfect the technologies that were developed and deployed in World War I. Furthermore, "[m]ilitary doctrine and tactics caught up with technological change in World War II ... The artillery, tanks, aircraft, and other machines of World War II were neither new in concept nor strangers to the battlefield. Rather they were simply more capable versions of earlier machines" (Hacker, 2005, p. 262). Enhanced air capabilities enabled new tactics such as the German Blitzkrieg, or lightening war, which combined technology and tactics to use speed as a means of causing mass disarray, confusion, and destruction, particularly in France (Frieser & Greenwood, 2005; Powaski, 2006). Because of the rapid pace of the attacks and their use of aircraft, these bombardments could happen anywhere at any time, which kept potential targets and the people who lived therein in a permanent state of heightened anxiety. Building on the success of these technologies, new developments occurred in the use of rocket technology, notably by the German's with their development of the V-2 which they used in raids on Britain (Haining, 2002). As both the blitzkrieg campaigns and the use of rockets suggest, Hitler had learned from the slow pace of trench warfare in World War I that if the war was fought using traditional tactics which were designed for traditional weaponry, then it gave the enemy time to develop new technologies and strategies to regain an advantage. By focusing on the speed side of the technological equation, which was unleashed with the enhanced use of machines in warfare, he hoped to catch the Allies off guard and crush them before they could develop successful counterstrategies. Indeed, this was the only way he could conceivably win the war, as he was significantly outnumbered and outspent by the Allies (see below).

Despite the disarray and speed of these offensive tactics, the Allies invested in the development of new defensive technologies specifically aimed to counter the speed advantage. For example, radio technologies were improved in the 1920s and adapted to military purposes in World War II. This permitted real-time communications between command posts and soldiers on the battlefield which enabled the transmission of enemy movements. In the 1930s several countries, including those on both sides of the conflict, used what was learned from radio research to develop radar capabilities that could warn of impending attacks and spot enemy transport beyond the range of the human eye, thus giving more advanced warning of these speed attacks. But it was not until the urgency of the War that the technology was rushed to be finished for use in battle (Mckinney, 2006a; 2006b; 2006c; 2006d; 2006e). As Painter (1922) diagnosed the situation, the technologies of war invented in World War I could not be un-invented. Even the countries who had argued for the banishment of certain technologies after World War I continued to refine them in the inter-war years, added them to their arsenals, and thought up new and more efficient ways to eliminate human lives in war.

Two of the most novel technological breakthroughs achieved in World War II had to do with the development of codes and ciphers—(and especially how to break them) which would pave the way for the post-war field of cybernetics—and advances made in chemistry and physics on the atomic structure. It was these developments that came to usher in a new technologically based logic which would come to totalize and dominate the post-war years in a way that rivals capital's guiding logic. The field of cryptoanalysis was notoriously led by the British scientist, Alan Turing[12] at Bletchley Park, who in 1941 sent the British Prime Minister Winston Churchill a letter "arguing that it was essential to give the highest priority to the recruitment of codebreakers and the provision of necessary equipment" (Hilton, 2000, p. 2) so that the Allies could break the German Naval Enigma code and the "German machine-cipher known in Bletchley as "FISH"" (Tutte, 2000, p. 9). In the 1930s the American mathematician Claude Shannon applied Boolean algebra to the problem of circuit simplification. By reducing the tasks of circuits and relays to a simple on/off binary, in which "the symbol 0 (zero) [is] used to represent the hindrance of a closed circuit, and the symbol 1 (unity) to represent the hindrance of an open circuit" (Shannon, 1940, p. 4), Shannon had invented a simplified mathematical language with widespread practical applicability for use in solving a variety of technical problems. With this binary language Shannon not only paved the way for faster data processing fulfilling the speed requirement, but he also laid the foundation for a language that could stand in as a substitute for any other language thereby simplifying its machinic reproduction and alteration. The British codebreakers were able to use this simplification to their advantage as they developed more sophisticated codebreaking machines to support the Allied efforts. The work done by these cryptanalysts paved the way for the post-war fields of cybernetics and computer science, introducing modern society to its information age (Pickering, 2010; Kline, 2015).

The other series of novel technologies developed during this time emerged from advances in chemistry and physics. They arguably had the most dramatic impact on World War II and subsequent impact on the mind of modern subjects in the post-war years, at least manifestly as opposed to the more latent impact that the cryptographic sciences had on the public in those years. These advances culminated in the development of the nuclear bomb

---

12 Turing's tragic life highlights the massive contradictions that dominated the West and its ideological conceptions of itself. While Hitler was busy sending Jews, homosexuals, Romani, Jehovah's Witnesses, and other groups he labeled as degenerate to the concentration camps, in England homosexuality was illegal and violators of the law were subjected to harsh punishments. In 1951, Turing was arrested for homosexual activities; in 1952 he was tried and "forced to accept injections of oestrogen;" in 1953 "as a known homosexual he fell into the new category of security risk" and although his research was instrumental in helping the Allies win the War, he was no longer allowed to "continue the secret work he had previously been doing;" in 1954, with his reputation destroyed by the British government, he committed suicide by "cyanide poisoning" (Hodges, 2004, p. 7).

with America's Manhattan Project. Although the Americans were the first to successfully develop the bomb, "[n]uclear fission was discovered accidentally in Nazi Germany on December 21st, 1938, nine months before the beginning of the Second World War" by the chemists "Otto Hanh and Fritz Strassmann, working at the Kaiser Wilhelm Institute for Physical Chemistry in Dahlem" (Rhodes, 2004, p. 17). Their discovery heightened fears that the Germans would use this knowledge to fashion more devastating weapons. The experiment was explained theoretically by the physicists Lise Meitner and O. R. Frisch (1939) in a paper they published in the scientific journal *Nature* in 1939. Then the Danish physicist, Niels Bohr, confirmed the experiments and immediately governments began investing money for scientists to learn how to control this process. Both the American theoretical physicist, J. Robert Oppenheimer, and the Italian-American physicist, Enrico Fermi, immediately realized the potential for this knowledge to be used in the manufacture of an incredible bomb (Weiner & Hart, 1972; Kevels, 1979). In 1943, R. Serber (1943) gave a series of lectures that outlined how this knowledge could be practically implemented as a military weapon, the conditions for its development, and the multiple kinds of damage the bomb would be expected to produce. With the groundwork laid, and fears that Germany was producing a bomb that would give them an unparalleled advantage in the war, President Roosevelt ordered the creation of several commissions to estimate the feasibility of the project, the necessary budget to get it off the ground and, using those findings, he quickly apportioned the funds for developing the bomb in secret (Hewlett & Anderson Jr., 1962).

Oppenheimer and Fermi were crucial to the next stages: they would come to be known as the father of the bomb and the architect of the nuclear age, respectively. Oppenheimer became the head of the Los Alamos Laboratory in 1943 and was chiefly responsible for developing the bomb for America and Fermi became head of the Argonne National Laboratory in 1946 where he developed the first nuclear reactor, both under the auspices of the Manhattan Project. On July 16, 1945, the first successful nuclear bomb was tested, and America won the race. Germany had already unconditionally surrendered in May, after a dual offensive by the Allies in the West and the Soviets in the East succeeded in capturing Berlin, but Japan refused the terms of surrender which prolonged the fight. Wanting revenge for the attacks on Pearl Harbor and a quick submission of the Japanese forces, President Truman, who had taken over after Roosevelt's death in April of that year, ordered the military to drop atomic bombs on Hiroshima and Nagasaki (Reed, 2014). Estimates of the death toll from just these two bombs surpassed 210,000 people, mostly civilians, who were wiped out in an instant (Yamamura, 2013). Devastated and humiliated for being the recipients of the single largest military attack on a civilian population in the history of our species, Japan surrendered.

The third and fourth amplifications—the death tolls and financial costs—could only fully be assessed after Japan surrendered and the hostilities of World War II ended. At the start of the war, the Axis powers outnumbered the Allies armed forces 6,262,000 to 5,480,000, but by the end of the war the Allies outnumbered the Axis with 28,620,000 to 15,560,000 combatants (Harrison, 1998, p. 14). It is difficult to accurately estimate the number of those who died, including the victims of the Holocaust and lives lost to related diseases and famines, but most have it somewhere in the range of 70 to 80 million total lives lost as a direct consequence of the War. The Soviets and the Chinese suffered the heaviest losses, with the Soviets losing an estimated 26–27 million lives (Ellman & Maksudov, 1994) and "a conservative estimate would put the total human casualties in China directly caused by the war of 1937–1945[13] at between 15,000,000 and 20,000,000" (Ho, 1959, p. 252). The Jewish casualties in the German extermination camps are generally estimated around 6,000,000, but recent scholarship conducted by Geoffrey P. Megargee (2009) in association with the Unites States Holocaust museum suggests the number could be more than double those earlier estimates. While there will never be a completely accurate accounting of all of those who suffered during the Holocaust, the number is surely larger than it would have been because of the West's reluctance to allow many Jewish refugees to enter their countries (Wyman, 1984; Hamerow, 2008).

The financial costs pale in comparison to the vast devastation of human life and the environment,[14] which can hardly be calculated in their historical and planetary impact, but nonetheless they were astronomical. The U.S. alone, is estimated to have spent $296 billion on the war, which in 2011 dollars translates to $4.104 trillion, nearly double the total *global* costs of World War I (Daggett, 2010). "[T]he Soviet estimate of the total war costs is 189 billion rubles, which exceeds total capital investment in the Soviet economy from 1928–1941" (Millar & Linz, 1978, p. 959), and which in current dollars also runs into the trillions and surpassed the total costs of World War I. When adding the total number of taxes collected in Nazi Germany during the war years and the total debt

---

13 This includes the casualties from the Second Sino-Japanese War, which predated the official declaration of World War II by about two years.
14 There is relatively little literature on the environmental costs of World War II but, as environmental concerns have taken a greater place in the collective conscious, over the last few years scholars have paid more attention to this dimension. See for example Gutmann, 2015.

they incurred over those years, Germany spent some 700,300,000,000 marks from 1938–1945 (Lindholm, 1947); which again, in current dollars eclipses the total costs of World War I but is far less than the combined Allied expenses. The combined GDP of the Allied nations when compared to the Axis powers, puts this in some perspective. The Allies GDP was about double the Axis powers GDP prior to the war, starting in 1938, but by 1945 it was five times the size (Harrison, 1998, p. 10). While this is in some ways indicative of the increased productive capabilities of the Allied powers—especially those who escaped fighting within their own borders—it also reflects the increased global wealth transfer to the Allied nations in the west that was at the center of Du Bois's diagnosis of the causes of World War I. Likewise, the view that this massive increase in GDP during the war years was a net positive for the West highlights how the psychosocial wellbeing of modern society has little to do with advancing the logic of capital and the promotion of business interests. The contradiction that all too often remains unexamined here, is that when the economy is the de facto measure of a society's health, business as usual can progress even when the people are suffering from mass anxiety and trauma.

That contradiction played out—as it continues to play out today—in sociology during the war years. Critical diagnoses of World War II were conspicuously sparse in mainstream academic sociology and few saw it as a sociological duty to call out these contradictions and expose the pathological structure of modern society. In Europe academic output was understandably limited by the war, but in America the attitude was often precisely that one should maintain business as usual; again, as during World War I, taking the war as a European problem and not a problem of modern society itself. Even when there was talk about how the war was reconfiguring society and therefore necessitated a reconfiguration of sociology, it was often taken as a chance to undermine the role of sociological critique rather than promote it. During the war the two leading sociology journals in America—*American Journal of Sociology* (AJS) and *American Sociological Review* (ASR)—made relatively few contributions to the social analysis of the war and paid scant attention to the impact it had on modern society writ large or on sociology. Beyond the rare article that alluded to the war in AJS, they published three issues that dealt with the topic of war[15]—(1) on the various scientific approaches to war, its typologies, and its intersections, (2) on morale, and (3) on the impact and effect of war on a variety of social institutions—and one which offered social

---

15 (1) Volume 46, No. 4, Jan., 1941; (2) Volume 47, No. 3, Nov., 1941; (3) Volume 48, No. 3, Nov., 1942.

forecasting perspectives on post-war life.[16] Most of the articles in these issues likewise studied the war as if it were an epiphenomenon of modern society but not causally linked to its structuring logic. The only article in AJS during those years that directly confronted the concrete material conditions that contributed to the war was an article by the sociologist Hans Gerth (1940), in which he offered a Weberian critique of the Nazi party. What set this apart, was that Gerth was expelled from Germany by the Nazi's and had immigrated to the United States before the outbreak of the war, therefore, because he was raised in alternative material conditions his thought was more attuned to the necessity of using a critical perspective in sociology to diagnose the situation than his American counterparts.

Likewise, in ASR, beyond a few scattered articles that looked at the effects of the war on select American populations, on the rare occasion that the subject of war was broached it was often done in the abstract with little or no direct engagement with the concrete material circumstances of the time. The most sustained acknowledgement of the war's effect on sociology in ASR was in a series of short statements on national affairs taken from a panel held on the topic and subsequently published in the April 1942 issue. Therein, the authors offered little variation in their views, overwhelmingly operating from a standpoint of methodological nationalism while issuing calls to maintain objectivity and value-neutrality despite the emotions provoked by current events. Carl C. Taylor (1942) in his summary statement, at the hint of sociological identity crisis expressed by members on the panel, advised that sociologists "forget their profession for the time being and do their part as ordinary citizens" (p. 158). He took this attitude to the extreme with a dismissive commentary and paternalistic advice that belittled the science and its practitioners, writing that

1. There is a great desire, almost anxiety, among sociologists to be useful in the present defense and war activities.
2. Many of them feel pretty helpless and others feel frustrated in the situation.
3. There is not much feasibility or practicality in many of their suggestions and little possibility that very many of them will be given opportunities to present their suggestions at the levels where administrative action occurs ...

There are probably two pretty basic causes for this plight of the sociologist. He has insisted on dealing with universes with the phenomena of

---

16 Volume 49, No. 5, Mar., 1944.

which he has no immediate, personal, and practical acquaintance. Worse yet, he has belittled the knowledge, understanding, and professional stature of persons who do have acquaintance with fields of phenomena that are not important in public and national affairs. Knowing only universes of immense scope in time and space, he would have to be Director of National Morale or Consultant to the President to be satisfied or to feel intellectually at home ...

The sociologist may be asked to help with some important jobs if he has previously demonstrated that he can deal with minor components of the situations which he would like to influence. (p. 158, emphasis added)

The message to sociologists was clear: abandon critical lines of inquiry or be subjected to a public campaign that would dismiss the sociological science outright. Modern society was under the control of the business class and its interests and would remain so. Critiques that pointed out the structural flaws of a modern society which placed capital at the center of its historical trajectory over planetary and human needs would be belittled as the ravings of out-of-touch (and worse, pseudoscientific) fools. Not even the reality of a second world war was enough to force a reckoning with the capital-aligned principle of instrumental rationality; so it continued to dominate mainstream sociologists in America whose thought reflected the dominant mode of thinking perpetuated by the material conditions they lived in and they overwhelmingly chose to save face with the ruling class by adopting this attitude as their own and by serving as roadblocks to the critical perspective within the discipline.

American exceptionalism was the tacit perspective of mainstream sociology which imagined that America's brand of modern society was somehow superior to and more advanced than the European types. Although the war's impact was felt in the American psychosocial landscape, it did not play out in terms of an American identity crisis (as it frequently did in Europe), rather it showed itself in the further spread of a managerial logic in which the costs and benefits of the war had to be weighed against the political and economic strategies of the American business class and integrated into their political economic program to maintain an obedient and pacified population willing to support their directives and maintain the direction they had charted for American society. Unprepared, inept, or unwilling, the bottom line was that American sociology failed to perform when there was the most pressing social need for it and, as a result, it continued to have minimal impact and relevance in the lives of the peoples of the modern world.

As the inter-War years effected the balance of global power, so too did the post-War years, as America, now armed with bomb and again escaping the war with its industrial base largely unscathed, was now firmly considered a global superpower. The publisher of Time and Time/Life magazine, Henry Luce, wrote in 1941 that with America rushing in to save the world again from the grasp of tyranny, it was time "to create the first great American Century" (Luce, [1941] 1994, p. 11). American exceptionalism was not just an attitude it was a strategic plan for the world order. In 1945, the head of the Office of Scientific Research and Development, Vannevar Bush (1945), outlined plans for a new direction for America, one guided by science and built around the concept of the information society which would take the wartime military advances in informational technologies and find ways to implement them in the social fabric of everyday life. American "administration officials fashioned plans for a series of supranational institutions and rules to govern the postwar world under American direction" (Eckes Jr. & Zeiler, 2003, p. 121). However, rather than solely relying on military might to get their way, thus recreating the colonialist project, America propped up the bomb as a horror that they never wished to use again, so long as the world followed their economic plans. They offered a carrot but carried a big stick. Information, particularly as it related to economic affairs, would become the new weapon of the post-war years. America's economic plan was built on a desire to modernize the world in the manner of imperialism (an accusation that they levied against the Soviets, but vehemently denied as their own intention), so that while all would supposedly benefit from the American way of life, America would benefit the most as global profits would now flow into her coffers. The United Nations replaced the defunct and failed project of the League of Nations to handle the political side of things. This was supplemented economically with an agreement made at Bretton Woods in 1944, that instituted the creation of the International Monetary Fund (IMF) "and the International Bank for Reconstruction and Development (commonly referred to as the World Bank" (Eckes Jr. & Zeiler, 2003, p. 124). Then Secretary of State, George Marshall, developed a plan for using these and other resources to rebuild Europe to America's benefit. "Officially known as the European Recovery Program (ERP), the Marshall Plan dispensed over $13 billion between 1948 and 1952 to Western European countries" (Wood, R. E., 1986, p. 29) which was supplemented with loans from the new international financial institutions. In exchange for accepting this aid from America, Western Europe was now dependent on American interests and fell under her long shadow.

As the British Prime Minister Winston Churchill would frame it, however, the darker shadow was cast by the Soviet Union who, devastated as she was, would become the primary ideologically "othered" force in the Western conscious. After the war, the alliance between the Western and Eastern powers broke apart and the world was carved into two sections, one under the domination of America and one under that of the Soviet Union. Pre-war tensions and disagreements between the capitalist and the communist models quickly reignited. Fascism was broken but not completely expelled from the world, instead Western governments reverted to their pre-War tactics and tolerated its presence so long as it gave allegiance to the logic of capital. Anxieties over communism were stoked by the American Diplomat George F. Kennan (1946) who wrote to Secretary Marshall and President Truman that the Soviet government "is actually a conspiracy within a conspiracy" lacking "an objective picture of the outside world." This led him to recommend an anti-Soviet stance in foreign policy and the maintenance of a strong military might to counter the Soviet's perceived ambitions in the world. There would be no more peace time and war time, maintenance of a war footing was now the de facto strategy of Western modern societies. As such, Kennan's tone set Truman's post-war foreign policy and motivated the adoption of the Marshall Plan as a means to gain influence over Western Europe. Churchill (1946) reinforced this approach in a speech, saying that "an iron curtain has descended across the Continent" with those on the eastern side becoming subject "to Soviet influence" which was a great cause of anxiety in a Europe decimated by war and suspicious of Soviet intentions. Furthermore, "[i]n front of the iron curtain which lies across Europe are other causes of anxiety" such as the still communist leaning Italy and the dilapidated France.

While England was fully willing to align itself with the new American world order, in a country like France answering the question was more of a challenge. They had to decide whose thumb they would rather live under, the Americans or the Soviets. The Soviets having suffered the largest economic and human losses in the war were in no way ready to take on the United States in terms of influence, but with the new position of power the U.S. found itself in, the French thinker Bataille pondered a more ambivalent stance:

> If the threat of war causes the United States to commit the major part of the excess to military manufactures, it will be useless to still speak of peaceful evolution: In actual fact, war is bound to occur. *Mankind will move peacefully toward a general resolution of its problems only if this threat causes the U.S. to assign a large share of the excess—deliberately and without return—to raising the global standard of living, economic activity thus giving the surplus energy produced an outlet other than war* ... It is true that the USSR is putting America through a difficult trial. But what would this world be like if the USSR were not there to wake it up, test it and force it to "change"? ([1949] 1991, p. 187)

In the post-War years the world was on a new footing, and although the Nazi's affront had been defeated, the anxieties of the world were only climbing. As with World War I, the weaponry and the technologies of World War II could not be dismantled to live on only in the history books. The power that was previously seen as belonging only to divinity was now materialized in atomic and nuclear bombs. By 1949, the Soviets had successfully tested their own version of the bomb which despite the decimated state of their country gave them an enormous power to keep the forces of the West at bay. Those who had gained this power now held the world in their grip and everyone lived in a state of constant anxiety as these technological forces, over which they could exercise no individual control, flourished all around them. Politically, American democracy was still rife with contradictions as the plight of racial minorities was still largely ignored and the Soviet solution under Stalin had turned into a closed form of totalitarianism that had little if anything in common with Marx's vision of a communist society. Economically in the post-War years, America and the West followed a model that split the difference between Bataille's presumed alternatives. On the one hand, they continued to invest heavily in the military-university-industrial complex to refine the weapons of war that they accused the Soviets of developing, which in turn made the Soviets follow a similar strategy. On the other hand, they resumed the project of creating a consumerist society as a means to pacify the discontent of the laboring classes who continued to toil under the regime of capital.

With sobered senses the world could not pretend as if the horrors of World War II were not wholly made possible by modern innovations, and therefore the potential for their use remained ever-present. Modern society could therefore not be anything else but a society of permanent anxiety as it continued its destructive tendencies at the same time that it preached an ideological vision of humanitarian goals. The contradictions ran side by side as the mental image of the West butted up against the material reality of everyday life in those societies. And yet, the guiding logics of the modernization project were not reconsidered, instead that project became the fundamental guiding principle of the American Century as the country refused yet again to critically evaluate and confront the darkest sides of modern life and the ways that they had shaped them. A deep unease set in on the Western mind, as people who had lived through the World Wars knew that there was something deeply flawed about the organizational principles guiding the trajectory of the West, and yet, as the consumerist society became more implemented, their lives achieved a level of modest comfort which they were not willing to abandon. Anxiety led to complicity, complicity led to repression, and repression led to feigned shock as the symptoms of alienation, anomie, and the spirit of capital continued to appear and tragically harm those who fell between the cracks.

For the mass society of the West, it was easier to pretend that those who suffered did so as a result of their own failings. This was especially easy to do as those who suffered the most were minority populations and therefore already maintained an "othered" position in America. Any notion of the social as a positive means of transforming society began to fade away from the mass consciousness of capital's subjects. Life for them would now begin to be totalized anew as their individuality was corrupted by consumerism and their characteristic uniqueness was reduced to informational content for the mass cultural project of maintaining a willing and obedient class of laborers to support the still unscathed agenda of the elite classes. The critical mind was in danger of losing the fight to keep the light on a better future for all lit, and the project initiated by the classics was in need of a revitalization to account for these changed material circumstances.

CHAPTER 4

Critical Methods 2

1   The Psychosocial Origins of the Frankfurt School

The early classics of the critical method were the first to formulate rigorous research agendas whose core aim was to peruse and interrogate the structural contradictions that constituted the foundational logics of modern life. At the time they were writing, the totalizing logic of religion had loosened its grip on the collective consciousness enough for bourgeois individualism to emerge, but by the time of industrialization the individual was still only in its historical infancy. Before this organization of the self could spread to the masses and mature, the classical theorists recognized that the bourgeois use of capital was having a comparable effect to religion in structuring the minds of the masses which limited the potential of individualism and the development of the self. The effect of religious alienation—that is, viewing material life as a profane prelude to an authentic and sacred life that would begin once one awoke to a spiritual existence after being cleansed through death—was that it served as a tool of domination over the masses. Religious and political elites could push psychological levers by appealing to the totalizing function of religious narratives and the loss of reward with threats of eternal damnation to control and pacify the masses if, and when, they might threaten the power structure. Losing this control mechanism and embracing a mode of social organization that promoted the individualism of all, would mean relinquishing the means of controlling the masses.

Despite egalitarianism being a guiding *ideological* principle of the bourgeois revolutions, material limitations were built into the structure of modern society to restrict the actualization of that principle to a select few. No longer chosen by "God", the elites now claimed that they earned their positions by virtue of their unique abilities as individuals. For the excluded others "alienation became part of the "program" of bourgeois society" (Dahms, 2011, p. 229), which was used to maintain the hierarchical organization of life with the bourgeoisie resting comfortably on top. It was true that the bourgeoisie possessed the means to cultivate unique forms of individuality that the laboring classes did not, but what they attributed to their own vanity was the result of social conditions which allowed their class to flourish while at the same time it repressed avenues for individual development in the laboring classes. After rational and empirical thought gained ground in the Enlightenment, liberal political programs took root in the West, and the scientific enterprise began to demonstrate its practical use-value, the disenchantment of reality advanced to the point that the collective conscious could not be relied on to self-police a belief system based on transcendental rewards. Foregoing the religious logic, which held off the reward until after death, the logic of capital was built on a rewards-based-system that effected the here-and-now of material life. Freedom for the masses meant freedom to sell one's labor power to the highest bidder and participate in the consumer market, not freedom to build a rich internal world as an expression of each person's unique individuality. The latter form of freedom was not abandoned as an ideal in the cultural conscious, but it was repressed and distorted, then produced and packaged as consumer commodities in forms deemed acceptable by the bourgeoisie that were then sold to the masses as a "reward" in exchange for the money they earned as obedient laborers. The problem was that the bourgeoisie did not sufficiently understand the totalizing nature of capital's logic and how they too would be caught up in the vortex of alienation, lose their own sense of individuality, and become dominated by their own machinations.

By the early 20th century, as technology became embedded in everyday life, a new totalizing logic appeared in modern societies that was a highly synthetic complement to capital. This new logic intensified the push of those societies in a mechanthropomorphic direction whereby "individualistic rationality [was] transformed into technological rationality" (Marcuse, 1998, p. 44). This is the power of the totalizing logic of information. The process of informationalization—that is, the reduction of reality to its informational content—began to emerge when the scientific use of reason became equated with a positivist stance on the world. In this model everything is transformed into data, a form which engenders itself to modern society's obsession with speed and, by reducing the friction of materiality—partially in analog form, and more completely in digital form—information is bound only by the speed of light. The project of converting all of reality into its informational double becomes one of the most powerful guiding principles of modern society; comparable in the power of its effect only to that of the commodification of reality, which is but one form of the informational transfiguration of material reality based on the notion of price. Examples of its effectiveness were evidenced in Taylor and Ford's regimes of "scientific management" and Shannon's binary language of zeros and ones, which reduced humans to their informational content by providing the tools for a rational computation of everyday life that corresponded to an increase in and evolution of alienation.

Under this logic, life could be approached as something free from the complicated emotions and struggles of a human species whose constitution was better suited to the world of nature than the artificial world of the modern machine, which demands an instrumental attitude and the application of rational thought to all spheres of life for it to function at its most efficient levels. Even the bourgeoisie—for whom individualism was a guiding ideal—could not escape the effects of this logic and its powerful synthesis with capital. Their minds too became just as cannibalized by this process as those they sought to control with their psychosocial strategies of domination. Under the sway of these two totalizing logics—capital and information—all modern life began to assume a highly administered quality to it in the 20th century. Narratives of the future that could be imagined arising out of this logic lost their broad humanist appeal, totalized as they were by the dystopian futures that represented the destinations of a society that followed commodification and informationalization to their logical endpoints. For the critically attuned mind, the present appeared more and more to be caught in an atemporal stasis with the whole of planetary life trapped in a 'casing as hard as steel'.

The existential dread of the proletarian masses began to creep into the conscious minds of the bourgeoisie who could not escape the structural effects of the system they built. As the minds and bodies of the laboring classes were ravaged by the system that demanded the exchange of their labor-power for the commodities they relied on for survival, their temporal horizon shrunk and the anxieties of the present overpowered and penetrated thought on future alternatives to this mode of life. In the West, the system of capital had successfully built an industrial basis for a society that could meet the needs of all. However, at the same time it had structured the minds of its subjects by providing pre-approved and readymade forms of commodified individualism for their consumption. These forms lacked the necessary mental preparations needed to find a social path that could lead beyond the course dictated by their present material conditions. Attempts to stimulate the proletarian consciousness were either stifled by the lack of an industrial base and authoritarian ideologies, as in the East, or the capitalist market place where they had to compete with other ideas as well as the mass cultural industry which produced useful entertainment vehicles for distracting the mind away from the aches and pains of a day spent laboring, as in the West. Those who had the time and energy to turn their mental faculties to the critique of society and a rigorous examination of the contradictions of modern life were the same bourgeoisie who now felt the amplification of the administered world in the tightening of the noose around their own necks, one they had built to control the masses, as it restricted even their avenues for finding and living a life built on the notion of individual freedom.

The most prominent members of the first generation of the Frankfurt School, as those who worked for and associated with the *Institut für Sozialforschung* came to be known, were from precisely this kind of psychosocial background. They were predominantly composed of German men born of Jewish descent, whose fathers had in the same years as Freud enjoyed the new social conditions that were made possible to those of Jewish ancestry, which opened new opportunities for their children to explore the fruits of the bourgeois lifestyle. Of the many affiliated members and associated scholars to the Institute, those who made the most lasting impact on the Frankfurt School's legacy and were the most dedicated to the project of updating and applying the critical method of the classics to the changed historical circumstances of their time were the social philosophers Max Horkheimer, Herbert Marcuse, and Theodor W. Adorno. The discussion in this chapter will primarily focus on related aspects of their contributions to the critical method, but other members and associated scholars of the first generation included: the cultural critic Walter Benjamin, the psychoanalyst Erich Fromm, the economist Henryk Grossman, the legal scholar and political scientist Otto Kirchheimer, the sociologist Leo Lowenthal, the lawyer and political scientist Franz Neumann, the social scientist, and Horkheimer's childhood friend, Friedrich Pollock, and the social scientist Karl August Wittfogel.

Max Horkheimer was born in 1895 to a "conservative" family of "firm believers in the Jewish religion" (Wiggershaus, [1986] 1994, p. 41). His father, Moritz, owned several textile factories and by World War I had "established himself among the ranks of [Stuttgart's] millionaires" (Abromeit, 2011, p. 19). As a young man, Max was groomed to be a manager in his father's factories, but it was the world of aesthetic pleasures and literature that ignited his passions. He found a sympathetic and lifelong friend in Pollock, whose father was also a factory owner "but had turned away from Judaism and had brought up his son accordingly" (Wiggershaus, [1986] 1994, p. 42). Max and Friedrich developed an intense relationship based on their mutually constructed views of the deficiency of bourgeois life and the effects that it had on deadening the inner world of the individual. The young Horkheimer expressed his feelings about how the logic of capital had come to totalize the minds and bodies of its human subjects by writing fictional works that explored these effects on his characters. In one such example, analyzed in Abromeit's biography, he channels his feelings into his protagonists who have a "hatred of bourgeois society [which he] depicts ... as a system of needs that prevents the upper class from pursuing any ideals beyond material wealth or dubious fame, and forces the lower classes into a brutal struggle for existence" (2011, p. 27). In 1913, aided by Pollock's influence and a failed affair with his distant cousin, Suze Neumeier—which took the form of an intellectual betrayal of Horkheimer's ideals—Max moved away from religion, but the fundamental questions that religion claimed to answer would continue to impact his thinking. In his most intellectually productive years he would write that religion was "synthetic, artificial, [and] manipulatory" (Horkheimer, 1978, p. 123). In his later years, however, he would come to ask the Durkheimian question[1]: "Isn't religion always needed because the earth remains a place of horror even if society were as it ought to be?" (p. 181). Of course, Horkheimer did not share Durkheim's positivist faith in the future, but he did recognize the social role that religion continued to play in people's lives even after its totalizing bond had ruptured in the Enlightenment. As a budding scholar, Horkheimer's ambitions were fueled by a desire to understand the human experience in its totality, especially as it related to the ways that the species artificially constructed and derived their meaning and directionality in modern life.

Horkheimer avoided an early military draft in World War I by working in his father's factory which was converted to manufacture goods for the war effort, but by 1916 he could no longer avoid the call. In his pre-war fictional writings, he expressed a strong disapproval of the War. When he was forced to enlist, he was purposefully undisciplined and suffered from poor health that got poorer as the war raged on. This had the desired effect in that he was never sent to the front line, however, in 1918 he was declared "unfit for further military service … and committed to the sanitarium Neu-Wittelsbach, in Munich, where he would remain until shortly before the end of the war" (Abromeit, 2011, p. 33). In 1919, with the war over, Horkheimer and Pollock decided to pursue their university education together in Frankfurt. Max studied psychology, philosophy, and economics, and defended his dissertation in 1923 on *The Antinomy of Teleological Judgement*, under the philosopher, Hans Cornelius, and then his *Habilitation*[2] in 1925 on *Kant's 'Critique of Judgement' as a Connecting Link between Theoretical and Practical Philosophy* (Wiggershaus, [1986] 1994, p. 46). Once he had paid homage to the traditional philosophical approaches in this early work, as required by his advisors, he was then free to begin anew an

---

1 In *The Elementary Forms of Religious Life* ([1912] 1995), Durkheim's last major sociological work, he writes: "[T]here is something eternal in religion that is destined to outlive the succession of particular symbols in which religious thought has clothed itself … If today we have some difficulty imagining what the feasts and ceremonies of the future will be, it is because we are going through a period of transition and moral mediocrity. The great things of the past that excited our fathers no longer arouse the same zeal among us … In short, the former gods are growing old or dying, and others have not been born … But that state of uncertainty and confused anxiety cannot last forever … There are no immoral gospels, and there is no reason to believe that humanity is incapable of conceiving new ones in the future" (p. 429–430).

2 A second dissertation that is a requirement of German academics wishing to enter the professoriate.

exploration of the more critical aspects of philosophical discourse. These he found in Hegel, Marx, and Freud, moving beyond the ossified boundaries of contemporary philosophy. When he assumed the directorship of the Institute, some five years later, it was their work which provided the framework for the research program that he would come to implement, one based on a "consideration of the totality of social relations; a commitment to dialectical and materialist methods; the theoretical importance of cultural and intellectual life; a focus on group—rather than individual—emancipation; as well as a critique of deterministic (i.e. vulgar) Marxism and Hegelian idealism" (Kautzer, 2017, p. 51). While Horkheimer was brilliant in his own right, he was perhaps most skilled in drawing likeminded scholars to join him at the Institute and work under his direction by helping them to recognize both the necessity of his critical project and that it could not be undertaken by a single individual but required an intellectual division of labor to assess the many dimensions of modern society.

One of the most important recruits to join Horkheimer's Institute was Herbert Marcuse. Born in 1898 in Berlin, in what has been described as "a typical German upper-middle-class" environment (Kellner, 1984, p. 13), Marcuse's background had many parallels to Horkheimer's own experiences as a youth. Although his father was not as wealthy as Horkheimer's, Marcuse's father was an owner in a construction company which allowed him to provide a modestly bourgeois lifestyle to Herbert. Like Horkheimer, Marcuse expressed that "he never felt any acute alienation because of his Jewish origins" (p. 13), suggesting that at least in these early years in Germany the power of capital had partially eclipsed the question of religious identity in terms of shaping the lifeworld of the individual and its social reception. He was also conscripted into the German military during World War I, and did not see action on the front lines, instead, as his grandson Harold Marcuse (1997) described it, he spent the war "wiping horses' asses for infantry in Berlin." The war prompted Marcuse to become politically active and "in 1917 he joined the Social Democratic Party (SPD) as a protest against the war and the society that produced it" (Kellner, 1984, p. 14), but he quit the following year over disagreements with their policies that opened the door to military interests.

In 1919, Marcuse resumed his university education, studying German literature, philosophy, and political economy. In 1922, in Freiberg, he completed a dissertation influenced by the work of Hegel and Lukács, called *The German Artist-Novel*, under the direction of the literature professor, Philip Witkop. Rather than move directly into work on his *Habilitation*, Marcuse's father bought him a share in an antiquarian book business in Berlin where he worked and continued his intellectual pursuits away from academia. During this time he read Heidegger's *Being and Time* ([1927] 2010) and was most impressed with how it "took everyday forms of alienation as its starting-point" and clarified "the question of authentic human existence" (Wiggershaus, [1986] 1994, p. 98). This prompted him to return to academia in 1928, to work on his *Habilitation* under the direction of Heidegger, on the topic of *Hegel's Ontology and the Theory of Historicity*. But two events prevented Marcuse from finishing it there: the first was the publication of Marx's early manuscripts in 1932 which he found to offer a far more concrete approach to philosophy than Heidegger's model, and the second was his horror at Heidegger's expression of public support for the Nazi project in 1933. While he had not especially been shaped by his Jewish heritage as a youth, he would recount that the social conditions in Germany at that time made it unrealistic for either a Jew or a Marxist to work within the academy (Kātz, 1982). Under these conditions Marcuse began to turn away from Heidegger's influence and enhance his reading of Hegel through Marx, thus earning him an invitation to join the Institute in 1933.

The youngest member of this Frankfurt trinity, was Theodor Wiesengrund-Adorno, born in Frankfurt am Main in 1903 to a father who inherited a wine business that had been passed down through the generations since its founding in 1822 (Claussen, [2003] 2008). Like Marx's father, Adorno's father converted from Judaism to Protestantism to assimilate into German culture, but his mother was a Catholic. As a boy he was especially influenced by his mother and his aunt, both professional musicians, who encouraged his education as a classical musician. He was too young to be drafted into the war effort, but found himself opposed to it, writing that his childhood uses of "[f]oreign words constituted little cells of resistance to the nationalism of World War I. The pressure to think along proscribed lines forced resistance into deviant and harmless paths" (Adorno, [1958] 1991, pp. 186–187). This also marks Adorno's early critical inclinations and his lifelong resistance to the social structuring of thought along pre-established lines guided and approved by bourgeois society. He recalls being criticized for his use of High German and foreign words as a youth, but he would later explain why he was drawn to the variance of language to attain precision in his expressions, which he saw as a mode of resistance:

> Vague expression permits the hearer to imagine whatever suits him and what he already thinks in any case. Rigorous formulation demands unequivocal comprehension, conceptual effort, to which people are deliberately discouraged, and imposes on them in advance of any content a suspension of all received opinions, and thus an isolation, that they violently resist. Only what they do not first need to understand, they consider understandable; only the word coined by commerce, and

> really alienated, touches them as familiar. Few things contribute so much to the demoralization of intellectuals. Those who would escape it must recognize the advocates of communicability as traitors to what they communicate.
>
> ADORNO, [1951] 2005, p. 101

Although as a youth he did not articulate the desire he felt to resist the structuration of his mind in accordance with German society, the seeds for this attitude were evident in the actions and interests the young Adorno pursued. His most passionate interest was in the avant-garde twelve-tone techniques and atonal compositions of Alban Berg and Arnold Schoenberg, with the former training him in musical composition.

As a complement to his musical studies, Adorno's philosophical interests were encouraged by a friendship he developed with Siegfried Krakauer, a man 14-years his senior, who introduced him to the work of Kant, and later to that of György Lukács and Ernst Bloch. It was Lukács's work that would provide "a form of philosophical thinking about history" that would inspire Adorno's "ideas on the philosophy of music and musical progress" (Wiggershaus, [1986] 1994, p. 81), while Bloch's influence would serve as a reminder of the need to keep open the horizon of utopia as a mode of resisting the temptation "to be constrained by what is authoritarian and repressive" (Adorno, [1958] 1991, p. 214). Much as Horkheimer and Marcuse found little support for their critical interests within the German academy, these influences on Adorno served as a distraction from his academic studies in Frankfurt, but they also provoked him to complete his dissertation at a hellfire pace in 1924 so that he could retain more of a focus on his critical interests. Like Horkheimer, Adorno's dissertation was also completed under the direction of Cornelius. It was accepted under the title, *The Transcendence of the Material and Noematic in Husserl's Phenomenology*. However philosophical this early work was, Adorno's real energies were devoted to music criticism and aesthetics, and he published "about a hundred articles" in that vein between "1921–32" (Wiggershaus, [1986] 1994, p. 70). Adorno returned to philosophy with his *Habilitation*, which he completed under the direction of the theologian and philosopher, Paul Tillich, in 1933, on the topic *The Construction of the Aesthetic in Kierkegaard*.[3] Since Horkheimer was Cornelius's assistant prior to his taking over the Institute, he and Adorno had been acquainted since 1922 and their influence on each other had already begun to take shape.

---

3 Published as Adorno, T. W., *Kierkegaard: Construction of the Aesthetic* ([1962] 1989).

Plans to establish the *Institut für Sozialforschung* in Frankfurt began to materialize in 1923, thanks to a generous endowment secured by Felix Weil, whose father, a Buenos Aires based German-Argentine named Hermann, made a fortune as a grain trader exporting crops from Latin America to Europe in the early 1900s. As a youth Felix was sent to Germany for his education where he took an interest in Marxism and was influenced by communist politics, but he had a "desire for an institutionalization of Marxist discussion beyond the confines both of middle-class academia and of the ideological narrow-mindedness of the Communist Party" and his father had ambitions "to go down in Frankfurt's history as a great benefactor" (Wiggershaus, [1986] 1994, p. 16). With this combination of circumstances Felix convinced his father to fund the creation of the Institute in Frankfurt. Initially the Institute was set to be headed by the sociologist Kurt Albert Gerlach, but his death in 1922 from medical complications prevented this from materializing. From 1924 until 1929 the Institute was directed by the political economist Carl Grünberg who set it on a research track guided by the principles of a non-dogmatic but fairly orthodox approach to Marxism. The early work did not stray far from traditional Marxist studies of labor and capital and was less involved in the development of new theoretical material to explain current phenomena than it was toward conducting empirical studies guided by Marx's theories. In 1930 Max Horkheimer was announced as the new director of the institute. He officially assumed the position in January of 1931 and changed its orientation toward "the history of social philosophy to put its current situation in perspective" (Jay, 1973, p. 25). In wanting to revitalize the theoretical project of the classics, Horkheimer was less interested in pursuing an orthodox approach to Marxism, which he saw as misunderstanding the historical moment, and was more guided by the influence of Western Marxism. This was a tradition that synthesized Marx's perspectives with those of Weber, such as in the work of György Lukács ([1923] 1971), and it emphasized Marx's philosophical thought as central to understanding his economic contributions, as in the work of Karl Korsch ([1923] 2012). This perspective also proved attractive to Marcuse and Adorno, with the former officially joining the Institute in 1933, and Adorno maintaining an intellectual link to the institute during his exile at Oxford, but only officially joining his colleagues some years later in America after the Institute relocated there to escape Hitler's Germany.

The similarity of the psychosocial backgrounds of Horkheimer, Marcuse, and Adorno, led to their long and productive intellectual relationships. They were guided by a shared commitment to the critical method. Once in America they would develop this program and method under the name Critical Theory, after essays by Horkheimer and Marcuse appeared in 1937 that outlined the project as a formalization of the methodological and theoretical approaches of the early classics of modern social thought. Their combined critiques of bourgeois life, a condemnation of war, a love of literature and the importance of aesthetics, as well as a common Jewish ancestry, and a desire to engage in rigorous socio-philosophical investigations of modern society, made for a unique and rare opportunity to produce social theory that was at its core a *social* enterprise sustained by intellectual debates, friendship, and respect that persisted despite inevitable disagreements. This friendship led to an especially powerful bond between Horkheimer and Adorno that lasted until Adorno's death in 1969.

Although they had already felt the effects that bourgeois society, the logic of capital, and instrumental and technological rationality had on disintegrating the inner world of the individual, when the rise of the Nazi regime necessitated that the Institute relocate to America during World War II, these social injustices that were flourishing in their native home compounded and amplified their sensitivity to the repressive tendencies of modern society. Their exile from Germany was a key factor in their intellectual development, providing two distinct explanations for why they so relentlessly pursued their critique of modern society in the face of such hostile conditions. First, they were forced to assume the "otherness" of their Jewish identities. These identities had not played a very large role in their psychosocial development as youths, but were, under Nazi fascism, socially imposed upon them. Since they had dedicated their lives to this project which was motivated by their recognition that one could not live an authentic existence in modern society as it was currently configured, when the limited sense of self that they had fashioned in the given material conditions was denied by the social milieu, they must have felt incredibly intense feelings of alienation, anomie, and anxiety. However, rather than being paralyzed by that anxiety, their socioanalytic dispositions provided the critical sensibility to recognize its warning and take action, first by moving the Institute to America to preserve the project, and second by recognizing that a society which has reached its critical[4] stage is one that is the most desperately in need of critical theory.

---

4 Etymologically the word "critical" descends from Latin into English. Dating from "1580s, "censorious, inclined to find fault," from critic + al (1). Sense of "important or essential for determining" is from c. 1600, originally in medicine. Meaning "of the nature of a crisis, in a condition of extreme doubt or danger" is from 1660s; that of "involving judgment as to the truth or merit of something" is from 1640s; that of "having the knowledge, ability, or discernment to pass judgment" is from 1640s. Meaning "pertaining to criticism" is from 1741". It is a warning cry of a crisis at the point of its catastrophic determination. It indicates an ambiguity as to the future directionality of the subject on whom the diagnosis "critical" has

As World War I had demonstrated, and as the ominous portents of World War II were suggesting, it was not just Germany that was in a state of crisis. Although it offered greater overt levels of expression, which were necessary for critical theory to avoid becoming mere ideology, America and the West were similarly facing a critical situation, as was the Soviet Union. Therefore, since the concrete socio-historical situations of Germany, America, and the Soviet Union each represented a unique form of modern society, each would require its own historical theorization which would then need to be compared to understand how and why this nexus of modern societies was facing a seemingly permanent state of crisis. This leads to the second reason why their exile in America aided their project. As Dahms argues, there was a

> *practical* influence in the comparative historical subtext of critical theory ... [that] resulted from the fact that the members of the Institute for Social Research in Frankfurt, subsequent to its reestablishment in New York in 1934, had primary experiences with at least two different incarnations of modern society, experiences that enabled them to apply a stereoscopic perspective on the organization and functioning of modern society as characterized by various types of "similarities" and "differences" ... in highly specific, and frequently no less problematic, and paradoxical ways [as the classics]. (2017b, p. 166)

What was missing from the mainstream practices of social science was the historical dimension, which recognized that modern society was dynamic and subject to continuous change as capital and technology came to bear on it with their fundamental logics which depended on either continuous growth or generational advance. What was missing from mainstream philosophy was an attention to the actual material conditions of the world, which philosophers were only too eager to ignore as they contented themselves with mere abstractions. What was missing from sociology was an accounting of the effect that the concrete material conditions have on how that science is practiced. By formulating a radical approach to comparative historical research that recognized the necessary synthesis of philosophy and social science, and being especially suited to perform this comparison, the Frankfurt School was able to, at various points, compare the socio-historical conditions that had given rise to the fascist nightmare in Germany, the distorted vision of communism in the

---

been applied but stresses the serious nature of the diagnosis. The term critical, like the affect "anxiety", is intended to serve as a warning to action.

Soviet Union, and the decadent and uncritical mindlessness of democracy and capital in America. Furthermore, they provided a model for understanding how "individuals' efforts to construct meaningful life-histories depend[s] on their ability to navigate modern society as a field of tension that may be inherently unreconcilable" (Dahms, 2017b, p. 177), thus they returned to the questions of the classics in linking the construction of the self to that of society and examined how the horizon of the possible changes alongside the evolution of the actual material circumstances. To better understand how they updated the critical method to their time at the height of international tensions, crises of capital, and World War, and what it would mean for the project of socioanalysis in our current socio-historical nexus, an examination of their foundational essays is required.

2   From the Critical Method to Critical Theory and Negative Dialectics

Once Horkheimer assumed the directorship of the Institute he used the inaugural address to outline how his research program was rooted in the critical method. He defined his vision of the Institute's direction as oriented toward an "interpretation of the vicissitudes of human fate," both those of the individual and the community, which "can only be understood in the context of human social life: with the state, law, economy, religion—in short, with the entire material and intellectual culture" as its object (Horkheimer, 1993, p. 1). Foundational to the project was "the question of the connection between the economic life of society, the psychical development of individuals, and the changes in the realm of culture in the narrower sense" (p. 11). While the first two aspects of that question harken to the projects of Marx and Freud, the later part was a nod to Weber's project and reminded his audience of the dynamic nature of modern society and the necessity to track the ways that culture continuously reshapes and recodes the psychosocial dimension. These questions became ever more pressing in an age of dynamic technological embeddedness coupled with the conscious awareness that modern advances continued to produce conditions that led to massively destructive wars on a global scale. Both conditions amplified the anxieties of life in bourgeois society in ways that had a profound effect on the continuing disintegration of the social and therefore on avenues for individual growth that were under a sustained attack from the totalizing logic of capital. Rather than simply prescribe solutions to the affective conditions that he experienced, as the Soviet's, the communists, and the orthodox Marxists were inclined to do, Horkheimer's call recognized that as a precondition for any attempt to reconcile the contradictions of modern society one would first have to understand how those contradictions emerged from the concrete material circumstances of the given modern society and how attempts at reconciliation could be thwarted and warped by those same circumstances under whose gravitational pull they were formulated. Short of that, it would be pure ideological naivety to expect that any practical action taken to adjust the contours of modern society would not end with the reemergence of new contradictions that would have just as a dark a hue as those in the current configuration. To achieve the scope of this ambitious project, Horkheimer instituted a division of labor amongst the Institute's members with each concentrating on a distinct realm of the modern experience.

The methodological problem he faced was that he could not simply task different scholars to pursue studies in their own fields, following their own methods and theories, and then piece them together after the fact in a logically consistent and coherent whole. Neither could philosophy and sociology—the disciplines that ostensibly claimed to be concerned with questions of the "good life"—answer these questions on their own. Philosophy, on the one hand, as it was practiced in the academy, was too rigid and abstract, contenting itself with its own discourse at the expense of ignoring the sociohistorical conditions in which it was produced. Sociology, on the other hand, in its academic guise "has nothing to say about the degree of reality or about the value of these phenomena" that it studies because it lacks the conceptual language to do so (p. 8). Therefore, for Horkheimer, only through a combination of the two "as a theoretical undertaking oriented to the general, the "essential" is capable of giving particular studies animating impulses, and at the same time remains open enough to let itself be influenced and changed by these concrete studies" (p. 9). What he proposed then is a critical method that responds to the dynamic nature of modern society by recognizing that sociology needs philosophy to orient its conceptual language while philosophy needs sociology to keep it historically relevant and prevent its concepts from becoming reified, thus placing the two in a dialectic. Therefore, members of the Institute would have to agree to give up a level of their individual autonomy to him as the director so that they could be guided by a central theoretical vision that would ground their social research.

Theoretically there was precedent for Horkheimer's project, as it was distilled from the animating impulse of the classics. Horkheimer explained it as "nothing but a reformulation—on the basis of a new problem constellation, consistent with the methods at our disposal and with the level of our knowledge—of the old question concerning the connection of particular existence and universal Reason, of reality and Idea, of life and Spirit" (pp. 11–12).

What was novel about Horkheimer's approach was that it proposed a social theoretical project that was itself *social*, insofar as the Institute would make use of the division of labor to channel the social power of its members to create an intertextual body of literature on the central problems of modern society that was more advanced than any one member could accomplish on their own. In other words, Horkheimer's vision shared a basic alignment with the early classics in terms of the questions they would ask, however, rather than taking the answers provided by the classics and applying or putting them to the test empirically, he recognized that there was a new matrix of social problems that was substantially more complex than what the classics experienced precisely because the economic system had not ceased its development and in that dynamic process modern society had reached a new stage that demanded new efforts and methodological strategies to theorize it in its complexity.

Horkheimer's address was followed some months later by a second inaugural address provided by Adorno that refined the critical diagnosis of the mainstream models of philosophy and sociology that dominated the academy. In this address Adorno critiques and links the Hegelian idea "that the power of thought is sufficient to grasp the totality of the real" and the Heideggerian notion "that being itself is appropriate to thought and available to it" (Adorno, T. W., [1931] 2000, p. 24). Adorno's claim is that the history of philosophy has itself demonstrated the inability of reason to develop "the concept of reality ... from out of itself" (p. 25). What Hegel had done was place reason as something that was already validated in and of itself, elevating it above reality as the means of examining and distilling reality into meaningful knowledge. Adorno, however, recognized that even statements made through the application of reason in thought "are inextricably bound to historical problems and the history of those problems, and are not to be resolved independently" (pp. 25–26). In other words, reason on its own cannot be separated from the effects that the concrete socio-historical circumstances exert over the determination of what constitutes objective reason. Heidegger's attempt, then, to move from objective being as exemplifying universal reason to the materiality of subjective being employing reason, appeared on the surface to provide a model to circumvent the failure of establishing a pure mode of objective reason. However, this model, which Adorno says is borrowed from Kierkegaard, has been unable to produce anything other than despair in the project because it cannot ground it in anything other than an "inner-subjective" space that negates attempts to move back to objective claims about reality. Rather than being able to produce objective and concrete claims as to the totality of reality, inner-subjective reason can only move "into transcendence which remains an inauthentic and empty act of thought, itself subjective, and which finds its highest determination in the paradox that here the subjective mind must sacrifice itself and retain belief instead, the contents of which, accidental for subjectivity, derive solely from the Biblical word" (pp. 27–28). What this distills down to then, is that the use of subjective reason to establish social judgements on objective reality must ultimately fall back on a form of religious alienation whereby the ultimate question of the totality can only be surmised through death, meaning that the question of being cannot be resolved from the standpoint of being any more than the totality can be resolved from the use of reason as if it were not itself subject to the effects of the historical totality.

Sociology for its part has therefore ignored the philosophical questions that seek to ground its claims on objective reality. It finds their output to be circular with the answers destroying the questions that were asked by failing to conclusively answer them or by demonstrating the logical impossibility of doing so given the restraints of material historical conditions. Thus, it falls back on the theological mode of premodern thought or a general relativism. In the quest to model itself after the modern natural sciences, sociology as the science of modern society, therefore, aligned itself with the pragmatic and positivist schools of thought which believe that the only value of science is to produce knowledge that is useful in the here-and-now by adequately reconstructing the materiality of the here-and-now in its informational form. But these notions of value and usefulness are themselves historically structured by the totalizing logics that shape society, meaning that in the current nexus of space-time they are derived from the interests of capital (which defines use-value) and technological rationality (which governs the constitution and use of informational content). If, however, the critical mind determines that the current configuration of society is not appropriate to the goals and values of the humans who compose it, then, by only producing knowledge (or information, as the case may be) that is deemed to have a use-value in accordance with the terms dictated by that society, it can produce nothing but affirmations of that society which further the very logics that produce the social conditions that distort and prevent the actualization of a values based life in the first place. Therefore, sociology, which is supposed to discover the natural laws of society upon which it can construct the good life, sets itself up as a logical impossibility and negates its own aims with its theories and methods.

The metaphor that Adorno uses to describe this situation of the mainstream variants of philosophical and sociological discourse is as follows:

> [W]hile the philosopher is like an architect who presents and develops the blueprint of the house, the sociologist is like the cat burglar who climbs the walls from outside and takes what he can reach ... For the house, this

big house, has long since decayed in its foundations and threatens not only to destroy all those inside, but to cause all the things to vanish which are stored within it, much of which is irreplaceable. If the cat burglar steals these things, these singular, indeed often half-forgotten things, he does a good deed, provided that they are only rescued; he will scarcely hold onto them for long, since they are for him only of scant worth. (p. 35)

The effect is that sociology scavenges philosophy to create endless empirical classifications in an informational model but, because their "knowledge" content is philosophically questionable, sociology lacks the ability to generalize from those classifications to the point that rational judgements could be derived from them, meaning that the worth of these classifications has little individual value and instead has value only for maintaining the social structure on its current trajectory. What this means is that sociology has become a machinic science that furthers the mechanthropomorfic trajectory of species-being, distorting claims that it is in fact a human science. And this is because sociology ignores the philosophical work that would need to be done to sort through the illogical elements of its discourse and because philosophy ignores the sociological work that would historicize its truth claims by placing them in a dialectic relationship. Therefore, mainstream approaches are trapped in 19th century models of thought because they refuse to be limited and guided by each other to the harm of both, as they are too wrapped up in their own projects to recognize the necessity of their being placed in a dialectical tension which would extend their usefulness in the dynamic of concrete material change. "The interaction between philosophy and sociological research that [Adorno] vociferously called for was to be achieved through 'dialectical communication'" (Müller-Doohm, 2005, p. 138). If they were to enter this dialectical relationship, then it would require acknowledging that the task of philosophy is "to interpret unintentional reality" (Adorno, T. W., [1931] 2000, p. 32), that is, to take the research performed by sociology and provide the interpretive frame for understanding "unintentional truth, rather than that truth appeared in history as intention" (p. 33). What this means is that at the current juncture of social organization, the classifications of sociology appear as historical truths, that is, as the rational intention of history; however, by applying philosophical interpretation to them we are able to see that they speak an unintentional truth that stands in critical opposition to the supposedly intentional rationality that guides modern society under the sway of its totalizing logics.

This approach would not have as its goal the production of use-values for the current system, rather it would provide flashes of understanding that could, in a historically dynamic society, negate the truth of the given reality by interrogating the truth of the false reality that gave rise to the question. At the same time, it would negate the answer because it would reveal that the circumstances produced the wrong question in the first place, thus bringing the material circumstances and their effect on thought into sharper focus.

> Materialist interpretation [along the lines that Adorno lays out] is valid for what is unintended, what can be broken down into its smallest elements, out of whose experimental arrangement an unexpected solution arises fortuitously. This is the Freudian process of dream analysis: a reading that decodes signs, or the transformation of codes into text ... The riddle does not encode a hidden reality that forms the basis of it (or is hidden behind it) and which must be uncovered; rather it calls for a solution that illuminates and dissolves 'the enigma in a flash'. Thus the riddle disappears with its solution.
> DAHMER, 2012, p. 100

The value in this project is that it would maintain the historical relevancy of philosophy without sacrificing itself to the modern logics, at the same time as it would allow sociology to maintain a commitment to illuminating the contradictions between human values and the modern system that warps them. The primary objective of this method, therefore, is not to develop theoretical systems for changing modern society—if this were possible, then it could only become possible if this method was successful in preparing the necessary preconditions, and only then if this aligned with a changed set of material circumstances—rather its object is to keep alive the importance of the need for a world in which the contradictions of values and facts of the given reality could be overcome if and when the conditions are ever ripe for the implementation of a new structural logic built on shared human values.

Programmatically, Adorno's and Horkheimer's addresses share the scope of the Institute's project, but Adorno's address lays out the problems of the tension between philosophy and sociology to a higher degree than Horkheimer's, while Horkheimer's does more to emphasize the areas of concrete material reality that are in need of critical interventions because of the effect they have on structuring individuals' minds in ways that conform to the totalizing logics. Furthermore, where Horkheimer's solution proposes a more interdisciplinary approach that blurs the lines between sociology and philosophy in the union of social philosophy, Adorno emphasizes the need for disciplinary boundaries so that each discipline can perform the task appropriate to its function according to a division of labor, but one that keeps them tied to each other within a dialectical relationship.

By setting up the problems with the current models of inquiry, outlining the Institute's goals and substantive areas of inquiry, they successfully differentiated their critical research agenda from the mainstream approaches of philosophy and sociology. Hitler's rise to power, however, impressed upon them the urgent necessity of providing a new model of research whose focus was specifically oriented to investigating the contradictions of modern society which were having devastatingly destructive consequences. As modern societies claimed to represent the best way forward for humanity, the Frankfurt School members became even more sensitive to how those societies kept boiling over in highly regressive ways when, after experiencing the traumas of World War I firsthand, they now had to flee their home in Germany and the social milieu indicated another modern war, more horrifying than the last, on the immediate horizon. Therefore, "[i]n the context of ... monopoly capitalism ... communist orthodoxy" and fascist authoritarianism, the Frankfurt School "attempted to rediscover Marx-the-dialectician, thereby recovering the concept of reason, which could refute instrumental ends-means rationality that was being used to support the reproduction process of both capitalism *and* bureaucratic socialism" (Schroyer, 1973, pp. 132-133). In the mid-1930s, now operating in an America that on constitutional grounds would ostensibly tolerate their critiques (at least more so than Germany or the Soviet Union), Horkheimer and Marcuse set their sights on formalizing precisely how their proposed model of the critical method would proceed as a guiding force of this research program. In 1937 they each published an essay which outlined their method under the name of Critical Theory. Horkheimer's essay is primarily oriented toward providing a model of critical theory for sociology, whereas Marcuse's essay pushes the same agenda in conversation with philosophy.

In Horkheimer's essay he again makes use of the comparative method to draw a distinction between the kind of theorizing he proposes and the traditional mode of theorizing that dominates mainstream approaches to sociology. "Theory for most researchers is the sum-total of propositions about a subject," Horkheimer writes, "the propositions being so linked with each other that a few are basic and the rest derive from these ... Theory is stored up knowledge, put in a form that makes it useful for the closest possible description of facts" ([1968] 1972, p. 188). This is the way that the natural sciences conceived of theory to communicate the store of accumulated knowledge to future practitioners of the science who wish to advance that knowledge and add their contributions to the general archive of the discipline. Theory serves as a guide for scientists to know which problems have been resolved, where further development is needed, and if in the resulting developments the fundamental principles need refinement or reconceptualization. Therefore, "the general goal of all theory is a universal systematic science, not limited to any particular subject matter but embracing all possible objects" (pp. 188–189). Although this model has success in the natural sciences, Horkheimer argues that it was misappropriated by the social sciences, leading to the production of traditional theory that is "concerned only with some descriptive analysis of a problem or phenomenon" (Thompson, M. J., 2017, p. 6); in other words, a furtherance of the project of categorization, or in the case of sociology, a social taxonomy. For example, with the implementation of Roosevelt's New Deal policies there was an "increased demand for statistical research" (Rawls, 2018, p. 2) as a form of informational social taxonomy for use in public policy. In the lead up to World War II the shift continued in American sociology "toward quantitative methodologies and theoretical collectivism" (Martindale, 1976, p. 141), but it was the social transformation of the war itself that reframed sociology and stigmatized qualitative methods and individuals working as theorists, thereby "create[ing] new conditions for theory and method that were regressive; making sociology less scientific and less creative" (Rawls, 2018, p. 16). Presidents of the American Sociological Association began a concerted effort in the post-war years to "tailor research *to meet the criteria of other sciences* that were more respected and better funded" (p. 17) so that their departments could gain a greater institutional footing. By the outbreak of World War II, sociology in Germany had been dissolved in the university systems. After the war it was subject to the same positivistic trends as American sociology (Lepsius & Vale, 1983), although there was more resistance and debate over this model in Germany thanks to the Frankfurt School's interventions than there was in American sociology where they only had a marginal status within sociology (Adorno, et al., [1969] 1976).

This positivistic approach to sociology—which relies on and promotes traditional theory—assumes that it can take information and use it to induce the natural laws of society, as the natural sciences have discovered the laws of nature, but "[s]ociety must always have already developed before its general rules can be formulated" (The Frankfurt Institute of Social Research, [1956] 1972, p. 4) and since it has not, such a project is based on a paradoxical assumption dooming it to failure. In other words, positivistic approaches to sociology and traditional theory fail to account for the implications of treating society as an object like those objects of the natural world which scientists can study in a laboratory by controlling all the constant and predetermined variables; that is to say that sociology is "resting on the paradoxical presupposition that research efforts should be directed at the refinement of concepts and methods that are inherently *static*, in the interest of illuminating a reality that is conspicuously *dynamic*" (Dahms, 2007, p. 192). On this ground, traditional theory relies on a philosophical contradiction which it ignores as it continues to promote the practice of an affirmative sociology that aligns with the institutional pressures to conform to the guiding logics of capital and information, under the aegis of technological rationality.

Critical theory, on the other hand, "constructs a developing picture of society as a whole, an existential judgment with a historical dimension" (Horkheimer, 1972, p. 239). The practice is "marked by tension" (p. 208) because critical theorists are cognizant of the fact that the subject who engages in critical thinking is "a definite individual in his real relation to other individuals and groups, in his conflict with a particular class, and, finally, in the resultant web of relationships with the social totality and nature" (p. 211). In other words, the self and society are co-constructed and contingent, meaning that in modern societies all thoughts are placed in tension between how the thinking subject would think under conditions of authentic freedom, subjectivity, and individuality, and the way that they are compelled to think in modern societies that structure individual consciousness by means of social domination. This means that what we think about and how we think about it are the result of living in society, and modern capitalist society is one that does not provide us with the social conditions needed to develop genuine emancipated thinking. So critical theory must operate at two levels simultaneously, first it must explain the dynamic nature of modern society and then it must perform "the normative evaluation of what made the object of investigation problematic" (Thompson, M. J., 2017, p. 6). In other words, sociological research must not only explain and code social phenomena, but it must understand how we are structured to think about the process of explaining and coding social phenomena and what impact our mental structuring has on what we do with it. "The Frankfurt School's critical reflections on science are thus primarily concerned with relating the object of knowledge to the constitutive activity of the subject within a historical context" (Schroyer, 1973, p. 134). With each new understanding of social phenomena, the social scientist must then see how previously understood social phenomena impacted this new understanding and how this new understanding impacted and impacts the discovery and use of its knowledge content within the socio-historical conditions that it was produced.

Two comparative ways of thinking about this paradoxical situation, that mainstream sociology has tended to ignore, have been provided by the sociologists, Niklas Luhmann (1994) and Harry F. Dahms (2008b). Luhmann (1994) asks us to question not only "What is the case?"—that is, what are we observing when we study a given social phenomenon—but also "What lies behind it?"—that is, what is structuring our perspective and what are we missing or not accounting for in the ways that we are observing the given phenomenon. To do both would involve a "formal congruency between society's and sociology's modes of communication" (p 138), thus requiring what Luhmann refers to as second-order observation.[5] The effects of implementing this communicative model on theory specifically, and sociology more generally, would be one where

> [e]ach concept that enters such a theory of modern society must be changed according to the specifications of this theory, while, at the same time, permitting maximum resonance with society. Such a theory would mirror or represent nothing. It would not be formed according to the alleged "nature" or "essence" of its subject-matter. But such a sociology would be a model of society in society that informs us about its uniqueness. It would offer novel possibilities for observation, independent from those in the function systems and everyday communication [which are warped by the domination of modern systems whose truth value is immanent to its own functions]. (p. 138)

Dahms (2008b) emphasizes the need to account for the space and time in which research is conducted to understand the effects that not only modernity as a whole has on our thinking, but that pockets of modernity, such as the distinctions between different features of modern societies that are characteristically unique to each but perhaps not unique to all of modern society, have on structuring our approach to research and therefore on how we perceive and interpret the phenomenon under observation. He calls for a renewal "of the "classical" commitment to analytical efforts that are, in fact, oriented toward a kind of *basic research* that is directed at illuminating the constitutional principle of a form of social organization that is built on and around contradictions, and that continues to produce and feed off of alienation, anomie, and the Protestant ethic, at an increasingly intensifying pace" (p. 49). In Dahms's approach, he more fully captures the dark side of modern society by emphasizing the symptoms it produces which condition our normative outlook and approaches to research. For Dahms this means that critical theory "must direct our labor at revealing the causes and consequences of modern society's inherent irreconcilability—on the assumption that if we should make it to the proverbial bottom, we might be in the position to conceive of social life in ways that are not determined by immersion in the gravitational pull of modernity's inability to allow for a qualitatively superior constitution of societal existence" (p. 49). Dahms's call for a model of basic research, echoes both Horkheimer's

---

5 See also Luhmann, 1993.

and Luhmann's projects, because they all recognize that by breaking out of modern society's dominant communication systems and refusing to allow research to only be structured in accordance with the goals of the totalizing logics, it would produce research that could have no immediate practical use-value to modern society because the constitutional logic of modern society is such that it negates attempts to restructure it according to differing sets of value-laden principles and therefore is not concerned with its own contradictory mode of organization. However, such research would provide a greater illumination of the contradictions of modern society and what we are facing in everyday life than current models, thus making it useful for the individual, if not modern society as such in its current configuration.

Given that even critical theory results from thoughts produced within the social domination of capital, thinking through the resultant contradictions, the limitations of this social system, and what the preconditions for their overcoming might be, are not sufficient for the project of critical theory. The material limitations that constrain thought are also, therefore, precisely what Horkheimer ([1968] 1972) saw as the conditions that necessitate that critical theorists "take refuge in utopian fantasy" (p. 211). This refuge, with no material foundation for its actualization in the current socio-historical context and thus existing and persisting in thought alone, must always be subject to its own auto-critique so that it challenges actually existing reality with every step while developing a rigorous distinction between the actual, the possible and the probable. For this reason, critical theory cannot be a static practice lest the dynamic system of capital coopt it. Critical theory is in constant risk of being coopted by capital because capital continuously transforms itself as it attempts to transcend its barriers and, in that quest, it is agnostic to values, thus allowing it to see in even the most critical thought that runs counter to capital an avenue to develop new markets that neuter critical thought by closing off its avenues of possibility; for instance, capital embraces the selling of books that critique the commodity form while also assuming the form of a commodity in the market of books. The critique is necessarily weakened by being forced into a material contradiction.

As capital shifts its perspectives it continuously distorts consciousness and reality until it has made affirmative thoughts out of, or at least neutered the thrust of, those that were once critical. Critical theory, therefore, responds to the dynamism of modern societies by placing the critique of its own assumptions front and center in its practice so that it can reload and recode its arsenal time and again in the face of the continuous evolution of social domination. Because the institutional logic of science as it is practiced in the academy rewards scientists who propose pragmatic solutions to current problems, "[t]he application, even the understanding, of these and other concepts in the critical mode of thought," necessary for the practice of critical theory, "demand activity and effort, an exercise of will power, in the knowing subject" (p. 230). Especially since critical theorists depend on academic jobs for their pay, their bucking of the institutional logics with their research is bound to be met with continued resistance. Being committed to the practice of critical theory will therefore amplify anxiety within the practitioner, because the practice demands nothing short of placing oneself in a precarious economic and social situation as an individual thinking outside the given framework of society, while simultaneously being physically bound to that framework. Given the trend toward a general lack of support for critical studies in the academy today and the lack of market demand for works dedicated to the development of critical thought, such a research program is not without its existential risks. Critical theorists, therefore, assume a position in which they maintain this anxiety in their practice to keep it alive, refusing, on the one hand, to channel it toward actions that displace the real object of anxiety onto socially approved outlets in exchange for a greater chance to make money or to experience small and temporary bursts of ecstasy (such as, winning professional awards, large book contracts, prestigious postings at elite universities, public recognition, etc.) or, on the other hand, to repress the sources of this anxiety and live in a state of cognitive dissonance, despite the personal sacrifices and discomforts that such a commitment is likely to cause. Their minds must, therefore, be content to take refuge in radical utopia as an idea of future possibility, even as material avenues for its actualization are overwhelmingly foreclosed. This is what sustains their practice as they continue to shine a light for others who may begin to feel this anxiety within themselves and look for ways to understand its sources.

Herbert Marcuse ([1968] 2009) echoed these themes in his complementary essay, also appearing in 1937, when he credited critical theory with uncovering "the responsibility of economic conditions for the totality of the established world and ... the social framework in which reality was organized" (p. 99). Like Horkheimer, Marcuse recognized that critical theory challenges the legitimacy of all modes of thinking which assume a purity untouched by material conditions. What critical theory stresses is that these practices take place within history, meaning that they must be understood and studied in the Weberian sense to ascertain "the causes of their being historically so and not otherwise" (Weber, 1949, p. 72). If philosophy was right in assigning reason as "the ultimate and most general grounds of Being" (Marcuse, [1968] 2009, p. 100) and "established [it] as a critical tribunal" that depends on the individual to exercise it, then reason "contains the concept of freedom as well" (p. 101) because if the individual is not free, then reason must appear in the form of something externally imposed upon Being. Since the empirical world is dominated by economic concerns and logics, it "appears to make reason dependent" (p. 103) on those concerns and logics because they ontologize the subjects of modern society and assign them with a mode of being proper to their functions, not to their own internal construction as an authentic mode of being.

When philosophy accorded reason as the means of achieving authentic being, and linked it to freedom, it appeared as if philosophy had reached its limit in providing the tools for achieving the good life. This is because the course of modern bourgeois society in the 19th century was, albeit unequally and to a limited degree, actually advancing conditions of freedom for greater numbers of people and providing avenues for them to experience an increased level of happiness. For example, many industrial technologies did lessen the physical, if not the mental, burdens of labor and many consumer technologies provided greater access to the spaces of the globe, as well as the means to stay in touch with friends and family who were spatially separated. The result of these historical developments during the time of industrialization was that philosophy distanced itself from the concerns of material reality, believing that it had already provided the necessary tools and pointed the way for the fulfillment of its project. However, since the conditions in the 20th century assumed a repressive form which negated the general progress toward individual freedom, as liberal societies maintain is their goal, conditions for the actualization of the philosophical project were no longer logically given within the historical unfolding of material reality. Philosophy, therefore, found itself with answers that no longer reflected the given conditions of fascist European countries, totalitarian regimes in the East, and democratic countries subsumed by the logic of capital in the West, which were indulging their "authoritarian barbarity" and separating "the previous reality of reason from the form intended by theory" (p. 116). Instead of creating the conditions for free individuals to use their reason and develop an authentic state of being, these modes of social organization began to sacrifice the individual to their logic, warp the social, and make culture vanish "to the point where studying and comprehending it is no longer a source of pride, but of sorrow" (p. 116). For Marcuse, then, philosophy must work to either preserve the categories of "[r]eason, mind, morality, knowledge, and happiness," since they "are not only the categories of bourgeois philosophy, but concerns of mankind," or it must 'derive them anew' (p. 108). In other words, philosophy cannot simply turn back to its foundational thought and believe that it communicates transhistorical truths, rather it must continuously refresh the categories that it has deemed to be of the highest importance in the quest for the good life against the storm of history lest they be washed from the collective consciousness or relegated to the dusty archive where they are reduced to their mere informational content as historical data points. Critical theory keeps those past hopes alive, without hoping for a return to that past and all the problems it dragged along with it.

Whether scientific practice or philosophical inquiry, neither can be performed in an isolated bubble outside of the space, time, and influence of modern society's totalizing forces. However, acknowledging this will place certain uncomfortable truths in the minds of scientists and philosophers—like how their seemingly disconnected practices exist in a larger social framework where they can and are often used to amplify unequal power relations and social domination. Even practices that claim otherwise, like critical theory, must be alert to the possibility that they can actually amplify and intensify social inequality and injustice by laying bare the logic of these processes for those who may use them to extend exploitation and domination. It is, for example, worth considering what capital learned from Marx's exposé of its internal contradictions and how that knowledge has been used in ways to counter the internal contradictions; as, for example, with Keynes' strategies for saving capitalism from its production and unemployment crises by encouraging social conditions to support the creation of a consumer class. The role of critical theory is therefore only to make "explicit what was always the foundation of [philosophy's] categories: the demand that through the abolition of previously existing material conditions of existence the totality of human relations be liberated" (Marcuse, p. 107). Not wanting to face up to this logical limitation of the critical project has led many academics to practice either a naïve form of critical theory that mistakes critical inquiry with the making of ideological judgements that do not account for how the logic of capital distorts them or a domesticated form that neuters the practice of its most critical and revealing concepts to avoid a direct confrontation with the reality they help to shape and reinforce (Thompson, M. J., 2016).

Many have, because of their anxieties over the social constellation of modern society and the painful symptoms that society produces, sought to channel those anxieties in available ways (political action, protest, etc.) to ease the pain, but this often requires either ignoring or refusing to accept that these actions run counter to the logic of critical thought precisely because they extend the logic of capital, sacrificing momentary easing of social domination for the extension of its most insidious forms. For example, modern societies have often cultivated the temptation to buy into a form of either scientific or technological utopianism, sensing in the generational advances and the deep technological embeddedness of society a movement that is not visible in the social, but which, in that movement, provides a sense that it is society itself moving forward and progressing. What those who succumb to this temptation miss is how science and technology are so entwined with the logic of capital that they ignore "the association of those men who bring about the transformation" of society (Marcuse, [1968] 2009, p. 115); that is, those who have the reserves of capital are designing these technologies not for social progress but to grow their own capital reserves and social power. As occurred in the early 20th century when technology was wrestled away from the inventors and placed in the hands of the financiers, "science and technology depends on them," that is, on those who worship the logic of capital, meaning that these avenues of change "cannot serve a priori as a conceptual model for critical theory" (p. 115) because they represent a furtherance of the logic of capital, not a negation of its structuring power. Although the future is highly likely to be immanently dictated by the paths of capital and technology, these are paths that must be submitted to ideology critique so that they are not mistaken as paths to freedom. Therefore, the critical theorist must above all else submit her own beliefs to auto-critique, so that she is not caught up in the ideological vortex caused by technical vertigo.[6]

Being critical of the approved modes of liberal political action, does not mean that critical theorists ignore the future, condemn praxis, or turn a blind eye to the immanent suffering of dominated peoples around the globe. Rather, by following the logic of the division of labor and in recognizing the severe restraints placed on individuals' time in modern societies, critical theorists are best suited to treat their practice of macro-critique and the development of critical theories of society as the most important form of praxis they can do in

---

6 Another example is the most readily available form of political action in Western democratic societies: voting, a practice that heavily relies on the ideological belief that this is a necessary and sufficient way of engaging in and supporting democratic ideals. In the American context, with a two-party system in which both parties solidly support the extension of the logic of capital, this has often meant assuming a perspective that it is better to vote for the lesser of two evils. After the election of 2016, which saw Donald Trump elected to office, many liberals turned this into a battle cry ("Blue no matter who"), condemning those who could not bring themselves to vote for the neoliberal policies of Hillary Clinton. This continued in the 2020 election with the Democratic party pursing a strategy which courted the "moderate" (pro-capital) Republican vote and shamed the "progressive" (ostensibly anti-capitalist) vote. There are, admittedly, many pragmatic and moral reasons why people may have decided that it was better to vote against the authoritarian and neo-fascist policies of Trump when there was a choice between that or the furtherance of a liberal status quo. Some members of the Frankfurt School even worked for the Office of Strategic Services, the forerunner to the CIA, during World War II . However, for a critical perspective that argues against supporting the vote as a form of political action in a two-party America, see Du Bois, [1956] 2002.

the given conditions. To the extent that critical theory emphasizes the materiality of history, it concerns itself "with the past—precisely insofar as it is concerned with the future" (Marcuse, [1968] 2009, p. 116). And, despite the many world crises raging in 2020 which are apt to produce a form of tunnel vision, it is not necessarily the immediate future of the next generation or that of the present that is most concerning. With resource depletion, climate change, and global political and economic crises threatening our survival, and the whole human species (beyond the divisions of religion, ethnicity, race, gender, sexual orientation, or national identity) subjected to varying degrees of alienation, anomie, the Protestant ethic, and repression, the concern is over the long-term future of the planet and life itself. It is this concern with the future that leads critical theory to challenge modes of science and philosophy that do not stress the importance of rationally thinking through how the effects of their practices develop historically and why science and technology are not the definitive vehicles of utopia.

Since critical theory has the normative goal of an authentic existence for all in a rational and just society, it concerns itself with why history has developed in ways that prevent the realization of this goal. When Marcuse, again echoing Horkheimer's words above, wrote that "[i]n order to retain what is not yet present as a goal in the present, phantasy is required" (p. 113), he meant that to prevent the goal of a better society from being snuffed out, the flame of its desire must be continuously cultivated in the mind. However, from a relativized standpoint, modern societies do not cultivate critical minds that can deploy their own reason, that have a sociological imagination, or that are inclined to engage in critical thought. Critical theory, therefore, must seek out and work toward the development of a particular kind of mind; one that is willing to engage in intellectually challenging work despite all signs warning against such an endeavor. Given that a rupture with the past birthed something as radical as modernity—with its aspirations for liberty, equality, and fraternity—it triggered this normative desire and logically illustrated its historical possibility. The rarity of such a moment in our species' history, when the oppressed could even for a moment peek beneath the veil of domination and see that a different reality is possible, serves as the justification for why critical theory refuses to let the thought die and sees its maintenance as the ultimate form of praxis, among others that are available within more limited space-time horizons. Even if the socio-historical conditions are such that critical theory cannot cultivate the conditions to actualize these goals materially in present circumstances, it continues to seek out willing minds to act as the material presence of this ideal by keeping it alive in thought until such a time when they might bear fruit.

Critical theory acts in full knowledge that it is developed and practiced in a reality that is a contradiction of unreconciled norms and values. Like the system of modernity that it diagnoses, critical theory therefore also appears as a contradiction. On the one hand, it insists on being ground in material reality, but it does not do this to the exclusion of the ideal. Rather, critical theory places the contradictions between the material and the ideal in dialectical tension and continuously sharpens them on each other, testing the waters to see if the ideal can ever penetrate the material or if the material will snuff out the last breath of the ideal. Recognizing that the material is the limit of possibility for the ideal, critical theory advances via the logic of negation so that it can discard categories and concepts that have either been turned into affirmative ones or have lost the essence of possibility as it reloads the concepts that are still needed within the current socio-historical conditions. This constant reloading is what takes the critical method of the classics and turns it into a dynamic practice of critical theory that is suitable to the continued diagnosis of a dynamic reality.

After working within the preceding model of critical theory for over 20 years, during the late 1950s[7] to the mid-1960s Adorno interrogated the critical method again to refine and reload it in a form he deemed more appropriate to the material conditions that arose in modern societies post-World War II. Although Adorno was clearly influenced by Marx in his dedication to dialectical thinking and he followed the central belief of critical theory that revolutionary change is of paramount importance if humanity is to build the social conditions necessary for the development of emancipated individuals, at no point did Adorno theorize the "concept of a collective revolutionary subject which might accomplish that change" (Buck-Morss, 1977, p. 24). When Marx wrote his theory of the proletariat as the class that would become the foundation of the social revolution, he was writing under a different set of material circumstances, ones that did not and could not reflect the material reality of 20th century. Marx's own commitment to the critical method would likely have made him one of the first to agree that this was the case.[8] However, the

---

[7] Some of the first seeds of this project appeared in "[t]he series of lectures Adorno delivered at the Johann Wolfgang Goethe University in Frankfurt in the summer semester of 1958," the topic, "an introduction to dialectics" (Adorno, T. W., [2010] 2017, p. xi).

[8] Shortly after Marx's death, in the 1883 preface to the Communist Manifesto, Engels already acknowledged a material change in this direction: "this struggle, however, has now reached a stage where the exploited and oppressed class (the proletariat) can no longer emancipate itself from the class which exploits and oppresses it (the bourgeoisie), without at the same time forever freeing the whole of society from exploitation, oppression and class struggles— this basic thought belongs solely and exclusively to Marx".

weakness of his class theory was that it offered a positive view of history that could easily be read in a deterministic manner, and without taking a holistic view of his work which would temper such claims, it opened the door for Lenin to latch on to those positive elements and reify the concepts.

Although Lenin claimed to condemn Hegel's idealism and, like Marx, conform to a materialist position, he snuck idealism in the backdoor of his theory and praxis when he circumnavigated the capitalist stage of development and attempted to create an ideal communist society out of the wrong material conditions. When Stalin took over, he had neither a logically consistent theory of communism,[9] nor a material reality that conformed to the theory of communism, so he largely abandoned the ideal and sought to affirm the material reality of the Soviet Union through pragmatic reform and authoritarian rule. With his death and Khrushchev's rise to power in 1953, attempts to theoretically justify the project all but vanished and revolutionary politics fully succumbed to a purely reactionary mode of politics. Adorno, having witnessed this history, assumed an approach that while retaining a Marxian orientation differed from Lenin and other orthodox Marxists, in that he based his arguments on an inner logic which held that "any philosophy ... lost its legitimacy when it overstepped the bounds of material experience and claimed metaphysical knowledge ... [and] that the criterion of truth was rational rather than pragmatic, and hence theory could not be subordinated to political or revolutionary goals" (Buck-Morss, 1977, p. 25). Failing to adhere to this logic led to positivist takes on reality that either sought to affirm the wrong material conditions, as with Stalin, or by rejecting reality elevated the ideal over the material thus making the philosophy descend into ideology, as with Lenin. For Adorno, the only logically consistent approach was one that negated the current material reality and the distorted concepts that reality produced of itself. By completing that task, Adorno argued that it would cut off positive theories that reality could use according to its own, flawed, principles. His refusal to produce positive theories based on ideals for the actualization of the revolution appeared even more justified after World War II, given how the applications of those positive theories were warped by material circumstances that emulated the totalizing logic of capital and the mode of thinking it produced along the lines of instrumental and technological rationality rather than leading to genuine emancipatory conditions in either capitalist or communist societies.

---

9   The "Socialism in One Country" policy was a clear break with Marx and Lenin's theories of communism.

In the early 20th century when capital faced its most serious challenges to date, with World War I and economic depression, the proletarian consciousness was not ripe for a social revolution. After the atrocities of World War II, the super-powers of the United States and the Soviet Union became locked in an ideological battle with neither side representing any strategy for movement toward an emancipatory society, meaning that the material conditions were still not right to produce the consciousness needed to actualize a class-based revolution. The German philosopher, Peter Sloterdijk ([1983] 1987), offered the following diagnosis of the material contradictions in post-War societies:

> What in the current world situation is a conflict within the system presents itself in an absurd way as a conflict *between* two systems. At the same time, this externalized conflict between the systems has become the main fetter to the liberation of human productivity. The so-called system conflict takes place between two mystified modifiers. By means of a paranoid politics of armament, two real illusory opponents force themselves to maintain an imaginary system difference solidified through self-mystification. In this way, a socialism that does not want to be capitalism and a capitalism that does not want to be socialism paralyze each other. Moreover, the conflict confronts a socialism that practices more exploitation than capitalism (in order to hinder the latter) with a capitalism that is more socialist than socialism (in order to hinder the latter). (p. 247)

As evidenced by the ensuing arms race that continued to soak up the surplus capital, even at the expense of social needs during a time of relative peace and prosperity, both the United States and the Soviet Union were far more concerned with propping up their own regimes on the back of military technology to maintain their global influence than they were with paving a better way forward for their own citizens. Because of Lenin's theoretical missteps and Stalin's pragmatism, not only did the Soviet Union pervert Marx's idea of communism to the detriments of their people, but once capitalist societies had advanced their industrial bases to the point where they could evolve into communism, actually existing communism in the Soviet Union and China had little if any appeal for the Western masses.

It was not a hard sell to convince the Western labor class that capitalism and its consumer society was the superior model, despite the prevailing symptoms of alienation, anomie, the Protestant ethic, and repression that they suffered from. It was, in their immediate self-interest, better to eat fast food and watch advertisements for consumer goods disguised as television entertainment than dealing with the Gulag or mass famine. Likewise, the Soviet Union could point to the capitalist model with all its deficiencies—as whole segments of the population fell through the economic cracks and were subjected to brutal racism—and convince its masses of the failures of capitalism. Rather than either system engaging in a self-reflective and critical practice that would expose the internal contradictions of their systems, they shaped and guided their societies around the obvious failures of the other side, ignoring their own shortcomings. Capitalism, in the post-War years, began to implement far more social welfare programs than its logic would have suggested. Likewise, Soviet-style socialism implemented far more projects oriented to the exploitative logic of capital to keep pace with the arms race between the two warring sides and to pacify the demands of the masses just enough to prevent wide-spread social unrest. As a result, there were far more similarities than differences between the two models in the post-War years, but because of their antagonistic stances the differences were exploited within the collective consciousness to demonize alternative models.

Not only did positive attempts to theorize a way out of the present circumstances rely on a philosophical contradiction, but they were at best prone to failure and at worst distorted by the totalizing logics into blueprints for enhancing and furthering social domination rather than lessening it. What made the dialectic attractive as a method for thinking within material conditions that warped and distorted thought, was that for Adorno it was a method for understanding how thought is distorted because it contains "a double movement, a movement of the objective concept, on the one hand, and a movement of the knowing subject, on the other" (Adorno, [2010] 2017, p. 24). Since it was all too clear from the concrete material circumstances that society and consciousness were both lacking in the necessary development, any successful attempt to reconcile the differences between the social and the individual must operate at both levels simultaneously to understand their impacts on each other. Furthermore, the dialectic could link thought to material reality because it is both "a method of thought" and "a specific structure which belongs to the things themselves" (p. 1) and it is through the alliance of the two that there emerges a "movement of the concept" (p. 9). The key take-away is that there is a distinction between the concept and the thing itself, and this is what the dialectic exposes. If capitalist or communist societies would confront the contradictions between the claims of their models and the concrete material reality of their societies, the gap would become evident and to rectify the contradictions in thought the concept would have to change. If the concept, then, would no longer retain the ideological claims that were not to be found in the present material reality, then the poverty of the ideas would become like a splinter in the mind of the people who would feel the living contradiction of their lives within their given society. Since history has no reason inherent to it, aggravating consciousness is not a positive guarantee that material circumstances would change. Nor does it guarantee that the circle will close in a final synthesis of the totality ushering in the end of history. Following this model does, however, enable a reshaping of the individual mind which could impact the collective consciousness, and a critical collective consciousness is a necessary precondition to any material change at the level of social structure.

Adorno could not, however, simply suggest that we rely on the dialectical method as practiced by Hegel or Marx, because "there is actually an inner affinity between dialectic and positivism" (Adorno, [2010] 2017, p. 117), in that, the dialectic, like the positive sciences, takes the immediately given as the objective fact of reality. Because of that common ground the divisions between the two methodological approaches must be clearly delineated so that dialectic does not collapse into positivism. Adorno explains, "in contrast" to positivistic "thinking," the dialectic "is expressly self-reflective in character" (p. 124) but "if we fail to reflect closely on these things [that are exposed by dialectical thought], we experience an ever stronger tendency to project what in reality is due to such objective circumstances precisely upon personal factors, upon the characteristics of particular human beings or particular groups of human beings" (p. 123). This is precisely the temptation that Marx warned against in the 'Preface to the First Edition' of *Capital Volume 1* ([1867] 1990), when he wrote:

> Individuals are dealt with here only insofar as they are the personifications of economic categories, the bearers of particular class-relations and interests. My standpoint, from which the development of the economic formation of society is viewed as a process of natural history, can less than any other make the individual responsible for relations whose creature he remains, socially speaking, however much he may subjectively raise himself above them. (p. 92)

Although Marx was not sympathetic to either the bourgeoisie as a class or individual members of it, he recognized that actions taken against specific individuals or groups would have no impact on the structures that allowed those individuals or groups to appear in the first place. This was a lesson that modern society refused to learn, as in World War I, when it assigned blame on specific groups and then recreated the conditions that gave rise to World War II. Again in World War II the atrocities were placed squarely on the back of the Nazi's as a group and Hitler as an individual, but this ignored the structural causes that allowed someone like Hitler to gain power and find a willing group to participate in his agenda thereby leaving the door open for fascism to rise again.[10] Since critical theory had already conclusively demonstrated that there can be no positive theory of a dynamic society, the tendency that emerged in post-war sociology with the project of social taxonomy was the creation of seemingly more manageable groups within society, neatly organized as information in a clear statistical language, to the detriment of the study of the structures of society and for the purposes of taking action precisely within acceptable scales of reality that left the social untouched and maintained the status quo of modern society. So, although it is possible to take individual or even limited group action (for example, under the pluralist model of democracy) on certain objects that are problematic within society, Marx and Adorno's point is that we must not give into the positivist temptation of thinking that these actions will impact what has caused that problematic object to exist in the first place. Action taken in this manner can occlude the social structures that are actually responsible for the problem which is merely manifesting itself within the group or individual context as an object upon which action can be taken. If those structural causes are ignored then the action taken upon the object can actually do more to extend the social domination of the structural apparatuses, even if the object disappears or is reconfigured. Since Marx never got around to writing his book on the dialectical method, which might have cleared up some of the issues with his own theories and their relation to praxis, there was a need to spell out a form of negative dialectic whose object was to prevent the lapse into positivism.

Adorno began working on this project in earnest starting in 1960 with a series of lectures on related themes. The argument was most fully developed in his 1965/66 lecture course on negative dialectics ([2003] 2008) and the 1966 publication of his book with the title *Negative Dialectics* ([1966] 2007). In his lectures he provides a straightforward answer as to what the method of negative dialectics refers to, and that is, "a dialectics not of identity but of *non-identity*" ([2003] 2008, p. 6). The purpose is not to unify thought and being in some final synthesis, as is the aim of the Hegelian dialectic, rather it

---

10 Horkheimer, for example, wrote an essay condemning the 1961 trial in Israel of Adolph Eichmann, one of the primary architects of the Holocaust. By attempting to place the blame on the individual, guilty of horrific crimes as he was, the attempt at justice failed to acknowledge the role of the structure that gave him the ability to commit his crimes. How, after all, Horkheimer asks, could justice be delivered on one man for the deaths that ran into the millions? "The very idea that Eichmann could "atone" for his deeds according to a human standard and the sentence of a human judge is a mockery of the sacrifice the Jews made, a gruesome and grotesque mockery ... The trial is a repetition: Eichmann will cause harm again" (Horkheimer, [1974] 2012, pp. 121-123).

is to articulate "the divergence of concept and thing, subject and object, and their unreconciled state" (p. 6) to discover what is non-identical in the two. Contradiction therefore takes center-stage in Adorno's thought, but it is "the contradiction in things themselves, contradiction *in* the concept, not contradiction *between* concepts" (p. 7) that interests him, just as it was the internal contradictions of capital that interested Marx. Adorno writes that "to think is to identify" and "[c]ontradiction is nonidentity under the aspect of identity" ([1966] 2007, p. 5). In other words, there is a tendency to think that our thoughts represent a unity with the objects upon which we apply our cognitive functions, that is, we think that when we think, we are thinking the objects as they really are. However, thinking must necessarily rely on concepts, that is, on abstractions from the material objects which come to exist as thought images that we use to identify material things. When those thought images are submitted to dialectical thought and compared to the materiality of the object from which they emanated, a gap appears in which there are, on the one hand, non-identical remainders left over in the object which are not contained in its abstraction as a conceptual thought image, and on the other hand, aspects of the concept which deviate from the materiality of the object. Another way of stating this is that conceptual ideals which are not materially present in the object persist in thought alone (which is the domain of the ideal, since the very notion of the ideal is that it transcends the material), thus distorting the concept from precisely what is supposed to be represented in thought.

The standard philosophical take of dialectics is that by advancing thought in this manner it can arrive at the truth of the object and therefore the truth of thought, however, Adorno warns against this form of identity thinking. He writes that although the philosophy of identity claims "that the true and the false can both be directly read off of the true" this proposition lacks validity because "the false, that which should not be the case, is *in fact* the standard of itself: that the false, namely that which is not itself in the first instance … that this falseness proclaims itself in what we might call a certain immediacy, and this immediacy of the false, this *falsum*, is the *index sui atque veri*"[11] ([2003] 2008, p. 29). Thinking what is false is therefore what constitutes a mode of thinking that is true, since the truth of an object cannot be reconciled in thought by forming a perfect dynamic synthesis between the two. What remains, and what can be validly established, is the non-identical, that which is false in the concept and in the application of the concept to the object; this is what Adorno considers "right thinking" (p. 29). This is precisely where Marx

---

11   Latin to English: the judge of both itself and present truth.

fails in his project because his dialectics lead him to the temptation of speculative thought in which he thinks that there is a truth in the "absolute potential to the productive energies of human beings and their extension in technology" (p. 96) thereby allowing the concept to contain more than is present in its material form. By moving beyond the point in which he exposes the falsity of capital that emerges from within its internal contradictions, Marx leaves this crack for positivistic thought to reemerge within his own speculative thought; in thinking the non-identical, it is exposed.

Since, for Adorno, there can be no talk of method without accounting for the content of the object it is applied to, this call for negative dialectics has a historical necessity to it that arises out of the conditions that he was writing in. Having witnessed the most devastating history of humanity in the 20th century, with two world wars and a social drive toward the development of technological rather than social rationality, the material circumstances in the 1960s pointed to a profound lack of critical self-reflection, even among those who preached radical and revolutionary politics. Their failures to transform society betrayed their ignorance, especially in the American context with its claims to various freedoms in the constitution, that "there is horror because there is no freedom yet" ([1966] 2007, p. 218). While the lack of freedom was visible on the surface of authoritarian countries, the consumer society in America and the West had hidden the ways that it restricted individual freedom with activities that on the surface appeared to promote freedom of choice and expression; of which the protest movements in the 1960s were but only one such manifestation. Without a negative dialectic that exposed the falsity of modern society's claims, the individual may at times "oppose himself to society as an independent being" but in such a context the result is that "[t]emporarily, the individual looms above the blind social context, but in his windowless isolation he only helps so much more to reproduce that context" (p. 219). "To dominate this conditioning, consciousness must render it transparent" (p. 220), which means that the only avenue of breaking out of this cycle cannot be found in pragmatic actions taken within the bounds of the lacking society. Rather subjects must, in realizing that they live within a lacking society, come to see how they have been produced as a lacking subject, and in interrogating the lack of self and of society, and the falsity of the world that denies such a lack exists, they will keep alive the concepts upon which a non-lacking society would necessarily have to be based.

Therefore, if avenues of action are foreclosed by the concrete historical circumstances in which we find ourselves, Adorno's lesson is to remember that thought remains. It remains a need of humanity to think through their conditions, to understand the dynamic between their self and their society, and to engage in a form of labor that progresses even by the smallest degree toward that good life that philosophy imagined all those years ago. The wish that underlies this need is precisely for the freedom needed to use reason and develop ourselves as authentic beings: true to ourselves and in harmony with our societies. Obviously wishes cannot be fulfilled in thought alone, but neither can the fulfillment of this wish be the result of actions taken in sociohistorical contexts that prevent the right thought from flourishing within the masses. Thought is, therefore, only the necessary precondition for any attempt to realize a different form of reality. Limiting the critical method to a negative dialectics that exposes the falsity of our condition can have the effect of confusing negative thought with pessimism, "[f]or suffering is the weight of objective realities bearing down on the individual" (Adorno, [2003] 2008, p. 110) and suffering can degrade negative thought into pessimism. But pessimism demands "passive contemplation" of this reality, whereas "[t]he effort implicit in the concept of thought, as a counterpart to passive contemplation, is itself this very negativity, a revolt against any demand that it should defer passively to every immediately given" (p. 112). This is the radical impetus behind Adorno's critical method, to resist the way the world conditions positive thought through the use of negative dialectics as a way to train the right kind of thought in the wrong world. The result is a method that produces thought which is of no use to the system but is of use for individuals living and coping with that system. By continuously proceeding via negation, it makes negative dialectics as dynamic as the world in which it must function, forever resisting its totalizing logics since it recognizes that even its own insights must be continuously subject to negation as our concepts and our material reality continue to transform under the conditions produced by the totalizing logics of capital and information. In that transformation the identical and the non-identical are moving targets and tracking them is what gives purpose to critical thought and keeps the project alive.

3   A Lesson for Socioanalysis: Anxiety and the Social Vicissitudes of Technology and War in Mass Society

Positive, traditional, theory serves a function in modern society because it mimics the totalizing logics of capital and information, providing tools that reinforce the structures of social domination. Ideal theories of social change are warped and negated by those logics; feeble protests that betray a lack of understanding of the irreconcilability of modern society with the values that it sublimates. Critical theory, including Adorno's model of negative dialectics, responds to these theoretical models by interrogating the psychosocial divide with a method for critiquing the material structure of society in its relation to the self, and the self in its relation to society, under the specificity of the given socio-historical conditions, so as to construct micrological ruptures in the totalizing logics at the point of their structuration of the subject's mind. The working hypothesis of this approach is that "[t]he separation of society and psyche is false consciousness; it perpetuates conceptually the split between the living subject and the objectivity that governs the subjects and yet derives from them" (Adorno, 1967, p. 2). This condition cannot, however, be solved on a purely methodological basis, since "[p]eople are incapable of recognizing themselves in society and society in themselves because they are alienated from each other and the totality" (p. 2).

Although C. Wright Mills ([1959] 2000) was right that a sociological imagination is needed for people to be able to link their personal lives to the given socio-historical circumstances, Adorno reminds us that our condition in modern society neither cultivates nor allows this kind of thinking to develop spontaneously. Likewise, modern society attempts to extinguish all sources of nourishment for this kind of thinking. That this mode of thinking remains present in certain pockets of intellectual resistance, despite modern society's attempts to annihilate it, is because it broke through social domination and entered the store of human knowledge at a time when society was going through a transition from the totalizing logic of religion to that of capital. This mode of thinking has been kept alive, handed down in thought and maintained in the archive of knowledge, ever since. There was no such rupture when the totalizing logic of capital gave birth to the totalizing logic of information. If a rupture had taken place then it would have opened new avenues for individual and social emancipation, rather these logics achieved a seamless transition and synthesis that is highly compatible with the project of negating critical thought. Under the combined effects of the totalizing logics of capital and information, alienation has been amplified, as the sources of this condition have compounded their effects. The result is that attempts to continue the transfer of this critical knowledge to new generations now faces an increasing number of structural and institutional barriers.

Therefore, it is a mistake to assume that sociological data will 'speak for itself' in our current socio-historical circumstances. This is especially so because the structure of everyday life works against individual attempts to cultivate the kind of mentality needed to theorize the links between sociological data and our personal lives on our own. The critical mode of thinking must be transmitted in a personal manner that requires the presence of individuals involved in the hand-off of this knowledge. Only in an intimate setting that recreates the conversational mode of knowledge transfer which allows for an affective transference between the parties can there be any success in maintaining this mode of thought. Sociology has, however, largely shirked its responsibility to do this; in large part because sociological knowledge transfer in the university system has been subjected to instrumental and technological rationality to such a degree that it has streamlined the process in a consumer-based model. "Intellect's true concern is a negation of reification" (Horkheimer & Adorno, [1947] 2002, p. xvii). Transferring critical knowledge to negate that reification is often a painful process as, if it is successful, then those ossified ideas of self and society break apart and the subject cannot retain the idea that they had of themselves or of their society prior to the initiation of this transfer without engaging in a severe form of cognitive dissonance.

For those who seek to be professional sociologists, the case is not much better. They are often trained under the guidance of affirmative methods and theories so that they can plug themselves into the totalizing logics as professional members of an elite class who seamlessly integrate their lives into the capitalist system. For the most part they receive little to no training in critical methods and critical theories, which means that they too are often unable to activate their own sociological imaginations, let alone serve in a capacity to cultivate this mentality within the masses who they claim to represent with their data. It is true that alienation cannot be reversed by either the application of the "right" methodology or the "right" theory, however, in their combination they can aid in the identification of how the totalizing logics of modern society imprint themselves on the structure of the subject's mind, so that we can identify the social roots of alienation and the contradictions of living in a lacking society while maintaining the utopian dream in the refuge of critical minds. What we are left with if we successfully activate our sociological imagination and engage in dialectical thought is a fragmented self in a fragmented society, the real challenge then lies in building something new out of those pieces that remain.

Critical socioanalysis, is the combination of critical theory and critical method, with the aim of transmitting this model of thought by providing the space and time for people who want to learn how to think about the contradictions they experience in their everyday lives as they struggle to understand the social origins of their alienation, anomie, Protestant ethic, repressive tendencies, and pressure to act and think in accordance with technological rationality. We may think of this knowledge as a code: once it is integrated in the mind of the subject it begins the process of reprograming their mode of thinking to better differentiate their psychological makeup from the social world and thus have a better grasp on the interconnections between the two. The socioanalytic session does not impose the code of the analyst onto the mind of the patient, as this would make the session nothing but another means of compounding alienation and extending social domination. Rather, in the socioanalytic session the socioanalyst serves as the object of the desire to alter one's coding; it is believed by the analysand that the socioanalyst possesses the "right" code and that they will share it with the analysand at the appropriate time. The socioanalyst, however, knows that they do not and cannot possess the "right" code, but that only within the process of the analysand's self-analysis will they come to learn how to write and execute their own code and see how structural forces will regulate and limit the success of that process. The socioanalyst is there to provide the tools for this self-coding. Since we can never escape the coding of modern society, which is externally imposed upon us, this internal coding must become an activity that requires constant attention and care, proceeding via negation so that it is always recoding its own program in ways that continue to counter the programing of current and likely new totalizing logics that will emerge as history continues its march. By engaging in this activity, it allows one to partially reclaim a mode of thinking that both responds and stands in opposition to the present material conditions rather than the alternative of either embracing alienation or living in ignorance of the gravity social structures exert over our lives to the detriment of our affective and material condition. Critical socioanalysis is not, however, a synthesis of psychology and sociology that builds a harmonious model of scaling concepts. Rather it recognizes the fractal nature of these scales and articulates the conceptual differences between psychological and sociological concepts and causes, for the purpose of gaining a greater understanding of the truth of the false consciousness that maintains their division.

 Attempts to find conceptual unity between psychological and sociological frames of analysis are guided by the problem of explaining individual motivations that either converge or diverge with social factors and conditioning. However, such attempts too often see the division as one of disciplinary boundaries that can be solved by finding the right conceptual framework to unite the two in theory. "This is possible only on the assumption that the divergence of sociology and psychology can be overcome independently of the real nature of their object" (Adorno, 1967, pp. 1–2). These lines of research were pushed in the U.S. by newly founded governmental organizations after World War II because there was a sharp rise in mental illnesses. That it became political signals a recognition that there was a social problem, but there was no desire to enact social reform, so efforts were implemented to treat this problem psychologically. Government and big business interests saw an avenue for encouraging a psychologically-oriented social psychology, one which would acknowledge the social problem but ultimately allow them to guide science in a way that it could ignore the social contradictions and through the use of psychopharmacological treatments and mental conditioning further the logic of capital (Scull, 2010; 2011). Sociologically-oriented social psychology also took notice of these problems, however, it received far fewer resources to dedicate to studying them. Rather than unifying and strengthening social psychology the two models diverged in their study of the psychosocial. The guiding orientation of their practice depended on whether it was housed in a sociology department or a psychology department.

Each discipline emphasized the importance of their foundational theories of causality and tacitly gave a nod to the other discipline to expand the range of their generalizations. For example, beginning in the 1950s, the sociologist Talcott Parsons began to construct one of the most rigorous models of this conceptual unity. However, he made his allegiances clear when he wrote: "Though it is logically possible to treat a single individual in isolation from others, there is every reason to believe that this case is not of important empirical significance. All concrete action is in this sense social, including psychopathological behavior" (Parsons, 1950, p. 371), thereby subsuming the psychological concepts into a sociological framework in precisely the same manner that psychology was doing to sociology. Parsons relied on a systems theory framework that treats "the *organization* of the personality as a system" (Parsons, 1964, p. 79) which he believed would place "psychoanalytic theory in such terms that direct and detailed articulation with the theory of social systems" would be possible (p. 110). By proceeding in this manner, he was able to construct a solid framework for analyzing the interconnections and interpenetrations of the physiological, psychic, social, and cultural systems—which he formalized in his AGIL model (Parsons, [1951] 1991; 1971)—but his theory reified the concepts to such an extent that while it had significant descriptive power, the model lacked the critical insight of placing the concepts in tension with their material forms and thus had limited diagnostic abilities. Despite the brilliance of its internal logic, Parsons's model of psychoanalytically informed sociology was largely abandoned in the social sciences, as was the project of finding a synthesis of the two. On the one hand, this happened because psychoanalysis was marginalized by psychology when it adopted the biomedical model and sociological institutions did not want to be the bearers of a cast-off method that no longer had prestige in its own disciplinary area (Chancer & Andrews, 2014). On the other hand, Parsons monumental influence in sociology also harmed the project because in the application of his model it became apparent that "societies problems, as they develop in concrete situations, do not fit into our academic categories" (Smelser & Wallerstein, 1998, p. 25), and if his attempt ended in failure, then it made younger scholars less likely to pick up the reigns.

Critical socioanalysis is also not an attempt to sociologize psychoanalysis by using psychoanalytic theories to explain social phenomena. It would be a mistake to think that the social is guided by the unconscious motivations of discrete individuals, particularly since society has reached the point where it has discovered that "the possibilities of choice available to the unconscious are so limited and perhaps constitutionally so meagre that the foremost interest-groups have no trouble in diverting them into a few chosen channels, with the help of well-tried psychological techniques that have long been in use in totalitarian and non-totalitarian countries alike" (Adorno, 1967, p. 7). What this means is that not only is the super-ego structured by society, but as the artificial conditions of modern society become our second-nature, the id is less structured by natural instincts and becomes more structured by this new model of "nature" and the mechanical instincts appropriate to its constitution. When "[a]dvertising" became the "elixir of life" (Horkheimer & Adorno, [1947] 2002, p. 131) for the culture industry, everyday life became infested with psychological techniques used to exert a greater degree of control over the unconscious mind to shape its desires in accordance with consumerist logic. Psychoanalysis was not left untouched by the effect of the societies in which it operates and so, it too, has largely become a practice that is guided by instrumental and technological rationality. "A technique intended to cure the instincts of their bourgeois distortions further subjects them to the distortions of emancipation" (Adorno, 1968, p. 10). That is, this model of psychoanalysis aims to convince the patient that they, as individuals, can overcome their alienation by means of their own efforts thereby providing a false sense of their 'self as emancipated' when this is a social impossibility. The effect of enhancing this mode of false consciousness is that "[i]t trains those it encourages to champion their drives to become useful members of the destructive whole" (p. 10). This trend was especially visible in the revisionist approaches to Freud's theories provided by Anna Freud and Heinz Hartmann, which came to dominate the American psychoanalytic scene with ego-psychology (Wallerstein, 2002), those of Alfred Adler and his individual psychology (Overholser, 2010), and those of Melanie Klein's object relations theory which came to dominate the British scene (Shapira, 2017). Attempts like these to sociologize psychoanalysis became "a means of [shaping] personal behave[iors] and attitudes to the *status quo* … [by] enhancing the semblance of a spurious identity of individual and society, of happiness and adaptation to omnipotent society … [And at best this becomes an ideology] which completely integrates the individual into an all-comprising organization that nevertheless remains as irrational as any psychological deficiencies of the individual ever were" (Adorno, 2018a, p. 642). Rather than being able to offer a "cure" for its patients' neuroses, these approaches "collaborate with the universal and long-standing practice of depriving men of love and happiness in favour of hard work and a healthy sex life. Happiness turns into something infantile and the cathartic method into an evil, hostile, inhuman thing" (Adorno, 1967, p. 8) thereby exacerbating the mechanthropomorphic transformation of species-being and our mental alignment with the totalizing logics.

With its focus on anxiety, critical socioanalysis interrogates the theories of the psychic world and those of the social world by placing them in tension during analysis. Its guiding hypothesis is that anxiety sits at the nexus of the psychosocial divide because in subjecting it to dialectical thought it reveals the distorted connections between the psychic and the social and the unresolved tensions that persist in that relationship which cause anxiety. Unlike anxiety that has strictly biological origins in the psychic apparatus, which upon locating its object can discharge its cathexis to achieve momentary relief, the anxiety that emanates from the constellation of modern society cannot be discharged upon its object. Anxiety of the kind that interests critical socioanalysis, is an affect that is subjectively experienced in the psychic realm of the individual mind, but is of societal origin, meaning that the object of anxiety is an intangible but objective force that has material consequences. As such, it cannot be reduced to either the psychological or the sociological dimension, any more than it can find a conceptual unity by harmonizing the approaches; both must be advanced simultaneously to expose how and why they fail to unify and, in that failure, locate between the cracks the sources of anxiety. As the critical psychoanalyst Joel Whitebook concluded in his study of psychoanalysis and critical theory, "[a] harmonious integrated self and life history, as envisioned in the classical bourgeois ideal of *Erfahrung*, are undoubtedly impossible today" (1995, p. 262), but in the fragments that are left of our lives, if we follow Adorno's line of thought, "consonance survives in atonal harmony" (Adorno, [1958] 1991, p. 248). While critical socioanalysis cannot promise to remove the social conditions that cause anxiety, it aims to explode the foundations that would see the psychic and social factors as in sync with each other—which would imply a material unity between the self and society that is not present even though they constitute each other—and out of the wreckage of society's approved concepts we can sort through the fragments and achieve something greater than the broken pieces of our lives in our mental reconstruction. Only by placing the concepts in tension with each other can the true nature of anxiety be revealed as a facet of the contradictory nature of life in modern society which will agitate the psychic dimension to reveal how it is at odds with the social constellation. In learning to locate and correctly identify the source of anxiety, the subject will gain possession of the ability to recognize, in a way that resists the totalizing logics of modern society, that their anxiety is the result of the contradiction between the concept that they have of the individual and the society that places limitations on their self-development; in Adorno's language, they will discover the non-identity of these concepts thereby helping them to better know the truth of what constitutes their anxiety in the falsehood of their thoughts.

The Frankfurt School theorists did not create a formal theory of anxiety, rather like the classics they explored other symptoms arising from the modern experience of which we could say that anxiety is a necessary byproduct, but two of their texts deserve brief mention. Fromm makes heavy use of the concept of anxiety in *Escape from Freedom* ([1941] 1969) but he does not give it substantive theoretical treatment. Instead he uses it in an everyday manner as one of several emotions arising from the modern experience. The closest attempt to theorize anxiety was in the essay "Anxiety and Politics," by Franz Neumann ([1957] 2017), but he admits that as a political scientist he is relying on "authorities from other disciplines" (p. 613) in his presentation of anxiety. He sees a clear conceptual link between anxiety and alienation, but he does not draw a distinction between fear and anxiety. Rather what most would call fear, he calls "true anxiety ... a reaction to concrete dangers" which he distinguishes from "neurotic anxiety ... produced by the ego, in order to avoid in advance even the remotest threat of danger" (p. 615). From these he distills three functions of anxiety: (1) "a warning role", (2) "a destructive effect", and (3) "a cathartic effect" (p. 616). The rest of the essay is dedicated to using the link between anxiety and alienation to explain "the affective identification ... of masses with leaders" (p. 618). Axel Honneth (2003) sees Neumann's reliance on Freud's notions of anxiety as a key weakness in his argument, but he completely ignores Adorno's critiques of psychoanalysis and advocates for a perspective provided by "psychoanalytic revisionism" (p. 253). This approach ultimately claims the primacy of the psychic origins of anxiety over social causes, ignoring the need to maintain a distinction between social and psychic concepts so as to place them in dialectical communication.

Despite their lack of theorizing anxiety, this affect was central to the mood, or in Heidegger's language, the 'attunement,' that the Frankfurt School diagnosed as emanating from modern societies. On the centrality of anxiety in everyday life, their position agrees with Heidegger's—who advanced the most philosophically nuanced conceptualization of anxiety in the 20th century—however, there is a clear distinction between Heidegger's causal arguments for where and how the mood of anxiety originates and therefore of the philosophical implications of maintaining such a position from that of critical theory.

To briefly sum up Heidegger's argument, we must begin with the distinction between fear and anxiety, which he rightly points out is either generally confused or lacks a clear distinction in their conceptual use. "Fear," Heidegger writes, "is always something encountered within the world, either with the kind of being of something at hand or something objectively present" ([1927] 2010, p. 136). That is, fear has a definite object and is experienced when in confronting that material object, the being who cares about being faces an immanent existential threat to its being. "By contrast, Heidegger argues that anxiety does not arise in the face of any definite possibility or entity in the world, but instead arises through their dissolution" (Dalton, 2011, p. 74). Anxiety is for Heidegger the general feeling of anxiousness that arises from "being-in-the-world" ([1927] 2010, p. 180) when that being is finally free to actualize itself as authentic Being. Anxiety reveals itself at the crux of the situation we find ourselves in, namely, that to 'realize the potential of Being, being must be free to choose and grasp itself' (p. 182), however, Being "exists in an inauthentic manner ... which amounts to *choosing not to choose itself*" (Magrini, 2006, p. 77). Because of his use of inner-subjective reason, Heidegger concludes that due to the indefinite nature of anxiety,

> [n]othing which is at hand and present within the world functions as that which anxiety is anxious about. The totality of relevance discovered within the world of things at hand and objectively present is completely without importance ... So if what anxiety is about exposes nothing, that is, the world as such, this means that *that about which anxiety is anxious is being-in-the-world itself*. ([1927] 2010, pp. 180-181)

In this way, Heidegger relies on a radical theory of existentialism that completely denies the effect of the objective material conditions on structuring the subject, instead placing the subject as the sole bearer of responsibility for their own anxiety because they have made the choice to flee in the face of freedom into the embrace of the prescribed roles assigned them by modern society, and in abandoning the development of their Being, it is the subject who is ultimately responsible for the attunement that they have with the world which is felt as a lack of freedom. The lack of historicizing his argument within the modern context denies the sociological knowledge that was produced by the classics. Rather than recognizing that the world has changed and that it structures subjectivity by limiting its freedom, Heidegger has disconnected this relationship of the subject to the objective world; the changes in subjectivity are the consequence of individual actions and free choice within the given socio-historical reality. Therefore, the reasons why there are no avenues to freedom is not the result of social domination, but rather because the subjects of modern society have made the choice to abandon the quest for freedom because it forces a confrontation with anxiety.

On the one hand, Heidegger is right in that he sees the necessity of understanding anxiety so that it can serve as a vehicle to achieving enlightened self-consciousness on the path to authentic being, and, that there is a tendency for people to turn away from freedom in the face of anxiety. On the other hand, Heidegger is wrong to think that this is not merely a precondition for embracing freedom, since he sees the subject as fleeing from freedom in society but fails to account for the social conditions that fail to provide the material basis for genuine freedom at the same time that they structure people in ways to make them retreat from freedom. Fromm argues this from the perspective of critical theory and provides a counter to Heidegger's argument by stressing that

> submission is not the only way of avoiding aloneness and anxiety. The other way, the only one which is productive and does not end in insoluble conflict, is that of *spontaneous relationship to man and nature*, a relationship that connects the individual with the world without eliminating his individuality. This kind of relationship—the foremost expressions of which are love and productive work—are rooted in the integration and strength of the total personality and are therefore subject to the very limits that exist for the growth of the self. ([1941] 1969, p. 29)

By combining the insights of philosophy and sociology, Fromm moves between the extremes to recognize and balance philosophical ideals within the given material reality.[12]

Failing to account for sociological knowledge is what made Heidegger's philosophical model entirely consistent with his embrace of the Nazi project in 1933. According to his own words, he "expected from National Socialism a spiritual renewal of life in its entirety" and once it was actualized for what it was, he saw his role as nothing more than a "political error" (Heidegger, 1998, p. 265). Although Marcuse pushed Heidegger to engage in a critical self-reflection, Heidegger refused, because to do so would be to accept a limit on his absolute individualism and would require admitting that his philosophical system was flawed in its reliance on the ideal of subjective reason. For him, it was enough that he had resigned his rectorship in protest of Nazi policies in 1934, and it was

---

12   Ultimately, however, Fromm's theoretical project, when taken as a whole, relies too heavily on the psychoanalytic dimension and the ability of the individual to make choices despite the limitations placed on her by the social dimension.

unnecessary to publicly distance his philosophy from Nazism. But "subjective reason can hardly avoid falling into cynical nihilism" (Horkheimer, [1947] 2013, p. 174) which is precisely what it did when Heidegger claimed that "[a]nxiety makes manifest the nothing" (Heidegger, [1929/1967] 1998, p. 88). The alignment between the core of his thought and the fascist project is here made evident. Heidegger either did not see, or refused to acknowledge, just how closely they aligned, which is why he could not publicly denounce his role without at the same time disavowing the entire foundations of his philosophical system. For if anxiety is only a reflection of the nothingness of our being, then it is also a reflection of the nothingness of the world. And if there is nothing, no psychosocial sense that is agitated by anxiety, just pure and empty being-in-the-world, then in our abandonment of everything it is true that we will find a sense of freedom, but this pure freedom is itself, nothing: an empty void in which the expression of liberated being is free to act without any restraint, for it has rejected all notions of psychosocial consequence and meaning in its abandonment of the world.[13] Because fascism offered a means of completely rejecting anxiety by fully embracing being-in-the-world and the celebration of all its horrors alongside all its ecstasies in the freedom of accepting whatever mode of being there was to be had in the pursuit of passions unrestrained by internal struggles, it was perfectly attuned to Heidegger's philosophical system. Even if he was opposed to the actions they took, for his philosophical thought passive resistance was sufficient and philosophy was above such material justifications.[14]

Despite the dangerous implications of Heidegger's concept of anxiety, it is a shame that the other major contribution to the study of anxiety during the mid-20th century, Rollo May's *The Meaning of Anxiety* ([1950] 2015),[15] only mentions Heidegger in passing and does not offer a critique of his model in

---

13 This is precisely the danger that Freud recognized could arise when the 'self' made a complete rupture with society, and why, while the individual must be defended against the structural forces that negate their path to individuality, so too, "civilization has to be defended against the individual" (Freud, S., [1927] 1961, p. 7).

14 This should make us recall Benjamin's diagnosis of fascism: "'Fiat ars—pereat mundus" [Let art flourish—and the world pass away], says fascism, expecting from war, as Marinetti admits, the artistic gratification of a sense perception altered by technology. This is evidently the culmination of *l'art pour l'art*. Humankind, which once, in Homer, was an object of contemplation for the Olympian gods, has now become one for itself" (2008, p. 42): a perspective evident in Heidegger's philosophy, which is supposed to stand on its own philosophical merit as if it could be divorced from the reality in which it was crafted.

15 The project has peripheral links to the Frankfurt School, as both Paul Tillich (Adorno's habilitation supervisor) and Erich Fromm are credited by May for their help in the project.

the sizable literature review which covers philosophical, biological, psychological, psychotherapeutic, and cultural interpretations of anxiety before offering a synthesis of these perspective for use in empirical studies on the affect. From May's survey of the literature, he synthesizes the various perspectives in "the following definition: *Anxiety is the apprehension cued off by a threat to some value that the individual holds essential to his existence as a personality*" (p. 189). By linking anxiety to personality, May is able to give an account for how anxiety appears to be without an object. He explains,

> Since anxiety attacks the foundation (core, essence) of the personality, the individual cannot "stand outside" the threat, cannot objectify it. Thereby, one is powerless to take steps to confront it. One cannot fight what one does not know ... It *is* "cosmic" in that it invades us totally, penetrating our whole subjective universe. We cannot stand outside it to objectify it. We cannot see it separately from ourselves, for the very perception with which we look will also be invaded by anxiety ... *[A]nxiety is objectless because it strikes at that basis of the psychological structure on which the perception of one's self as distinct from the world of objects occurs.* (p. 191)

Because of this May makes an interesting discovery that sheds far more light on Heidegger's project. There is an odd and contradictory aspect to anxiety in its relationship to the self. On the one hand, the person who is more self-aware, and has a more developed and richer inner world, is one who is more susceptible to feeling anxiety because they are sensitive to the social pressures for conformity with the masses. On the other hand, "mounting anxiety reduces self-awareness. In proportion to that increase in anxiety, the awareness of one's self as a subject related to objects in the external world is obscured" (p. 191). This would provide an explanation of the different reactions to anxiety that the Frankfurt School and Heidegger embodied in their works and actions. On the one hand, although the Frankfurt School experienced anxiety with the rise of Hitler in Germany because they felt the existential threat that he posed, they did not become paralyzed by this anxiety, rather they achieved a release of that cathexis by engaging in critical thinking which allowed them to maintain, through the use of dialectical thought, a dynamic understanding of the relationship between the self and society, the subject and the object. This allowed them to maintain a view that did not diminish one for the other but recognized the necessity of maintaining avenues of dialectical communication open between both. In that sense, dialectical thought is a protection against the ways that the isolated self within society is structured to think and provides a means of resisting the temptation of thinking with the grain of society. On the other hand, Heidegger's sense of self was so threatened by being-in-the-world that his anxiety overwhelmed him and severed the links between the subject and the object in precisely the manner that society had conditioned him to think. Fascism, and the Nazi project, therefore, does not represent an *overcoming* of anxiety, it is rather a material representation of becoming *overwhelmed* by anxiety! Fascism heightens the anxiety of modern reality so that it can function by overwhelming the thought of individuals to the point where they become completely disconnected from the social at the same moment that they are caught in its most powerful current. Maintaining a critical perspective on the link between the self and society is therefore a necessary protection against precisely these kinds of totalizing logics that pervade the modern experience.

Coming from psychology, it is to May's credit that he recognizes and incorporates the cultural dimension in his analysis of anxiety, as this is a more central concern of socioanalysis. He proposes that the "quantity of anxiety prevalent in the present circumstances arises from the fact that the values and standards underlying modern culture are themselves threatened" (p. 222). But his cure for this cultural anxiety falls back on an idealist theory of social change, and like all ideal theories of social change there is an element of truth in that if the world would embody them, then we would have a better world, however, they are fundamentally at odds with the given material reality of modern society. The approach he offers is to make a distinction between the notions of society and community, with society being something that is externally imposed upon the individual and community being something that emerges only when one relates "one's self to others affirmatively and responsibly" (p. 223). Society is therefore to be resisted while community is embraced. His solution for building a community requires an economic task and a psychological one, which ignores how modern society structures both. Economically, community is built when there is "an emphasis on social values and functions of work," psychologically, this requires "the individual's relating himself to others in love as well as creativity" (p. 223). However, he does not, because he cannot, tell us how to complete these tasks and although this foundation may be the basis for an ideal community, such a community cannot exist outside of society. The truth remains that the material reality constructed by the totalizing logics prevailing in our modern societies has no social basis for either building a society or a community based on shared values and love. What he has ignored here is the central lesson of critical theory, that before any such attempts can be made we first must meet the necessary precondition and this involves a movement in the concepts of self and society that can only occur as a result of engendering dialectical thought in the masses; although even if such a precondition were met, the possibility of this project would still face nearly insurmountable odds. Furthermore, this position is the result of denying anxiety an object. Socioanalysis maintains that anxiety only has the appearance of being objectless, because its object is intangible but objective with material consequences. It is precisely a failure to understand the social as an object, because as Durkheim tells us, it can only be understood in its effects and therefore it is the hypothetical object which makes sociology possible. Therefore, the object of anxiety, if not the social itself, is at least a byproduct of the social as it is constituted in modern society, which masks it, giving it the appearance of being an affect without an object. That it has an object does not imply that the subject can act on the object, or can achieve a cathartic release onto the object, but only by engaging in dialectical thought can its object and the contractions between the subject and this object be realized as what they really are.

Ultimately then May agrees that anxiety is not only a negative phenomenon that paralyzes individuals with the loss of self-identity when there is a disconnect of the self and society. Rather, as a result of his theoretical synthesis and empirical studies, he concludes that "the positive aspects of self-hood develop as the individual confronts, moves through, and overcomes anxiety-creating experiences" (p. 372). The difference between May's conclusion and that of socioanalysis is clear in the last point. The goal of psychological treatments of anxiety is ultimately to overcome anxiety and it does this by fixating on the psyche as an object over which the individual can exercise a measure of control. By attaching the anxiety upon an object, which is not the object of anxiety, it thereby dissipates the anxiety. But as Freud had already clearly demonstrated, this technique can only achieve limited success because the real object of anxiety gets repressed, and there is always a return of the repressed as anxiety is bound to manifest itself anew. Since socioanalysis would place the individual's anxiety in tension with the social, it is keenly aware that the conditions for overcoming anxiety are not possible in the current material reality, precisely because its object is dynamic, morphs as material conditions change, and reproduces itself in compounding layers of social domination. Therefore, socioanalysis only confronts and works through anxiety, without offering the promise of overcoming it.

As such, the Frankfurt School recognized that in addition to constructing critical theories of the psychic and social dimensions of the modern experience, so too they had to track the material changes of modern society to understand how and where these compounding layers of social domination appeared and what effects they had on reconstituting and reproducing the conditions that gave rise to anxiety. Keeping the classical tradition alive, they set out to explain how "[t]he individual is entirely nullified in face of the economic powers" which "are taking society's domination over nature to unimagined heights" (Horkheimer & Adorno, [1947] 2002, p. xvii). Due to the totalizing nature of the logic of capital, by the 1930s when they began their work in earnest, it was no longer possible to examine social domination by focusing primarily on the economic dimension. Capital is a totalizing logic, so it penetrated all the spheres of human action (politics, economics, culture, religion, family, education, etc.) and altered the coding of those spheres to its own ends, as such all those spheres need to be analyzed and their effects diagnosed.

By way of a brief review of this totalizing process, the classics had tracked how through the division of labor, the social became manifest. The bourgeoisie used their reserves of capital to mobilize the social for the purposes of expanding the range of capital's power. Securing the means of production, they purchased raw materials and labor-power to produce commodities which they exchanged on the market for a return of their capital investment and a portion of the surplus labor derived from the exploitation of human laborers. Production and exchange formed the social relations upon which modern society organized its principles. The proletarians primarily consumed commodities in exchange for the money they earned by the sale of their labor, and the bourgeoisie primarily exchanged commodities for a return of capital. Capital self-valorized and grew, while the commodities were realized in their use-value, meaning that at the end of the exchange process the proletarians were left with no money and no commodities, and had to sell their labor-power anew, while the bourgeoisie were left with larger amounts of capital and retained exclusive ownership of the means of production. The bourgeoisie reinvested a portion of that capital in making the means of production more efficient, which placed such stress on the division of labor that those who relied on their labor-power increasingly faced situations in which they no longer had relevant skills which made their labor-power redundant and unnecessary for capital. On this basis the inequalities of modern society became more and more apparent and ideological interventions were required to maintain the acquiescence of the laboring class. Since religion had worked so well as a totalizing logic in premodern times for controlling the masses, capital made use of the Protestant mentality to bind predominant notions of morality in Western societies with an ethical mandate that it was good to work and to limit one's consumption to the minimum necessary for its reproduction. This helped to structure the thoughts of the laboring class in a way that they too began to embody the spirit of capital even though they were not the ones who received the primary benefits of this system. Secular asceticism, thereby, made it a moral imperative for people to turn away from the temptations of the new system by emphasizing the means over the ends as a form of instrumental rationality. Since this ran against the natural instincts of a pleasure-seeking species, it required massive psychological repression for people to place limits on these instincts. With a psychological hold placed on the instinct toward pleasure, the only avenue for avoiding a pure state of unpleasure was death. Enlightenment thought that countered religious alienation was thereby subverted by the material conditions which could only be sustained if death represented something to look forward to as a release from the burdens of a life in which pleasure was denied.

Capital hit a point of crisis in the early 20th century, but its full resolution was delayed by World War I which extended the functioning of the system by soaking up some of the surplus capital and unemployed masses, but as this did not offer a permanent solution to the crisis the pre-war conditions returned with a vengeance in the interwar years. As Horkheimer described it in 1939, around the time that the Frankfurt School theorists began to track these changes in earnest, "the contradictions of technical progress have created a permanent economic crisis" ([1939] 1989, p. 77) and "[t]he same economic tendencies that create an ever higher productivity of labor through the mechanism of competition have suddenly turned into forces of social disorganization" (p. 78). The surplus reserves of capital had gotten too big and there were not enough productive outlets to either employ the whole of the labor class or place the surplus reserves of capital in motion to continue its self-valorization because, production technologies had become so efficient that they were eclipsing the actual rate of consumption (a rate that was limited, on the one hand by the ideology of secular asceticism, and on the other hand, by the material fact that laboring classes did not have enough money to consume more). A new psychological mindset had to be coded into the masses so that capital would be free to continue its spiraling growth. But there is a danger in recoding the social masses, especially if it causes a significant rupture between the previous coding and the new coding, as this can cause anomie in those who lag in this recoding process.

Since there was no desire among the bourgeoisie to abandon the totalizing logic of capital, they had to implement a new totalizing code that complemented this logic. In Germany, with nearly half of its population unemployed, it was easier to capitalize on their anxiety to mobilize them toward the fascist doctrine of war. This was not an anti-capitalist doctrine per se, but it was one that took the social power of the masses and continued to exploit it, only instead of the goal being strictly to grow capital for its own sake, it used that capital to fuel the war machine in a total war (i.e. to advance a totalizing logic of war as a complement to that of capital). It required a strong narrative force that countered bourgeois notions of individualism so that the labor base would continue to work on behalf of a project that did not directly benefit them as individuals. However, the pleasurable release that such a model offered was one oriented around the avoidance of unpleasure not the pursuit of pleasure, that is, the function of the death drive, which in achieving its ultimate goal in death, leads in the end to the peace of the nothingness. Hitler's model was therefore in opposition to the bourgeois model because the goal of the bourgeoisie is not to end up with nothing, but to end up with more of the something they already have, capital, and to gain capital they still needed a mass of living bodies willing to sell their labor-power. In the United States and the West, the solution was to unleash some of the pleasure principle by creating a consumer society that, while emphasizing individual consumption, would not allow for a rupture of the totalizing logic of capital that dissolved the individual, thus keeping the masses under social domination.

The totalizing logic of information was thereby deployed in the West to dissolve meaning in the individual experience while not denying the pleasures of individualized consumption. Lacking meaning in the construction of one's inner world depletes the value one finds in engaging in a project of self-development that requires so much personal work and effort. Therefore, capitalist societies had to find ways to convince the masses "that there is only one way of getting along in this world—that of giving up [our] hope of ultimate self-realization. This [we] can achieve only through imitation" (Horkheimer, [1947] 2013, p. 141). Imitation is the basis upon which technology spreads as it represents "the efficient, smooth, reasonable unfreedom which seems to have its roots in technical progress itself" (Marcuse, 2001, p. 37). If the Western bourgeoisie were to directly (that is, without capital) manipulate the labor class to mobilize the excess social power that was sitting dormant waiting for capital to unleash it, then they would have had to follow Hitler's method, which tied the ideological control of the masses too strongly to the actions of a single man, meaning that it could rise and fall with his successes and failures, and ultimately leave them with nothing. This did not represent long term thinking, which was nearly impossible in Germany at the time since it was overwhelmed by an anxiety which negated attempts to look beyond the present conditions. War was a perfect distraction from the lack of a long-term strategy, but the risks outweighed the benefits insofar as this strategy could backfire and the masses could easily turn against the visible manipulator of their power. This is not to say the West did not also use war to their benefit, but for them war was a means of reinforcing the ideology that underpinned the war, not an end in itself as a replacement or a superseding logic of the guiding principles of capital. Therefore, the death drive of the West had to be countered by loosening the grip on the pleasure principle just enough so that capital could ultimately survive and continue its expansion without losing its structuring power.

Technology not only embodied this logic, but it also assumed the form of the vehicle for carrying the new totalizing logic to the masses. It served a mediating function between the laboring classes and the capitalist class which hid the fact that "the basis on which technology is gaining power over society is the power of those whose economic position is strongest. Technical rationality today is the rationality of domination. It is the compulsive character of a society alienated from itself" (Horkheimer & Adorno, [1947] 2002, p. 95). Whereas the Nazi project provided war as an objective outlet for an anxiety that they wished to deny, Western liberal democracies further dissipated the object of anxiety by using "[c]ulture" to "infect … everything with sameness" (p. 94). The social could not become the solid object of anxiety, social domination infected every sphere of human action and no single object could be fixated on to achieve a cathartic release. Anxiety, therefore, became ever present, constantly appearing in overlapping spheres and each required its own conceptualization. Anxiety was scattered across the spheres of society as the force of the social followed capital in its becoming subsumed in all the structuring logics of modern society. Social progress was essentially negated, because attempts to discharge anxiety on singular objects were thwarted by countervailing trends in other spheres of social reproduction.[16] As technology imitated the human functions of labor and life itself, it reflected back upon that life an image of sameness which imitated itself. Mechanthropomorphism runs in two directions, first by making the machine imitate life, and then by making life imitate the machine, bringing the two into a symbiotic relationship with each internalizing the other in a feedback loop that further distorts the image each time it completes the cycle. "Under the impact of this apparatus, individualistic rationality has been transformed into technological rationality … This rationality establishes standards of judgement and fosters attitudes which make men ready to accept and even to introcept the dictates of the apparatus" (Marcuse, 1998, p. 44). To spread this technological mode of rationality, technology had to be used to create the conditions for its spread so that it could structure everyday life while being "on the alert to ensure that the simple reproduction of the mind does not lead on to the expansion of mind" (Horkheimer & Adorno, [1947] 2002, p. 100).

---

16   Thinking of the social in this manner can be quite maddening, as it requires thinking within contradiction. At the precise moment when the effects of the social are most clearly dominating life in modern societies, is also the moment that it seemingly disappears and takes on a mass existence which no longer has the historical characteristics of the social while at the same time retaining many of the structuring functions that we've come to associate with it.

Newspapers, radio and film, served as the primary transmitters of this cultural homogeneity used to advertise products and lifestyles, and promote their consumption as the repetition of the same in everyday life: individual consumption in a mass unit lacking any individual characteristics. By creating a mass identity, the bourgeoisie capitalized on the fact "that mass and individual are contradictory concepts and incompatible facts" (Marcuse, 1998, p. 53). Since the individual and the social derive from the same historical process, this transformation of both into a singular mass means that we must now contend, not only with the disappearance of the individual, but also with the disappearance of the social. Once television became a massive success in consumer societies in the 1950s, even the most intimate spaces of the lifeworld were cannibalized by the totalizing logic of capital working in conjunction with the totalizing logic of information.

> Entertainment is the prolongation of work under late capitalism. It is sought by those who want to escape the mechanized labor process so that they can cope with it again. At the same time, however, mechanization has such power over leisure and its happiness, determines so thoroughly the fabrication of entertainment commodities, that the off-duty worker can experience nothing but the after-images of the work process itself … The spectator need no thoughts of his own: the product prescribes each reaction … through signals.
> HORKHEIMER & ADORNO, [1947] 2002, p. 109

Television became the vehicle for transmitting the ideological conditioning for the massification of society. It did not rely on punitive tactics, such as those employed in workplaces which implemented Fordist and Taylorist versions of "scientific" management; rather it presented itself as a reward after a long day spent toiling, a time for rest and relaxation, a passive recoding. Not only did the introduction of the totalizing logic of information not reverse the dissipation of the individual caused by the totalizing logic of capital, but it seemed to have dissipated the social as well by eroding foundations for its actualization within the hands of a class whose interests diverged from those of the capitalist. At the same time, mass conformity made class distinctions somewhat meaningless as people came to identify, not in terms of their relation to the labor market, but in terms of consumption habits, which were massified and thus largely homogenized. Rich or poor, people owned a television. The size of the television might change, but the content it streamed was identical.

Marcuse wrote that "technics is itself the instrumentality of pacification" ([1964] 1991, p. 238). This is what technology reveals and "[t]echnology is a mode of revealing" (Heidegger, [1977] 2013, p. 13). This mode of revealing does not stop, it is a continuous process, which is why generational advance happens so rapidly with technology. Each generation reveals what the next generation could do. Within the human it reveals itself as an instrument of their own pacification and acceptance of their place in the mass society. Materially, technology announces itself as a means of pacifying our lives from the burdens of labor and the dangers of nature. Technology promises to free humanity from the struggles of daily survival. Under the sway of the logic of capital, however, technology becomes about its own language, information, and the totalizing process of informationalizing reality so that technology can expand itself. The individual that capital presents to technology, is not the individual of old, there is nothing left in the concept that German idealism developed (Lukes, 1971), rather this new individual is nothing but a data point. This idealization of the logic of capital is realized in technology because it also reveals that it can be used to bring about the mental pacification of humanity, making the informational and the material content of the concept coincide.

In the workplace the laboring class is reduced to mechanical functions, each part of their labor follows a script that enforces the sameness of the experience, thus allowing the experience to be informationalized. Whereas the division of labor originally gave rise to individuality, here in the routinization of instrumental rationality, technological rationality is born, and this mode has no need of the individual, it needs an obedient mass. Leaving work offers no release from this situation. On the drive to and from work the radio structures the mind by bombarding it with informational content. In the form of advertisements, it tells people who they should be, how they should act, what they should consume. Just as the signs that litter the roads and fill the field of vision with consumer options. When one returns to the domicile, what was for a time the last refuge of the individual, they are too exhausted mentally and physically from the demands of controlling their urges that would interfere with the repetitive behaviors that their jobs demand to engage in critical self-reflection and actively build their inner life-worlds in their "free" time. But television offers them a passive means to experience a simulation of the inner lifeworld that is denied to them in this world, and thus has mass appeal to the tired masses. Others have done the labor of writing approved stories that tell of lives that they are unable to have themselves; but these lives are no less empty of meaning because they too are the mere repetition of information, a formulaic broadcasting of sameness following set scripts. Through the simulation the viewer visits the countries of the world, has sex with the beautiful people, experiences the thrill of war, of the car chase, of the bank robbery, goes to space, meets aliens, peeks beneath the veil of the consumerism of the super-rich, no longer bound to a prohibition against luxury goods, and escapes the tedium of this repetitive lifestyle. Everyday life, like the simulated versions of it on television, becomes something unreal, completely and utterly alienated in every way. The simulation of life on television is so far removed from the actual life conditions that it cannot be comprehended as anything other than a simulation, so that even news stories of the continuing traumas and misery of everyday life also feel unreal, nothing more than a simulation made in a factory, a lifeless commodity that is used up upon consumption. Freedom is choosing the channel that has a genre which suits your imagination, this is the new individualism, so long as everyone sits in front of the screen and passively lets the coding of their minds take place.

In the 1960s when pockets of the masses felt the emptiness of this consumer lifestyle, when they looked and saw that they were actually doing nothing more than passive activities and the world was given to them in a mediated fashion, many attempted to rouse the social again, to heed the call of the revolution to agitate the conformity of post-war life. But there was no material basis upon which they could rouse the social, take action, and overturn what had already been totalized and dissipated, any more than there was a material basis for the faux-individualism of those who shouted in Western societies from positions of relative comfort that they demanded change. When Gil Scott-Heron (1971) wrote that "the revolution will not be televised," he meant that it would mean leaving the couch and going out into the world to actualize the change, but such actions were stripped of meaning under the totalizing logic of information, and the revolution became just another plot line to another movie, another show, another news report, another Tuesday night. Life resumed on Wednesday morning when inevitably they showed back up to punch the clock so that they could do it all over again. As Marcuse ([1964] 1991) diagnosed this situation, with no small sense of irony:

> The products [radio, television, film, etc.] indoctrinate and manipulate; they promote a false-consciousness which is immune against its own falsehood. And as these beneficial products become available to more individuals in more social classes, the indoctrination they carry ceases to be publicity; it becomes a way of life. It is a good way of life—much better than before—and as a good way of life, it militates against qualitative change. Thus emerges a pattern of *one-dimensional thought and behavior* in which ideas, aspirations, and objectives that, by their content, transcend the established universe of discourse and action are either repelled or reduced to terms of this universe. (p. 12)

This life is good because it continuously erases the bad by reprograming our minds in a dynamic way that orients them to focus on the rewards of obeying the dictates of this life, but in no way does it deliver the "good life" with any semblance of the philosophical implications contained in that idea. Critical thought demands that we focus on the bad and maintain it in our minds against the forces that seek to erase it with simulations of the good.

To the symptoms of alienation, anomie, the Protestant Ethic, and repression, the Frankfurt School adds, technological rationality and one-dimensional thought to our socioanalytic diagnostic toolkit. These symptoms are the result of the development of technology (both those of war and consumerism) and our adaptation of it into every aspect of our lives. Technology is not, however, inherently a bad and dangerous thing. It could also be the means for us achieving a peaceful and good life, free from the daily burdens of our toils, and therefore contains the possibility of our reclaiming individuality and rebuilding the social as a force of qualitative change. The totalizing logics of capital and information, however, negate that possibility in this material reality. What then are we left with? For Adorno, since we were dealing with totalizing forces that make claims of representing the whole of our experience, the only way that we could understand their full impact was with a mode of thought that also attempted to think the whole. "Dialectic," in his usage, "is a form of thought which speaks of constellation, of interconnection, of the whole, even while it cannot claim any confident grasp on such a whole, for indeed it has nothing simply at its disposal" that can counter the material reality which we must accept as the truth of our condition without acquiescing to it (Adorno, [2010] 2017, p. 219). "[A]ll we can do is emphasize the fragmentary character which is perhaps the only form that dialectical thought is possible today" (p. 219). In the face of the overwhelming power of the totalized world, critical socioanalysis sets itself the task of looking for those fragments, by working through anxiety at the point where it resides, at the psychosocial divide.

Conclusion to Part 2

Anxiety is not a catch all affect that explains the whole of the individual and the social experience of modern society, but it is a central carrier affect of the moods and symptoms produced within modern societies that are structured by the totalizing logics of capital and information. This is why a focus on anxiety gives us a window into the ways that the self and society are warped by the organization of modern society. It is not correct to say that anxiety serves as a bridge that crosses the psychosocial divide, rather it exposes the fact that there is no bridge, that despite the ways that the self and society constitute each other, they are disconnected in the most important of ways. In the 20th century there was a great acceleration of the productive technologies and the way of life in modern societies. Transportation and communication technology shrank the time-horizon of the globe. The time needed for critical self-reflection and the building of a rich inner world was negated by the demands to adapt to this reality on pain of being left behind in the past it rejected. As the political and economic landscapes evolved to accommodate the speed of these transformations, so too did the human species begin the process of remediating itself to better adapt to the new circumstances of their everyday life. Contemplative thought was reduced to reactionary thought, as people were compelled to make use of instrumental rationality to maintain a place within an economy that was no longer quite so dependent upon them. As the capitalist class invested in more efficient and faster technologies to gain a competitive edge in the market, the laboring class increasingly had to see and treat themselves as machines void of any "human essence" to gain a competitive advantage over each other in the market of labor-power. This instrumental rationality gave way to a technological rationality, a deadening of the human elements in favor of those machinic ones that were so much more valued in modern society according to its own terms. Alienated from society, the self was alienated from itself and in sensing this change but not knowing exactly what was happening, anxiety began to radiate from the self and society as neither was able to fix its bearings within the individual or the social. Without the ability to critically evaluate these changes, it felt as though the anxiety was without an object, it emanated from the very fact of living in a modern society that had dissipated the foundations of the individual and the social, while the self and society persisted.

Although modern society nurtured a heightened sense of anxiety, it was not all bad. There were many ecstasies to be found in the consumer society. The steam engine and the combustion engine opened new space-time horizons to the individual experience. Electricity was a wonder that recalled the magic of fairy tales. Inanimate objects came to life and performed tedious tasks without complaint. Music and news floated invisibly in the air and we could catch their sounds on the radio, bringing the world right into our most intimate spaces. Television brought us images of far off places and distracted us from the uncomfortable truth of our existence. For all the ecstasy a technological life promised, it still did not deliver the *good* life. Frustrations boiled over across the globe as people fought against the loss of meaning and the loss of opportunity to make something better for themselves than a life spent toiling for the basic necessities. When even opportunities for the mere reproduction of life began to dry up, there was little reason to cling to life, and war allowed people the means to embrace their death drives with an outlet for their nihilism. But war too, had changed. There was no intimacy, no greater human connection or mythic sense of heroism or honor to be found in battle. Efficiency and speed had mechanized the killing fields. Technology was not only for easing the burdens of life, it could also be engineered to end life. Guided by the principles of efficiency and speed, the technologies of war were perfected so that they could end life in the blink of an eye. The whole globe now sat under the technological gaze as the nuclear bomb cast its long shadow over the collective consciousness. Powers that had previously been reserved in myth for the gods, now rested in the hands of humanity and they showed a willingness to use it to preserve the system that their totalizing logics had built. If there was meaning to be found in premodern warfare, there was only ideology moving people like cogs in its modern form.

Working under these conditions the members of the first generation of the Frankfurt School set themselves the task of understanding why these contradictions persisted in modern society despite and because of the vast advances in scientific and technological rationality that fanned the flames of the material possibility that the world could be arranged differently. Their theoretical treatises were buttressed by empirical studies on prejudice, anti-Semitism, the family, personality structures, and more. They followed the classical modern social thinkers in recognizing that it was modernity itself—married as it is to the logic of capital and amplified by the logic of information—that undermines traditional norms and values and prevents progressive norms and values from taking root. But rather than retreating in thought to a conservative desire for the old ways of living that left social injustice unquestioned and unchallenged, or to a naïve progressivism, they took refuge in the phantasy that eventually the conditions could arise that would allow for the rational organization of modern societies, in which the productive forces could be used for the benefit of all. However, their research showed just how dissolved the foundations had become for the actualization of the individual and the social, as avenues for qualitative change disappeared. The technological system was built by humans, but it had recreated the species in its image.[1] The old concepts persisted in thought, just barely, but they no longer described the actual conditions of modern life. In what is perhaps the most famous work to come out of their collaborative efforts, in *Dialectic of Enlightenment* ([1947] 2002), Adorno and Horkheimer advanced the thesis that "[m]yth is already enlightenment, and enlightenment reverts to mythology" (p. xviii). In other words, modernity contains the seeds of its own undoing within itself, which becomes a problem when the culture industry encourages people to throw off the burden of thinking and its disruptive potential in exchange for the comfort and ease of the repetition of the same; even when that repetition visibly leads to the destruction of reason and freedom, thus of individuality. Myth is precisely that narrative force that denies the non-identity of concept and material reality and promotes their false unity. In Marcuse's *One-Dimensional Man* ([1964] 1991) he complemented this thesis when he concluded that the products of modern society "promote a false consciousness that is immune to falsehood" (p. 12). Because our modern lives do, in actuality, include more comforts than those of our pre-modern ancestors, we come to think that they represent the minimum acceptable standard of life, which "militates against qualitative change" that would threaten this moving threshold of acceptability. And as Erich Fromm ([1941] 1994) pointed out, with these artificial conditions taken as a new state of nature, people are often all too willing to exchange freedom for consumer goods and would rather bow to authoritarianism than risk the loss of these modern comforts.

The work then that remained for the social sciences and philosophy was to interrogate this new way of life and see what fragments of the old hopes remained; if not to actualize the ideal reality, then at least to ensure "that Auschwitz not happen again" (Adorno, 2005, p. 191). Critical theory and negative dialectics are models for thinking through the contradictions, but since they run against the grain of modern society they appeared as unreasonable and without use-value in a system guided by both the profit motive and by the coding of reality. Instrumental and technological rationality warped the perspective on the real and nurtured a false sense of reality, clouding out the very things people said were important to them as they actively worked against realizing those things. To engage in their practice demanded nothing less than

---

1   As I've described elsewhere: "Such are the successive phases of the human: [1] it understands itself as the reflection of a supernatural deity; [2] it understand the deity as the reflection of the self; [3] it understands the machine as the reflection of the self; [4] it understands itself as the reflection of the machine; [5] it is a purely artificial construct, a fractally ambiguous image erasing itself in the fictionalities of being" (Crombez, 2015, pp. 54-55).

confronting the anxiety of the world head on, not in the hope of overturning it, but in the hopes of reclaiming a mode of thought that could see the world for what it is. In the years that they engaged in this practice, they came to see just how regressive society had become and how many had embraced the robotic life and reveled in the passive acceptance of the given conditions. From 1969 to 1979, the decade in which Adorno, Horkheimer, and Marcuse passed away, they witnessed the world as it entered a quasi-stalemate scenario. The Cold War between America and the Soviet Union turned to fatal strategies that made everyone a hostage and a terrorist at the same time (Baudrillard, [1990] 2008). Under the weight of capital and the technology it relied on to continue its advance, life felt like it had reached a point of polar inertia (Virilio, [1990] 2000). It was moving forward at ever faster speeds, all the while it felt as if it was at a standstill, especially since in the West this speed was increasingly experienced from the couch where people were physically stationary, passively watching as images of the world flashed at the speed of light on the television screen. Everyone was holding their breath to see what would happen next, what new catastrophe would capture their attention in the screened images and potentially destroy their life. It was such an absurd condition under which to live that most could only relate to it in its utter unreality; that is, as unreal as a Hollywood blockbuster. But it was indeed *real* life in an *unreal* form. At the same time the rest of the world could do little more than wonder when they would get pulled into the standoff and pay the ultimate price as their resources were plundered by the machinic societies against whose onslaught they were powerless. Whether they wanted to change or not, they knew that adapting to these logics was their only hope of survival and one by one they acquiesced to its logic and modern society encapsulated the entirety of planet earth. Not everyone shares in the ecstasies of this life, but all share in the anxieties.

In this totalized world, critical theory and negative dialectic were needed more than ever. Instead, however, the pull toward negativity that Adorno warned against caused a reactionary movement that was geared toward looking for the positive in the systems that were already in place. After the first generation of the Frankfurt School passed, the "critical theory" of the second generation no longer represented the same motivational forces as their project. With the failures of the anti-capital revolutionary movements of the 1960s, and the actuality of a communism that had only the name in common with its concept, thoughts of rupturing the grip of capital evaporated. Their direct successors at the Institute, under the leadership of Jürgen Habermas, turned toward a practice of critical liberalism that sought to buttress liberal institutions as a way of slowing or perhaps reversing their descent into meaningless bureaucracies whose only function was to act on behalf of the totalizing logics.

This was a "critical" theory that had been totalized by the logic of capital, which is why it no longer tracked the logic of capital and why it reified the social in institutions that shared nothing in common with the concept of the social. In France, theory made a poststructural turn, and in many ways, this was closer in thought to the aims of critical theory, but they could not keep the past hopes alive and at the same time reject the past that had nurtured those hopes. They knew nothing of that world anymore, which only persisted in the archive as stories of a past that felt far further removed historically than the years suggested, and so they threw the baby out with the bathwater and became fixated on the present to the exclusion of the past and the future. Caught in the atemporal void of the present, they could do nothing but critique the totalizing logic of information and throw their heads back with laughter at the meaninglessness of this life that was fully absurd.

The material conditions that gave rise to and reproduced society's absurd stage, are the context for the next scenic landscape. If socioanalysis is possible, if there is any way to reclaim our anxiety, then we must confront the material conditions of the present and understand this journey into the absurdity of our current moment. Only on those grounds can we construct a framework for critical socioanalysis, which is perhaps precisely the kind of absurd practice that is needed to revitalize critical thought in an absurd world.

PART 3

Anxiety-Dreams of Posthuman Futures: Sorting through the Discourses of 21st Century Life

∴

Introduction to Part 3

When we try to make sense of the material reality of the 21st century it is easy to get overwhelmed. It evades most generalizations and is prone to partiality and fragmentation. The absurd is precisely that which is unreasonable and illogical, therefore it is what eludes "good" sense. The mechanthropomorphic, or posthuman, world that we now live in pushes the limitations of the human mind. It contains so much information about itself that tracking the general condition of society taxes the upper limits of thought in the attempt to maintain clarity between all the different concepts and forms that coexist in an ununified whole, that is to say, in paradoxical fashion; but it is precisely the contradictions of our existence which are necessary to grasp if we are to explain our current condition. I proceed on the grounds that we can at best scratch the surface level and perhaps locate some of the rhizomatic roots that operate beneath the surface. To grasp the scenic landscape of the 21st century and to develop a theoretical model for socioanalysis that operates within that landscape, we need to know why this is the case and to do that we need a better understanding of how changes in the material order rupture the continuity of social organization as history moves from one stage to the next, and the effects this has on thought itself.

Although their lives overlapped, the world of Marx differed from the world of Durkheim, Weber, and Freud, as much as their lives differed from those of Horkheimer, Marcuse, and Adorno, and as our lives differ from theirs. Yet we persist in calling all these worlds modern. These differences are the result of the material consequences of the totalizing logics and the effect that each of their modes of domination has on the structure of reality and the structure of the mind in the shift from one to the next. What Marx exposed as internal contradictions of the logic of capital became more visible externally by means of the amplification and intensification of that logic in material form (i.e. the continuous revolution in the means of production) as the totalizing logic of religion faded from dominance. As the bourgeois program was executed and ran through its functions, it continuously transformed material reality and in so doing, transformed thought. But thought lags material change. The process does not happen in one simultaneous moment, rather material change takes place and thought then attempts to catch up to those changes. Movement in both is the precondition for history moving from one stage to another.

Material change, however, does not wait for thought to catch up; it is a moving target for thought and it responds to whatever stage thought is at in determining its directionality. This means that material changes and the corresponding changes in thought do not have to align, and in fact, can never completely align in a dynamic reality. Their disjunction and disunity is precisely what makes the current condition such an absurdity. Thorstein Veblen (1919) described the modern approach to understanding this process as he experienced the transformation from the totalizing logic of religion to the totalizing logic of capital in the following terms:

> This unsettling discipline that is brought to bear on the workday experience is chiefly and most immediately the discipline exercised by the material conditions of life, the exigencies that best men in their everyday dealings with the material means of life; inasmuch as these material facts are insistent and uncompromising. And the scope and method of knowledge and belief which is forced on men in their everyday material conditions will unavoidably, by habitual use, extend to other matters as well; so as also to affect the scope and method of knowledge and belief in all that concerns those imponderable facts which lie outside the immediate range of material existence. It results that, in the further course of changing habituation, those imponderable relations, conventions, claims and perquisites, that make up the time-worn system of law and custom will unavoidably also be brought under review and will be revised and reorganized ... (p. 9)
>
> Given time and a sufficiently exacting run of experience, and it will follow necessarily that much the same standards of truth and finality will come to govern men's knowledge and valuation of facts throughout; whether the facts in question lie in the domain of material things or in the domain of those imponderable conventions and preconceptions that decide what is right and proper in human intercourse. It follows necessarily, because the same persons bent by the same discipline and habituation, take stock of both and are required to get along with both during the same lifetime. More or less rigorously the same scope and method of knowledge and valuation will control the thinking of the same individuals throughout; at least to the extent that any given article of faith and usage which is palpably at cross purposes with this main intellectual bent will soon begin to seem immaterial and irrelevant and will tend to become obsolete by neglect. (p. 10)

At the time, it seemed as if the logic of religion would not be able to stand up to the reshaping of the mind that economic discipline initiated or to the stampede of scientific knowledge that was demonstrating its practical value in the technological forms that buttressed economic discipline. Therefore, Veblen surmised that the domination of religious thought would give way under the domination of economic principles which were recoding thought to align with a new set of cultural goals.

However, before religion could be overcome—and, Durkheim excepted, it really only became obvious that it would not be in the course of the 20th century since religion persists as a fragmented logic within the group dynamics of society—a crisis of knowledge occurred that further interrupted the dynamic between material reality and thought in such a way that thought increasingly did not, because it could not, correspond to material reality. The persistence of religion in modern society is one of the unreasonable and illogical contradictions that marks the failure of thought to fully conform with material reality. Within the scientific process the seeds of the totalizing logic of information were planted, which enhanced these contradictions by amplifying the effects of alienation. Veblen sensed the edges of the material transformation that this new logic was initiating, but he did not and could not fully grasp the effects it would have on thought. However, he hypothesized that this represented a new rupture within the modern that would usher in a new set of material circumstances and thoughts that were decidedly different than those of the modern. He continues,

> But it has been only during the later decades of the modern era—during that time interval that might fairly be called the post-modern era—that this mechanistic conception of things has begun seriously to affect the current system of knowledge and belief; and it has not hitherto seriously taken effect except in technology and the material sciences. (p. 11)

Veblen was the first to hint at the possibility that there could be a "post-modern" despite the fact that the complete overturning of the old logics had not yet occurred within the mind. This means that the modern had not reached its end, even if it had concretely developed its foundations in the material world, because thought lags and in that lag new material interruptions arise that disrupt the continuity of thought just as interruptions arise that disrupt the trajectory of material reality. These disruptions primarily came in the form of technology, and as these technologies began to totalize the modern experience, they announced the presence of a postmodern reality that was beginning to eclipse the modern reality.

If the premodern was primarily dictated by the totalizing logic of religion with its roots in the discourse of theology, then the modern is primarily dictated by the totalizing logic of capital with its roots in the discourse of political economy, and this postmodern is primarily dictated by the totalizing logic of information with its roots in the discourse of science and technology. These discourses occur simultaneously while following their own logic, but at points they overlap in shared goals. The logic of capital was originally easier to track materially because its primary objective was the transformation of nature into the form of the commodity, and its secondary objective was to recode thought in a way that reinforced the primary objective. Once that secondary objective advanced under the logic of information, it became more difficult to track materially because the primary objective of information is the transformation of knowledge into its technical form—which largely takes place behind the scenes in specialized occupations and practices—and its secondary objective is to recode the mind to correspond to the artificial reality that capital constructed; but it disrupts its own process as we experience technical vertigo. As new technologies emerge, old technologies are abandoned, which often means resetting the informational logic to correspond with the newest technology so that its informational content takes precedent within the material conditions. The generational advance of technology is so rapid, that when coupled with the material lag of its implementation, by the time it is implemented it has already advanced to a new version which repeats the process; as a result, thought fragments as it loses track of its material referent which is constantly in flux and as people adopt different generations of technology at the same time. In other words, Veblen's diagnosis that "given time and a sufficiently exacting run of experience" thought will come to align with the new material conditions, is no longer valid. The fact that the logic of capital and that of information are highly synthetic, overlap, and interpenetrate each other, also makes it difficult to distinguish between the modern and the postmodern. Many flat up deny that such a distinction is either possible or logically consistent, preferring to see both as two sides of the same coin or as competing ideologies[1];

---

1   Taken to its logical end point, some have defined the modern as precisely what can never arrive, since it is either always partial, or it is a directionless becoming guided only by whatever is new so long as it is different than what came before. In other words, to be modern is to be a contradiction, it is to claim novelty in the present while being chained to the past. Therefore, Bruno Latour ([1991] 1993) could claim in one breath and with only a hint of irony that "we are modern" (p. 3) in a book that bears the title: *We Have Never Been Modern*. Latour's first point is that by living in decidedly different circumstances than the traditional mode of life we have identified ourselves as modern, and thus act as if we are representative of the modern. The second point acknowledges that the constitutional logic of the modern is unobtainable according to its own terms, but we ignore those contradictions in our claim of being modern, making it a claim that is simultaneously true and false. Ultimately Latour sees the postmodern as a "symptom" (p. 10) that must be understood as a result of this contradiction. Jürgen Habermas ([1985] 1987) recognizes this contradiction as well when he takes

however, there is value in marking out the ruptures in our material experience if we are to gain a grasp on how thought is likely to develop in reaction to those ruptures and what anxieties will arise in the overturning of the old ways as thought attempts to catch up with the new material forms.

The French psychoanalyst, Jacques Lacan, made a distinction that proves helpful for thinking through these ruptures and tracking how and where the totalizing logics primarily operate within society as a way for us to get a better grasp on anxiety. The Marxian model for the critical method, which was adjusted by the Frankfurt School, saw a distinction between the ideal and the material, which formed the basis for dialectical thought operating within modern conditions. For Lacan (1975) these correspond to the order of the Imaginary and the Real, but to these he adds the Symbolic order, which Fredric Jameson sees as necessary to understanding the postmodern and the "properly representational dialectic of the codes and capacities of individual languages or media" (1991, p. 54). Although the totalizing logics interact with all three orders, they each primarily correspond to one of these orders. The totalizing logic of religion primarily corresponded to the order of the Imagination; religion dominated Imagination as it subsumed ideals under its ultimate narrative of supernatural life. This was the cause of mental domination in premodern life. The totalizing logic of capital primarily corresponds to the order of the Real, as it dominates material reality by subsuming all of nature under its ultimate narrative of the commodity. This is the cause of social domination in modern life. The totalizing logic of information primarily corresponds to the order of the Symbolic, as it subsumes the languages and codes that attempt to mediate our understanding of the Imaginary and the Real. This is the cause of mass

---

a normative stance toward the modern which binds it to a specific set of enlightenment ideals. For Habermas (1983), modernity is 'an incomplete project' "formulated in the 18th century by the philosophers of the Enlightenment ... to develop objective science, universal morality and law, and autonomous art ... for the rational organization of everyday social life" (p. 9). By seeing the modern as a project, Habermas looks not to the logic of the modern as a philosophically reified concept, but to the normative dimension of the process of becoming modern, i.e. modernization; which he maintains as an ideal. Bound to a certain set of norms and oriented to a certain set of values, social actions can be judged as to whether they align with the modern project or not and Habermas needs this measuring stick for his project of critical liberalism. Since the postmodern critique of reality would deny him this measuring stick, he treats it as a "neoconservative" ideology that "welcome[s] the development of modern science, as long as this only goes beyond its sphere to carry forward technical progress, capitalistic growth and rational administration" (p. 14). Rather than reading the postmodern as a critique of the material conditions, Habermas abandons critical theory and engages in ideological warfare, with the modern as representative of the project of becoming human and the postmodern as representative of the project of becoming machine.

domination in postmodern life. In the attempt to make this Symbolic order more closely align with the Real, our language and codes began to assume the shape of information, which instead of achieving the goal of representing the Real as is, served to demonstrate just how wide of a gap exists between the real and its symbolic coding. Without affect and depth of meaning, information exposed the logical shortcomings of our language and codes when they are applied to material things. This translated into a crisis of knowledge, especially material knowledge, which is necessary to understanding our condition. It is in our understanding of this Symbolic order, revealed by applying dialectical thought to the totalizing logic of information, that we come to recognize that "there's an anxiety provoking apparition of an image which summarizes what we can call the revelation of that which is least penetrable in the real, of the essential object which isn't an object any longer, but this something faced with which all words cease and all categories fail, the object of anxiety *par excellence*" (Lacan, [1978] 1991, p. 164). In other words, the reason that anxiety appears without an object is because our words and concepts fail to fully capture the reality of our condition, they can no longer name the object and due to the shifting nature of the Real we cannot bring that object into the domain of information as it contradicts the nature of the Symbolic order. To understand the object of anxiety, that intangible but objective force with material consequences, we must confront that which is precisely coded in such a way as to evade all attempts to confront it.

While the demands of everyday life distract us, capital and information change us into beings that are no longer human. We are no longer creatures of the imagination, or creatures of the real, we come to see ourselves as creatures of the Symbolic order: that which is artificially constructed. The human was the basis for both the individual and the social, so whatever has a self and lives in a society, is today something that is decidedly posthuman, coded and recoded on the fly. The contradiction between our identification as humans living in a material reality that denies us all opportunities to actualize our humanity, is the framework in which socioanalysis must begin. What still distinguishes the human form from its technological other, is the claim to using knowledge in our thinking instead of mere information. But we must recall that mechanthropomorphism extends in two directions; history has shown that the first direction, that of making the human more machinelike is the more powerful of the two, but there is also the direction of making the machine more human. Resisting informationalization is a call to knowledge, to find meaning in the dialectic of the symbolic and the real through the use of language that is necessarily imprecise because it involves tone, modulation, and the transmission of affective notions that do not correspond to the machinic logics of capital and information. The utopian refuge of socioanalysis is meaning; it is a utopian refuge because the sources of meaning in material reality are drying up or becoming absurd. As such, the practice is based on the notion that maintaining knowledge and meaning in our reality is a good thing, in this way it keeps a past hope alive because that hope is based on the notion that critical knowledge is closer to the real than affirmative information. In keeping with critical theory, socioanalysis does not, however, express a desire to return to the past— knowledge was no closer to the real then than it is now—rather, it accepts the present as the objective material circumstances from which we must proceed. Within the current nexus of the contradictions of our concepts and material reality, which socioanalysis subjects to interrogation within analysis, it must look to where meaning, if any, persists and how to exploit the fragments of meaning that remain to bring them to the forefront of the mechanthropomorphic process. It asks if there is still meaning in the individual, in the social, in reason, in freedom, and in meaning itself; and, if there are fragments of meaning in these concepts, how do they correspond to material reality and is there any critical value to maintaining those fragments in thought? To answer these questions, critical socioanalysis must sort through the discourses of the symbolic order and interrogate their relationship to the real. In order to set up a theoretical method for socioanalysis that can accomplish this goal, let us first explore the scenic landscape of this postmodern condition in greater detail to gain a deeper understanding of the conditions that make socioanalysis necessary and from which it is produced.

CHAPTER 5

Scenic Landscape 3

1   The Postmodern Rupture in Modern Society

After World War II, questions about the directionality of modern society had to assume a global dimension. Missteps, as the two world wars proved, were costly, deadly, and psychologically and socially devastating. With science having demonstrated that it could compute the math behind harnessing the power of the atom, and then apply those calculations in the material form of the bomb, it appeared to have the capacity to solve some of the most challenging problems that humans had set for themselves. Governments and private corporations began to pour millions and then billions of dollars into scientific research.[1] But as the former administrator at Oak Ridge National Laboratory, the nuclear physicist Alvin M. Weinberg (2014) explained, the post-war problems were not of the same order as those tackled by the natural sciences. He said, "Nuclear energy offers a Faustian bargain. It offers the world an inexhaustible source of energy. But in return, it demands a vigilance and longevity of our social sciences to which we are quite unaccustomed." Although the effects of stimulating science with capital were immediately visible in the material and mental realms, and the Frankfurt School's diagnosis of technical rationality was derived from those effects, thought had to catch up to really grasp the full impact of these changes. That is, thought must think about itself to understand how it is changing, and this requires an analysis of the dialectic of the real and the symbolic.

Despite all its theories and methods, the social sciences were hardly equipped to handle the scope of the task, especially as they persisted in following the traditional model of theorizing. The science popularizer and astrophysicist, Neil deGrasse Tyson (2016) explained: "In science, when human behavior enters the equation, things go nonlinear. That's why Physics is easy and Sociology is hard." In the Cold War years, there were fewer attempts to theorize modern society as a totality, instead, much traditional theory focused on specific nation-states or groups within those states, rather than modern

---

[1] As of the writing of this, the most recent data made available by the National Science Foundation (2017) shows that from 1990 to 2016, total university research and development funding from federal, state and local, university, industry, and other sources grew from just under $30 billion per year to just over $70 billion per year.

society as such. The most comprehensive traditional theories produced in American sociology during those years, were focused primarily on America. I argue that the French theorists, the philosopher Jean-François Lyotard, the sociologist-cum-metaphysician Jean Baudrillard, and the urbanist Paul Virilio, were among the next generation to advance the critical method and attempt to understand the impact that modern forces, with America and the West as the primary drivers of modernization, had on the totality of the modern experience of reality. Their position in France was uniquely suited to critical thought as they occupied a position that was both within the core of advanced modern societies, and at the same time externally totalized by American doctrines, thereby allowing them to experience the tensions and contradictions to a greater degree than those in America as these processes recoded their own society from the outside. There is not the space here to explore each of their critical methods in full, but I will touch on their thought throughout this chapter as their perspectives help explain the novelty of the late 20th and early 21st centuries.

Since the Second World War, and the technologies that it and the subsequent Cold War produced which have proliferated in modern societies, we have been dealing with an artificial reality that is in equal measures made up of a tangled knot of the Imaginary, Real, and Symbolic orders. If the modern is precisely defined by its impermanence and its dynamism, then according to its own terms, the process of modernization, which is the carrier of specific, but not static, sets of norms and values, must itself be an embodiment of the contradiction between the modern's ideal form and its material manifestation. In other words, the relationship between how we see, experience, and envision the modern is tied to how the modern presents itself to us and how we are socialized as agents of modern society. With the spread of technology into the mass conscious, the modern appeared to finally be on a footing to escape the past, and yet, echoing Marx's words, it continued to drag the past with it in the form of continued domination. One of the first noticeable effects of this new kind of social organization was in how it shaped the social character of people in advanced modern societies on the cusp of demonstrating postmodern tendencies. Two landmark studies in traditional theory described these changes quite well.

First, the sociologists David Riesman, Nathan Glazer, and Reuel Denney ([1950] 1989), in their best-selling work *The Lonely Crowd*, offered a diagnosis of the ideal types of American character and how that character transformed in relation to material changes. The material change that they tracked was the S-curve of population growth, which is "an empirical description of what has happened in the West and in those parts of the world influenced by the West"

(p. 8) as people became increasingly urbanized. They track a series of three changes in social character that correspond to three points on the S-curve. Each of the social character typologies describes how the person who embodies that character type relates to their environment. Using demographic data to buttress their argument they trace the shift in population within modern societies and link those social conditions of life to different orientations of the self. When there is a high death-rate and a high birth-rate, population is low and is at the bottom of the S-curve. People born in this type of society are close knit. They embody a tradition-directed social character in which the "person feels the impact of his culture as a unit," they are expected to "behave in the approved way," and "the sanction for behavior tends to be the fear of being shamed" (p. 24). These are societies that are primarily under the influence of the totalizing logic of religion. When these societies possess sufficient technology to stabilize their access to resources, they enter a transitional phase and the death-rate decreases but the high birth-rate remains. Population experiences a boom and enters the upward slope of the S-curve. People born in this kind of society are more closely integrated in the familial unit. The growth of cities means that there is less integration because the increase in quantity of people impacts the quality of the relationships. As such, people in this society develop an inner-directed social character in which the "person has early incorporated a psychic gyroscope which is set going by his parents and can receive signals later on from other authorities who resemble his parents" (p. 24). These people obey their internal piloting and when they go off course it can produce feelings of guilt. These societies are primarily under the influence of the totalizing logic of capital and embody the Protestant ethic to a high degree as a remainder that carries over from the period of religious domination. Now living in a wholly alienated society, after a period of population boom, the death-rate remains low but the birth rate levels off. Population is high and is at the top of the S-curve. People born into this society are other-directed. They are "at home everywhere and nowhere, capable of a rapid if sometimes superficial intimacy with and response to everyone" (p. 25). They learn "to respond to signals from a far wider circle than is constituted by [their] parents" and "the border between the familiar and the strange" breaks down. Because their relationships are often superficial, "one prime psychological lever of the other-directed person is a diffuse *anxiety*" that acts "like a radar" (p. 25). This is the description of a society that is primarily under the influence of the totalizing logic of information. Living in a world in which the individual and the social have dissipated, we have never known a non-alienated life, thus we are alienated equally in our homes as we are in our surroundings, but we remain torn between persistent human feelings and the mechanthropomorphic pressures to act in a machinic manner. This is the description of the social character of those living in a mass society, a condition that persists today in an amplified form, with all the anxiety that comes along with that intensification.

The second landmark traditional theory that tracked these changes was the post-industrial society thesis advanced by the sociologist Daniel Bell ([1973] 1999).[2] He begins by dividing society in to three parts: (1) social structure, comprised of "the economy, technology, and the occupational system"; (2) the polity, which "regulates the distribution of power and adjudicates the conflicting claims and demands of individuals and groups"; and (3) culture, "the realm of expressive symbolism and meaning" (p. 12). As a mode of organization, Bell links the post-industrial society to the social structure where he sees five dimensions of change that differentiate it from industrial society:

1. Economic sector: the change from a good-producing to a service economy;
2. Occupational distribution: the pre-eminence of the professional and technical class;
3. Axial principle: the centrality of theoretical knowledge as the source of innovation and of policy formation for the society;
4. Future orientation: the control of technology and technological assessment;
5. Decision-making: the creation of a new "intellectual technology." (p. 14)

These changes in the social structure are not totalizing in the same way that industrialism was for early capital. Modern society cannot completely rid itself of its industrial base, as it requires this base to produce the goods it consumes. However, just as when Marx was writing, and industrialism had not fully replaced traditional modes of labor with a mechanized form, it was still appropriate to call the industrial model modern. So too, this post-industrial mode of organization represented a kind of break in advanced modern societies away from how they had been organized previously, which led many to come to call these societies postmodern. There was no end of labor in postindustrial societies, rather the form changed but wage-labor persisted indicating that this was still a mode of social domination and control of the masses.[3] Industrialism

---

2 The French sociologist Alain Touraine is generally credited with developing this thesis throughout the 1960s and publishing them as a complete work in 1969, under the title *The Post-Industrial Society: Tomorrow's Social History: Classes, Conflicts and Culture in the Programed Society* ([1969] 1971), but he receives no more than a passing mention in Bell's work (p. 39).

3 Since the jobs in post-industrial societies are not oriented to the production of material goods, professional jobs are largely bureaucratic and non-professional jobs are largely responsible for handling menial tasks that the professional class are too busy or too disinterested to do

was exported to countries who wished to begin the process of modernization in exchange for producing consumer goods for the West with the labor of their peoples. The West, in turn, assumed a managerial role over those societies that took the form of neo-colonization, by shaping a technocratic class to rationally orchestrate these new global links in a way that reinforced a mutual dependence of nation-states while maintaining a racialized and ethno-nationalist hierarchy. On the surface, the goal was that by linking national economies they would become codependent, meaning that economy would take the crown from politics and dictate political maneuvering in the hopes that this would prevent another world war. However, since America possessed the bomb and the most powerful military force the species had ever known, the threat of violence always hovered in the background, which allowed them to negotiate the terms of these relationships so that they were the primary beneficiary. This global economic matrix did not guaranty a better quality of life for the rest of the world, instead it funneled the resources and wealth toward a select few American hands and a few others of the global elite. Social logic was nowhere to be found in this economic logic, as the technocratic view was fully guided by economic and technical rationality.

In 1973, when Bell published his study, he was able to adequately describe the impact of the first three of his dimensions of change, but future orientation and decision-making could only be speculative affairs.[4] Rather than critically evaluating these, he falls on the side of an ideal image of the world, concluding that the post-industrial "reality is primarily the social world—neither nature nor things, only men—experienced through the

---

themselves. They share the quality of being machinic in that they involve repetitive tasks with little to no variation or opportunity to engage in creative thought. The anthropologist David Graeber (2018) has taken to calling these "bullshit jobs" because technology has advanced to the point that we can meet our material needs working far few hours per week than we do on average, but "[i]nstead technology has been marshaled, if anything, to figure out ways to make us all work more" (p. xvii).

4 As these changes were more fully realized materially, they still largely persisted at the level of a 'feeling' that was structured by these changes and was visible in the cultural realm. The German literary critic, Andreas Huyssen would in 1984, only commit to saying that "what appears on one level as the latest fad, advertising pitch and hollow spectacle is part of a slowly emerging cultural transformation in Western societies, a change in sensibility for which the term 'postmodernism' is actually, at least for now, wholly adequate. The nature and depth of that transformation are debatable, but transformation it is. I don't want to be misunderstood as claiming there is a wholesale paradigm shift of the cultural, social and economic orders … But in an important sector of our culture there is a noticeable shift in sensibility, practices and discourse formations which distinguishes a postmodern set of assumptions, experiences and propositions from that of the preceding period" (1984, p. 8).

reciprocal consciousness of self and other" and in this sense it "gives rise to a new Utopianism" (p. 488). Since the totalizing logic of information was still in its infancy, Bell would have had to apply a critical Marxian perspective on the logic of capital to this process and to the polity and culture to understand how in their interactions it would warp this vision. Although the post-industrial society thesis recognized that there had been a revolution in social relations, it did not mean that the continuous revolution of the means of production would not be applied to these new forms of labor. Bell's thought was still structured by a modern mentality, one that believed in a rational world of social progress which would usher in a qualitatively superior future. What Bell saw as humans working in a group dynamic, he translated as a shift from labor mediated by material technologies to a form of labor mediated by technological rationality, which was an internalized form of technology meaning that it retained the totalizing logic of information, but since this labor relied on human interaction he misread this as a social process and did not account for the fact that these were alienated beings performing these interactions. Since post-industrial work still relied on the division of labor and since it was the full embodiment of technological rationality, it represented a fertile new ground for technological development and the reintegration of machines into the labor process. In other words, the technocrats applied their own logic to themselves, thereby achieving an autopoiesis of the logic of capital in service of their own alienation. Therefore, instead of building a new social utopia, it in fact exacerbated the state of alienation and further transformed the world along mechanthropomorphic lines.

Since the end of World War II, the United States' national economic policies had been shaped by the Fordist/Keynesian mentality and Roosevelt's New Deal, which held that it was both good for capital and for global stability if there was "full employment" and "a strong underpinning to the social wage through expenditures covering social security, health care, education, housing, and the like" (Harvey, 1990, p. 135). Although the Fordist and Keynesian policies had the effect of creating a consumer class and ushering in a new standard of living, they relied to a high degree on the fact that America was the only advanced industrial economy that did not have to rebuild after the War. High exports allowed the American middle class to thrive in the manufacturing sector which served as the basis of the consumerist lifestyle. Once European markets were rebuilt, and Japan achieved modernization of their economy, international competition began to weaken the dollar, and the effects were felt throughout the global economy. "Between 1973 and 1983, double-digit inflation became commonplace in most Western economies" (Warlouzet, 2017, p. 18) which triggered several problems:

(1) "competitiveness decreases," which "can be offset by devaluation—however, devaluation fuels inflation and increases the cost of imports and of external borrowing"; (2) "a high inflation creates uncertainties for consumers and companies alike, leading to a postponement of investments" and higher unemployment rates which exacerbate the uncertainties; (3) "the combination of high inflation and rising unemployment renders stimulus plans more difficult to implement since they tend to fuel the former without diminishing the later"; (4) "inflation can have adverse social consequences if wages, pensions and welfare benefits are not indexed" which can lead to an "erosion of savings"; and (5) "in terms of international cooperation, strong differences in terms of inflation rates between neighboring countries … hamper trade and monetary cooperation" which causes political tensions. (pp. 18-20)

1973 saw the U.S. economy back in recession, proving that Fordist and Keynesian policies had not prevented the crises of capital, but had merely delayed them. Since the economy ran on fossil fuels that were imported from the middle East, when war broke out between Israel and a coalition of Arab nations in 1973, the OPEC nations, "led by Saudi Arabia, announced a five percent cut in the supply of oil," and "a further five percent reduction every month, until the United States stopped obstructing a comprehensive settlement of the Israel-Palestine conflict" (Mitchell, T., 2010, p. 190). This strategy hit the US hard because it directly impacted "[m]iddle class citizens" with "anxiety over the future availability of an essential commodity, and prices increased almost by the day" (p. 190). Since the automotive industry was one of the largest manufacturing sectors and employers in the U.S., jobs were also impacted by the increased presence of imported cars from Japan and Germany that offered better fuel efficiency.[5] As jobs were lost, prices increased and savings were depleted, 1973 also saw the first computerized credit card system and more American families began turning to the system of personal credit to extend their consumer lifestyle through debt-financing. The combination of these global issues "shook the capitalist world out of the suffocating torpor of 'stagflation' (stagnant output of goods and high inflation prices), and set in motion a whole set of processes that undermined the Fordist compromise" (Harvey, 1990, p. 145). Labor

---

5 "In 1972, all import brands combined held just a 13 percent share of the U.S. market. That shot up to a then-record 15.8 percent in 1975. They never looked back" (Treese, 2013). Although many foreign owned car companies manufacture the vehicles in the US today, from April 2017 to March 2018, the Japanese companies Toyota and Honda alone accounted for around 25% of the total cars sold in America (*The Wall Street Journal*, 2018).

became the new target of economic policies that demanded *"flexible accumulation"* (p. 147), that is, the ability to move away from measuring the health of the economy in terms of full employment—a goal that somewhat accounted for the social dimension—to measuring it by GDP and profits—a goal that aligned with the new totalizing logic of information. David Harvey argues that a major restructuring of labor was able to take place because of "time-space compression in the capitalist world" in which technology had advanced to such a degree that "the time horizons of ... decision making have shrunk" because of "satellite communication and declining transportation costs" (p. 147).

Transportation technologies had been steadily advancing since the invention of the automobile in the late 19th century and the invention of the airplane in the early 20th, but commercial aviation was a revolutionary development in the inter- and post-War years. The primary driver of the aviation industry was the military, who was the first to deploy this technology in practical ways. Although there was a sustained attempt to develop an aviation transportation system for public use in the inter-war years, "[t]he commercial industry was a minority component prior to 1957" (Collopy, 2004, p. 87). "By 1954, 13 of the 500 largest American firms derived most of their revenue from the design and manufacture of aircraft ... During this time, aerospace established its place as the largest employment sector in the U.S. economy, with 800,000 employees, and the sector with the highest percentage of technical jobs" (p. 90).[6] In the 1960s and 1970s the emphasis in manufacturing shifted from primarily military to commercial applications. In 1970, Boeing "introduced the 747 into service" and it had a "global reach, the ability to carry 400 passengers 6500 miles" (p. 91). With consumer demand, mass production, and heavily subsidized by government, the commercial aviation industry shrank the globe, allowing people to travel to nearly any destination on the planet in under a day.

Revolutionary as commercial aviation was, in terms of restructuring the economy and initiating a shift toward a postmodern culture, it had only a marginal impact. The two core developments that enabled satellite communications, rocket and computer technology, had the most monumental impact. Like aviation these developments also had their roots in the military and then later shifted to commercial applications. In October of 1957, the Soviet Union inaugurated the space age by launching the world's first artificial satellite, Sputnik I, into low Earth orbit. It was like something out of a science fiction novel (Verne, [1865/1869] 2011; Wells, [1901] 2017; Heinlein, [1947] 2004), as fantastic as the first humans setting sail across the ocean and leaving the sight of

---

6 See also Pattillo, 1998.

land. Since the USSR had successfully exploded a nuclear bomb in 1949, the fact that they now had rocket technology that could enter earth's orbit led to paranoia in the United States which suddenly felt vulnerable despite its possession of the bomb. The bomb could now potentially strike anywhere at any time.[7] More than the fears of nuclear war, the successful launch of Sputnik I "was the greatest defeat Eisenhower could have suffered" and it was interpreted as "a sharp slap to American pride" (McDougall, [1985] 1997, p. 133). When Sputnik II was launched the following month, Eisenhower could no longer downplay the significance of this event to the American public. The U.S. was quickly able to launch their own satellites, achieving success in January 1958, but this appeared in a wholly reactionary light to the Soviet's trailblazing victories. In April 1958, Eisenhower responded by submitting a bill to congress that would establish the National Aeronautics and Space Administration (NASA) to take the lead in competing with Soviet advances in space exploration. In May 1958, the Soviet's launched Sputnik III underlying the urgency of the bill, and by October 1, 1958, it passed both houses and NASA was formed. Less than any real desire to push science and develop intellectual curiosity, the issue NASA was tasked with was recovering national pride; in doing so, they moved the goal posts and made the space race about who could land on the moon first. The Soviet's managed to crash into the moon with their Luna 2 mission in 1959, but this "failure" demonstrated how precise the engineering calculations had to be to achieve success, which is why research had to go beyond rocket technology. NASA developed a ten-year plan and was to receive funding of over $1 billion per year, with most of it going toward basic research. The Kennedy administration funneled even more money at NASA in the hopes of speeding up the success of the mission, but by the late 1960s when the public learned that "the United States invested $146 billion dollars in government R&D from Sputnik to the moon landing" (McDougall, [1985] 1997, p. 437), the economics of pushing science for reasons of pride appeared in a dim light. The success of the moon landing, in 1969, shortly after Nixon's election, did reinvigorate a sense of American patriotism in the collective conscious, but when the economy entered recession in the

---

7 As David Krieger, the founder of the Nuclear Peace Foundation, saw it: "In the Nuclear age, a future for humanity can no longer be assumed" (Falk & Krieger, 2012, p. 3). But rather than provoking mass anxiety, survey data suggests that many Americans supported nuclear proliferation and the arms race with the Soviets (Kramer, Kalick, & Milburn, 1983). The problem was that the masses could not fully conceptualize the impact of the bomb as it triggered mass cognitive dissonance.

1970s the money spent on this initiative was deeply scrutinized and attacks against "big government" became a political rallying cry.[8]

To the masses, who had been conditioned to think in pragmatic terms using technical rationality, they could not see the full range of applications that all of NASA's basic research led to, they could only experience it in terms of its stated goal. Despite how historically monumental the moon landing was, people still had their eyes to the ground as they were struggling with the pressures of everyday life. Surveys conducted in 1969 showed that only those in excellent health supported the moon landing by a majority. As health and, therefore, the quality of life declined the majority of respondents disapproved of spending government money on what was essentially deemed a vanity project (Bainbridge, 2015, p. 19). There were no easy and direct links to be seen on the surface level from the space landing to economic growth, which in the collective conscious was deemed the measure of quality of life, as the visible technology that enabled this feat did not have immediate commercial applications. However, strides in computing were made behind the scenes that enabled much of the basic research NASA conducted to take place, and this technology pointed the way to a reshaping of the postindustrial economy.

The term "computer" was originally applied to people whose job was running calculations. The first modern machine computers were of the analog variety, a model "developed by Vannevar Bush and others at the Massachusetts Institute of Technology in the late 1920s" (O'Regan, 2012, p. 26). They were bulky and slow machines but could solve differential equations faster than by hand. Early digital computers made use of Shannon's (1940) binary language of 0's and 1's, which allowed for far more complicated programs and equations. But it was not until William Shockley, John Bardeen, and Walter Brattain, working at Bell Labs, developed the transistor in the early 1950s that digital computing really took off. In 1954, the first transistor-based computer, named TRADIC was invented at Bell Labs for the US Airforce. In 1959, IBM announced its 1401 model as the first digital computer designed for business, especially for the business of R&D. These early computers were massive, expensive, and produced an enormous amount of heat, so they required air conditioning which added to the cost of operating them. "[F]rom the mid-1950s to the mid-1970s, IBM" controlled "the mainframe computer market" (Campbell-Kelly & Aspray, 2013), but although they had success in selling computers to business, thereby creating the foundation for a postindustrial society, the users of these

---

8 For an account that incorporates the sociological dimension of the inauguration of the space age, see (Bainbridge, 1977).

computers had to be highly skilled and trained, so the effects were largely limited to the economic sphere. It was not until the company Xerox began to buy up small computer companies in the 1960s that innovation in computing began to seriously consider the consumer market. Xerox invented the Graphical User Interface (GUI), bit-mapped graphics, the word processor, postscript (to allow printing), Ethernet (which enabled computer networking), and object-oriented programming, as well as built the first personal computer (Watson, I., 2012, pp. 116–117). Xerox, however, struggled with cost, reliability, and software issues, leaving an opening in the market. In the mid-1970s several companies released personal computers, including IBM, Radio Shack, and Commodore International; again, they all had limited impact because of the high entry cost and buggy processing. Steve Jobs and Steve Wozniak made great strides in 1984, with the release Apple's Macintosh computer. It was less the result of their technological genius, than their marketing campaigns, which sought to make computing "cool," that accounts for their early success in creating a consumer base for the technology. Then Bill Gates and Paul Allen, who founded Microsoft in 1976, signaled a turn toward programing these new machines with software that really opened the door for the mass consumer market. From 1980 to 1984, there was a seven-fold growth in personal computer sales, from 1 million to 7 million units sold per year; by 2004 that number had grown to nearly 180 million units sold per year (Reimer, 2015). The spread of personal computing was the most significant "rupture" in the modern that initiated the postmodern society, as the cultural attitude was enhanced by a material recoding of the everyday experience.

The larger economic shift toward flexible accumulation made use of these technologies as they reshaped the production and consumer markets. With a technological base of satellite communications, a global transportation network that could move goods across land, sea, and air, and the introduction of personal computing, production could more easily be decentralized to take advantage of cheaper sources of labor in the modernizing countries of the "third world." Production became a far more flexible enterprise, as standardization in production enabled factories to move their base of operations and chase the cheapest labor, since it was no longer a necessity that manufacturing be done locally or with a "skilled" labor base. On the consumption side of things, there was less flexibility as many consumer technologies were expensive and had repeated use-value making them only occasional purchases, so there was not a constant demand for these goods, which meant that competition was fierce. "[T]o the degree that information and the ability to make swift decisions in a highly uncertain, ephemeral, and competitive environment became crucial to profits, the well organized corporation has marked competitive advantages over small business" (Harvey, 1990, p. 158). This translated into a wave of corporate consolidation during the late 1970s and 1980s, with large corporations gobbling up small businesses to shrink consumer choice and gain greater control over a limited market.[9]

Government had to be reshaped and its goals and policies reevaluated to support these business interests and convince the mass that it would be in their benefit to move away from the social safety net and target of full employment. The Reagan administration in the US, and the Thatcher administration in the UK, made a concerted effort in the 1980s to dismantle the last vestiges of the social in government and replace it with economic first policies. Their approach to governance came to be known as neoliberalism, which David Harvey defines as "a theory of political economic practices that proposes that human well-being can best be advanced by liberating individual entrepreneurial freedoms and skills within an institutional framework characterized by strong private property rights, free markets, and free trade" (2005, p. 2). Against what was deemed "big government" policies that followed Fordist and Keynesian mentalities, was this new model of the state and system of governance whose primary purpose was to create markets for business and militarily defend those markets without directly intervening in them. The beliefs that lay at the core of this model were that businesses possessed the information needed to respond to the dynamism of the economy and that "the state cannot possibly possess enough information to second-guess market signals" (p. 2). Being able to have that information and respond faster than anyone else is what gave a competitive edge in this new economy, and these new technologies provided the foundation for accelerating the transmission of this information. This in turn put pressure on developers to continuously innovate the speed of these devices, which would guaranty a rate of consumption that outpaced the life of the technology because its value was no longer linked to its lifespan, but to its speed capabilities, meaning that the generational advance of technology is what undermines its own lifespan and creates a continuous demand for the products on the consumer market.

If the computational speed of processing ever more complex information is what guides these new technological developments, then it is the information

---

9 "US companies spent $22 billion acquiring each other in 1977, but by 1981 that had risen to $82 billion, cresting in 1985 at an extraordinary $180 billion. Though mergers and acquisitions declined in 1987, in part of a response to the stock market crash the total value still stood at $165.8 billion for 2,052 transactions (according to W.T. Grimm, a merger consultant group. Yet in 1988 the merger mania kept going. In the United States merger deals worth more than $198 billion were completed in the first three-quarters of the year ..." (Harvey, 1990, p. 158).

itself that comes to shape the economy. Harvey doesn't make a clear distinction between knowledge and information, but he sees that "knowledge [better to say information] itself becomes a key commodity" in this new economy, "to be processed and sold to the highest bidder, under conditions that are themselves increasingly organized on a competitive basis" (1990, pp. 159-160). Since the competitive edge in this new economy is tied both to information and processing power of that information, this necessitated a "complete reorganization of the global financial system" so that it could find ways "to gain paper profits without troubling with actual production" (p. 163). On the one hand this involved the use of consumer credit financing, which extended consumption on the back of increased debt (Ritzer, 2007). The average American went from someone with a savings account to someone who now carried a credit card balance (Kus, 2015). In the world of finance, this meant a shift from investing in the means of production, which only had a slow and gradual return on investment, to a market based on financial speculation; that is, a new market that speculates on the basis of all this information. This provided a virtual outlet for the surplus capital that was only tenuously connected to the material world. Operating now under the totalizing logic of information, capital had found a way to circulate itself between computer networks, moving and self-valorizing by speculating on the material world. Actual transformations in the material world meant less than the flow of information, as the real speculation was tied less to actual material happenings than it was to what the next infusion of information might contain. In other words, it was a second-order speculation of information about information about the material world. Even if things looked bad from a material perspective, with the new way of defining governance and the new economic basis, the information economy could grow capital either way and make it appear as if it was a successful system on paper. For Harvey, these transformations account for what people label as postmodernism: "a climate of voodoo economics, of political image construction and deployment, and of a new social class formation" (p. 336). With a greater understanding of the material basis of this new form of society, in conclusion it is worth understanding the impact that these transformations had on thought, and to do that we must incorporate a more critical analysis.

Since the 1980s, the social character in postmodern societies is alienated in the experience of the self both in and out of the home and is prone to anxiety. With the social structure organized along postindustrial lines, jobs are highly dependent on those who can function in accordance with technical rationality or are willing to perform the service of menial tasks. The standard of living increases, but the rate of pay stagnates, and savings turn into debt. Technologies of business and domestic life are more seamlessly integrated, and the experience of reality is largely one mediated by the screen both at work and in free time. There is a spatiotemporal compression and the vast world shrinks as it is hyper-connected through networked communication and high-speed transportation. Governments abandon their role as guardians of the social and give priority to the economy, which is measured by positivistic information, not critical evaluation of the quality of life. While the numbers look good for the economy under these new measures, as it increasingly comes to rely on finance capital at the cost of diminished investments in productive capital, the disconnect between the information on reality and the materiality of reality widens. Information takes precedence, but that information is only ever about the given moment as it has no ability to predict long term trends with any meaningful accuracy since it is always subject to interruptions and since the whole point of it is to be as dynamic as possible, open to all possibilities; immanent speed replaces social forecasting and the future appears lost to the whirlwinds of the present.

If the modern rejects the past, then it is the postmodern that rejects the future, and in their combination our condition appears to be atemporal. Veblen's signaling of a post-modern arrived too soon to have any real impact, for collective thought had not caught up to material reality and technology had not yet been fully dispersed across all spheres of life. But it also speaks to the failures of mainstream sociology, as pointed out by the Frankfurt School, in the general disinterestedness of tracking and critically evaluating the effects of these material changes rather than just performing a social taxonomy of the given reality in service of this informational logic. In 1979, the year of Marcuse's death, the French philosopher, Jean-François Lyotard, proclaimed the arrival of *The Postmodern Condition* ([1979] 1984).[10] His "working hypothesis is that the status of knowledge is altered as societies enter what is known as the postindustrial age and cultures enter what is known as the postmodern age" (p. 3). Recognizing the temporal lag between material change and changes in thought, Lyotard draws a distinction between society, as the material conditions, and culture, as the domain of making meaning out of those conditions in thought; culture also has a material presence, but it is more about the symbolic than the real. The result of this break between society and culture is that "the general situation is one of temporal disjunction" (p. 3) or, what I refer to as atemporality.

---

10   In 1971, the Egyptian-American literary scholar, Ihab Hassan ([1971] 1982), used the term postmodernism to describe a literary style that was distinct from that of modernism. While he identified many of the differences between the two forms, he was writing literary theory, not social theory, so his work had negligible impact on the social sciences.

This atemporality, or temporal disjunction, is not just a description of the antagonism between modernist and postmodernist ideologies. Rather it is a description of a material reality that has for so long run out of sync with itself that it has ripped the social fabric and made the contradictions manifest in an absurdist way wherein thought increasingly denies material reality. As Lyotard describes it, the problem in this postmodern age is that there is a crisis of knowledge, which is a crisis of meaning. As technology enables an acceleration in the transformation of material circumstances, the time needed to think, to form knowledge, and glean meaning from it, is negated by economic demands. The logic of capital requires knowledge "to be produced in order to be consumed" and "consumed in order to be valorized in a new production" (p. 4). This transforms knowledge into information, because information is processed but knowledge requires reflection to discern the content of its meaning. "[R]eflection is ... thrust aside today ... because it is a waste of time ... Reflection requires that you watch out for occurrences, that you don't already know are happening. It leaves open the question: *Is it happening*?" (Lyotard, [1983] 1988, p. xv). For the modern, science was still the discourse for determining if an event was or was not happening; for the postmodern, science is the discourse of information production.

Through the scientific method, science set the rules for its discourse, but in embracing the informational model it abandoned the quest for philosophically legitimating its claims to knowledge. Lyotard performs a meta-analysis of the rules of science to demonstrate how its discourse is internally valid but is not externally legitimated. This stands in distinction to narrative knowledge, which is externally legitimated but lacks internal validity. Each has its own discourse that follows specific rules, which he calls, after Wittgenstein's (2009) 1921 use of the concept, language games. Narrative follows a temporal rhythm or beat that is culturally recognized and which is largely agnostic to the content as it can function under a plurality of different language games. Narratives are spoken, and others listen, and if those others recognize the narrative rhythm, they respond to it, and retell the narrative to establish its historical presence. Regardless of the truth content, narratives have a claim in and to reality by the presence of their performance. "Narratives," thereby, "define what has the right to be said and done in the culture in question, and since they are themselves a part of that culture, they are legitimated by the simple fact that they do what they do" (Lyotard, [1979] 1984, p. 23). Science, on the other hand, "requires that one language game, denotation, be retained and all others excluded" (p. 25). Since denotation requires the use of the most surface level meaning of the word, it excludes the feelings and ideas that the word would suggest in its cultural context. Since narrative relies on rhythm, it is precisely these feelings and ideas that are transmitted in its affective performance. Under the totalizing logic of information, science must flatten language so that it more directly corresponds to the informational form of material reality so that it may be subjected, by anyone, to "falsification." The moves in the language game of science are therefore not to be predicated on the sender or the receiver of its content, but on the content itself. If the content does not match up with the denotative rules, then it is rejected as invalid and jettisoned from the scientific discourse because it would require more than the transmission of information allows. Narrative is deemed legitimate if it persists in its retelling, therefore its legitimacy is granted externally from the discourse by the social handoff of the narrative in its inter-subjective telling, whereas science is deemed legitimate if its content is internally consistent, that is, in its objective presentation. What this means is that "[i]t is therefore impossible to judge the existence or validity of narrative knowledge on the basis of scientific knowledge and vice versa, the relevant criteria are different" (p. 26). What is true for both modes is that they have a focus on performative adherence to their own language games rather than to truth as such. The postmodern condition is therefore marked by a suspicion of "grand narratives" which claim to scientifically provide a narrative for the directionality of history but, cannot, according to the terms of its own language game, achieve both internal validity and external legitimacy. Therefore, once it became obvious that society had fallen under the totalizing logic of information, it became clear that there was a disconnect between visions of a qualitative superior future, which were bound to the narrative form, and the actual mechanisms by which material reality advanced, which was bound to the informational model claimed by science.

Lyotard, therefore, sees the modern and the postmodern as concurrent systems that "often coexist almost indiscernibly in the same piece, and yet they attest to a *différend* [a difference of opinion] within which the fate of thought has, for a long time, been played out, and will continue to play out—a differend between regret and experimentation" ([1988] 1993, p. 13). Narrative no longer seems able to point to the future in any material fashion, so it assumes a sense of nostalgia for the past when the enlightenment goals of the modern could still retain some relation to material processes. Information cannot point to the future either, as it is bound to the present, no matter how much speed or processing power we can technologically achieve, information will never move beyond the curvature of time; its very goal is to subsume space and time into their purely informational other. In this sense, postmodern society is an atemporal society. But, "[l]ife itself ... resists its reframing in terms of information" (Rose, 2007, p. 48). The postmodern cannot extinguish the modern; it drags it along as a dead weight, as they primarily operate in two distinct but interconnected fields. The modern is still primarily concerned with the real, this is still the domain of material capital and the production of material goods. The postmodern is, however, primarily concerned with the symbolic, this is the domain of information and the extension of capital into the virtual realm. The postmodern idealizes information because it is more fully manipulable, and subject to faster manipulations than the material, thereby allowing it to more easily transcend the traditional barriers of capital that ultimately are the greatest threat to its continued existence. In this sense, capital remains the definitive subject of history. It was the first functioning artificial intelligence, as it still largely controls the directionality of the totalizing logic of information (although this logic is not dependent on the logic of capital to carry out its objectives) according to its own goals, not those of biological life; it remains to be seen if the totalizing logic of information will sufficiently overcome capital in the way that capital overcame religion, since it is capital that maintains a material link which threatens both logics from continuing to execute their programs. What remains a characteristic of the "human" in the species, is that which is still concerned with the future, which is why anxiety persists. However, without a social avenue for achieving the future—because the present moves so quickly that all life can do is react to it—there is no material avenue for dissipating our anxiety. It is dissipated everywhere in simulated outlets that serve to distract life from its machinic functions and its mechanthropomorphic transformation, but it is reconstituted in the same moment by those very processes. If it can be said that there is any meaning to be found in this life, then the only meaning in postmodern life is the meaning that can be derived from the living contradiction of this absurd reality in which we have imploded the very basis upon which a harmonious life could begin.

2   We Are All Cyborgs Now: Life in the Mass

In 1960, inspired by NASA's initiative to land people on the moon and the sense of a common purpose that it gave to the scientific enterprise, Manfred E. Clynes, a research scientist in neurophysiology who was heavily influenced by music, and Nathan S. Kline, a clinical psychiatrist with a focus on psychopharmacological drugs, coauthored a paper in which they wrote:

> Space travel challenges mankind not only technologically but also spiritually, in that it invites man to take an active part in his own biological evolution ... The task of adapting man's body to any environment he may choose will be made easier by increased knowledge of homeostatic

functioning, the cybernetic aspects of which are just beginning to be understood and investigated. In the past evolution brought about the altering of bodily functions to suit different environments. Starting as of now, it will be possible to achieve this to some degree *without alteration of heredity* by suitable biochemical, physiological, and electronic modifications of man's existing modus vivendi (1960, p. 26) ... This self-regulation must function without the benefit of consciousness in order to cooperate with the body's own autonomous homeostatic controls. For the exogenously extended organizational complex functioning as an integrated homeostatic system unconsciously, we propose the term "Cyborg." The Cyborg deliberately incorporates exogenous components extending the self-regulatory control function of the organism in order to adapt it to new environments ... The purpose of the Cyborg, as well as his own homeostatic systems, is to provide an organizational system in which such robot-like problems are taken care of automatically and unconsciously, leaving man free to explore, to create, to think, and to feel. (p. 27)

Today science has succeeded in creating cyborgs, these human-machine hybrids, along some of the lines envisioned by Clynes and Kline.

As of 2015, "roughly one million people use insulin pumps" which for diabetics' act as a mechanical pancreas (American Diabetes Association, 2015). "Between 1993 and 2009, 2.9 million patients received a permanent pacemaker in the United States," to regulate their heartbeat (Greenspon, et al., 2012, p. 1541). Millions more have implanted defibrillators, artificial hip and knee joints, cochlear implants and hearing aids, and the neurostimulation market, which has already helped thousands of Parkinson's patients, is expected to be one of the fastest growing medical fields over the next few years for all sorts of nerve system ailments. Even wearing spectacles or contacts, taking analgesics and vaccines, and using prosthetic limbs and CPAP devices, are representative of a form of technological human enhancement; many do this without giving it a second thought. Yet traditional thought paradigms dominate the field with most either rejecting the notion of human enhancement on religious, utilitarian, or philosophical principles (Fukuyama, 2002, p. 88; Kaczynski, 2010), or defending these Neo-Darwinian principles as a moral and ethical imperative to improve lives using the tools of modern society (Harris, 2007, pp. 19-35; Enriquez & Gullans, 2016). Comparatively, relatively few have taken a critical approach that accounts for the anxiety that arises as these transformations occur under the sway of the totalizing logics of capital and information. Lacking such a perspective it is not possible to fully understand the consequences of enhancement on the social makeup of the species and society and the costs that each will bear during the transitional phase of perfecting this symbiosis (Rose, 2007; Fuller, 2011). Regardless of one's ideological disposition, and although science has not yet created the cyborg of the space age, there is no question that it has created cyborgs to adapt to our own environment and to overcome nature here on planet Earth.

That environment and what we think of as nature is changing. It is becoming something new, and science strongly suggests that these changes are inhospitable to life itself. "The Living Planet Index, which measures biodiversity abundance levels based on 14,152 monitored populations of 3,706 vertebrate species, shows a persistent downward trend. On average, monitored species population abundance declined by 58 percent between 1970 and 2012" (WWF, 2016, p. 12). Correlated to this, data provided by the National Oceanic and Atmospheric Administration (NOAA) shows that "[s]ince 1970, global surface temperature rose at an average rate of about 0.17°C (around 0.3° Fahrenheit) per decade—more than twice as fast as the 0.07°C per decade increase observed for the entire period of recorded observations (1880–2015). The average global temperature for 2016 was 0.94°C (1.69°F) above the 20th century average of 13.9°C (57.0°F), surpassing the previous record warmth of 2015 by 0.04°C (0.07°F)" (Dahlman, 2017). The accepted scientific estimate is that if temperature rise exceeds the 2°C threshold the consequences for life will be devastating. Over the same period, the United States Census Bureau (2002; 2018) estimates that the world's population was just north of 3.7 billion in 1970, and nearly 7.5 billion in 2018. If we take this information at face value, then on the surface it looks like the human species is thriving, while animal and plant life, as well as the planet itself, are dying.

The ecosystems that support life are flooded with the byproducts of modern life, many of which have harmful effects, even on the "human" species. They are impacting the air we breathe: "in 2013, 5.5 million premature deaths worldwide, or 1 in every 10 total deaths, were attributable to air pollution" making it "the fourth leading health risk worldwide" (The World Bank and Institute for Health Metrics and Evaluation, 2016, p. x). And the water we drink: "3 in 10 people worldwide lack access to safe, readily available water at home, and 6 in 10 lack safely managed sanitation" (WHO and UNICEF, 2017), "[a]s a result, every year, 361,000 children under 5 years of age die due to diarrhea related to poor sanitation and contaminated water, which are also linked to transmission of diseases such as cholera, dysentery, hepatitis A, and typhoid" (World Bank, 2018). And the soil in which we grow our food: "[g]lobally 24% of the land is degrading. About 1.5 billion people directly depend on these degrading areas" (United Nations, 2018). Since governments in postmodern societies under the sway of neoliberal ideology no longer speak the language of the social, human and environmental costs are deemed secondary to monetary impact, so many have attempted to put this in economic terms. In the United States alone, these changes "are currently causing, on average, $240 billion a year in economic losses, damages, and health costs" (Watson, McCarthy, & Hisas, 2017, p. 5). Despite these attempts to make an economic case for slowing down or reversing the metamorphosis of our environment, the United States—which as Baudrillard ([1986] 2010) says "is utopia achieved" (p. 83) because it "has concretely, technologically achieved this orgy of liberation, this orgy of indifference, disconnection, exhibition, and circulation" (p. 105) that simultaneously brings about "the anti-utopia of unreason, of deterritorialization, of the indeterminacy of language and the subject, of the neutralization of all values, of the death of culture" (p. 106)—is one of the most reluctant to join international efforts to combat these issues. Furthermore, many flat up deny that these changes have anything to do with modern society, and even among those who do admit that these changes are linked to industrialization and the consumerist lifestyle, they are reluctant or unable to take concrete steps to change their behaviors. Behavioral changes are themselves commodified and sold back as consumerist choice, thereby immediately restructuring attempts to deconstruct the link between the self and the society of environmental destruction. If this is the material case of our world, then what lies behind it?

From an individual or a social perspective these numbers are shockingly devastating and demand immediate steps to take rational action to counter them, but once crises enter the planetary scale, then they are truly a mass phenomenon and as such we must understand them from the perspective of the mass. "[T]he mass," Baudrillard tells us, "is characteristic of our modernity: a highly implosive phenomenon, unable to reduce itself for any traditional theory and practice, perhaps any theory at all" ([1978] 2007, p. 36). He continues, the mass "has no sociological "reality." It has nothing to do with any real population, body or specific social aggregate" (p. 5). What this means is that "[t]he masses as such do not exist: except that they are constantly invoked in discursive practices and in that sense quite real, in Philip K. Dick's crypto-Lacanian sense, as something that refuses to go away even if one stops believing in it" (de Zeeuw, 2014, p. 61). Without trying to sociologize this concept and give it a definition that it necessarily evades, we can think of the mass as arising in a situation when there is enough quantity there is a change in quality, and the mass arises when there is a dispersal of the social throughout all spheres of society marking its full absorption in the structural and totalizing logics. Rather than being able to rally the social forces to combat the planetary changes that threaten the continuation of life, "the masses are a mute referent," "[t]heir strength is *immediate*, in the present tense, and sufficient to itself" (p. 36). Being atemporal, always locked in a presentist perspective as a consequence of the logics of capital and information, the mass only has the "capacity to absorb and neutralize" material reality which it does in its symbolic practices that have suffered "the central collapse of meaning" (p. 36).

We can conceptualize the mass by taking an example from astronomy and the process of stellar fusion. It already fits as a good example of society sui generis, since a star is primarily made of hydrogen atoms (individuals) which at a certain mass become something more (the social). A star is something more than a bunch of hydrogen atoms, just as an atom on its own cannot be said to be a star. A feedback loop of opposing forces ignites the star's core and sets the atoms in motion. The greater the mass the greater the gravitational force, and that force in a star is so strong that the atoms are pulled together at incredible speeds, allowing them to collide and in rupturing their atomic bond in that collision, a massive amount of energy is released. The release of this tremendous atomic energy works against the gravitational pull and creates a feedback loop of forces pulling inward and forces exploding outward. Neither can serve its natural function without the other, both the star and the atoms that are caught in this feedback loop are at the mercy of each other, only surviving through a process of mutual reinforcement and destruction. But once enough of the hydrogen atoms have converted into helium and other heavier elements, the star can no longer maintain this delicate balance because the gravitational force is now dealing with objects that possess greater mass. It becomes "an opaque nebula whose growing density absorbs all the surrounding energy and light rays, to collapse finally under its own weight" (p. 36–37). The mass becomes "[a] black hole which engulfs the social" (p. 37). This is precisely what has occurred in the transition from what we think of as modern societies to that of postmodern societies.

The consequences of this new transformation have at best largely eluded mainstream sociology, or at worse it has ignored them, because it persists in relying on 19th century definitions of the individual and early 20th century definitions of the social.

> This is, therefore, exactly the reverse of a "sociological" understanding. Sociology can only depict the expansion of the social and its vicissitudes. It survives only on the positive and definitive hypothesis of the social. The reabsorption, the implosion of the social escapes it. The hypothesis of the death of the social is also that of its own death. (p. 37)

In relying on concepts that no longer align with material reality, sociology assumes that the rational presentation of information will keep the masses within reason, that is, when they are presented with the sober truth of their material conditions they will engage in deliberative communication which will moralize that information by infusing it with meaning in such a way that free individuals can mobilize the social and take concrete action to intervene in the material world. But this has not happened, and sociology has not theorized upon the basis of these new conditions in which the social has transformed into a mass, because ultimately to do so would be to admit once and for all that sociology is no longer a relevant science for understanding the totality of society under the present material conditions in which it operates. To work with the mass as a central concept would necessitate a new scientific practice. Short of creating this new practice, sociology can only, at best, examine the pockets of modern society that make up its social taxonomy, but in doing so, it ignores how these pockets are neutralized by the mass as they are pulled by the gravitational force of the totality. This is why Habermas' ([1981] 1984; [1981] 1989) theory of communicative action was dead on arrival; it had more in common with traditional theory in that it assumed a foundation for rational communication rather than taking the critical theory perspective that would have recognized the need to interrogate that foundation as a precondition to any theory of communicative action (Dahms, 2019). For Baudrillard ([1978] 2007), communicative action today is "nonsense," because "the masses scandalously resist this imperative of rational communication. They are given meaning: they want spectacle" (p. 40). This signals a rejection of "the "dialectic" of meaning" (p. 40) so all "react in their own way, by reducing all articulate discourse to a single irrational and baseless dimension, where signs lose their meaning and wither away in fascination: the spectacular" (p. 41). The inability or unwillingness of people, and especially social scientists, to think the horror that is the world-with-out-us, threatens to negate the continued reality of the world-with-us. This does not mean that there are not people who stand up and speak out and take action against these devastating planetary changes, but they ignore the reality of the mass which is why their rational communication does not penetrate the mass, meaning that what they perceive as rational is in fact irrational because it is directed at the wrong scale. The mass is indifferent to any and all actions, it absorbs and neutralizes them. Which is why, even those who do change their lifestyle (drive electric, eat vegan, recycle, etc.) have a negligible impact on occurrences at the planetary scale which can only be overcome by a force equal to or greater than its own mass. So, if the social has become the mass, then what does this have to do with cyborgs?

There is more than one way to make a cyborg. The processes envisioned by Clynes and Kline assumed that technology would be implanted to supplement the human, and that it would be unconsciously synthesized without changing the fundamental structure of what makes a human, human. But the unconscious and the conscious do not run as disconnected platforms in the mind, they are interlinked and changes to the structuring of the unconscious effects the functioning of the conscious mind. The technologization of modern society recoded the mind to act in accordance with technological rationality, that is, to become more machine-like. The symbolic codes in the unconscious are in postmodern societies structured by the language of machines, a language that runs beneath the surface of the technologies that we interact with like a machinic unconscious. A computer requires hardware, the real material base of the machine, or what we might think of as the exogenic components, and software, the symbolic coding of that machine, to run its programs, or what we may refer to as its endogenic components. Clynes and Kline focused on the hardware side of the process of building a cyborg, the exogeneous dimension, to the exclusion of the software side of things, the endogenous dimension. But the shift to postmodern society has resulted "in the transformation of the subject himself into a driving computer ... The vehicle thus becomes a bubble, the dashboard a console, and the landscape all around unfolds as a television screen" (Baudrillard, [1987] 2007, p. 20). The current mode of cyborgification follows more closely the aims outlined by the psychologist J. C. R. Licklider (1960), who in the same year as Clynes and Kline, wrote:

> The hope is that, in not too many years, human brains and computing machines will be coupled together very tightly, and the resulting partnership will think as no human brain has ever thought and process data in a way not approached by the information-handling machines we know today ... As a concept, man-computer symbiosis is different in an important way from what North (1954) has called "mechanically extended man" ... It seems likely that the contributions of human operators and equipment will blend so completely that it will be difficult to separate them neatly in analysis ... Men will set the goals and supply the motivations, of course, at least in the early years ...

Think of how this came to be in the terms that Deborah Lupton (1995) used to explain her relationship with a personal computer:

> I am face-to-face with my computer for far longer than I look into any human face. I don't have a name for my personal computer, nor do I ascribe it a gender ... However, I do have an emotional relationship with the computer ... I have experienced anger, panic, anxiety and frustration when my computer does not do what I want it to ... I live in fear that a power surge will short-circuit my computer ... or that the computer will be stolen ... A pen now feels strange, awkward and slow in my hand, compared to using a keyboard ... There is, for me, almost a seamless transition from thought to word on the screen ... While people in contemporary western societies rely on many other forms of technology during the course of their everyday lives ... the relationship we have with our PCs has characteristics that sets it apart from the many other technologies we use ... People [speak] of being left wondering what to do ... not knowing how to occupy themselves with no computer to work upon/with (pp. 97-98). The relationship between users and PCs is similar to that between lovers or close friends. An intimate relationship with others involves ambivalence: fear as well as pleasure ... Blurring the boundaries between self and other calls up abjection, the fear and horror of the unknown [i.e. anxiety], the indefinable ... Computer users, therefore, are both attracted towards the promises of cyberspace, in the utopian freedom from the flesh, its denial of the body, the opportunity to achieve a cyborgian seamlessness and to 'connect' with others, but are also threatened by its potential to engulf the self and expose one's vulnerability to the penetration of enemy others. (pp. 110-111)

The scene is one of being seduced by the computer, by the technological extension of the mind as it couples itself with the processing power of the machine in a virtual synthesis. At the same time when it is scrutinized there is something obscene about the coupling of the human and the machine. Just as within the sexual act, this coupling amplifies the disintegration of the self by making it become something other than what it was prior to the moment when we allow the other access in the act of penetrating and overpowering our self. Unlike the consensual sexual act, in which the effects are temporary and fade, the penetration of the computer is a continuous process with shorter and shorter windows of decoupling. For Freud the universal taboo was incestual relations, but many cultures also maintained a taboo against interspecies relations, and since the act of human-computer coupling is ubiquitous it has lost the mark of its obscenity and these taboos have faded in the coupling of humans with their technological progeny. The boundaries of the human self and its technological extension are dissolving, and in their coupling the self is remade, but without direction this remaking produces nothing so much as an overwhelming sense of anxiety.

Again, we can witness the material amplification of this process by tracing the evolution of technology. The computer was originally designed to be a calculating machine, but by the 1980s its potential as a means of communication began to be realized. With the success of codebreaking in World War II, sending and receiving military and other sensitive government communications from around the world took precedence in the Cold War as information became the most valuable commodity. Traditional means of communicating—post, telegram, and telephone—were too easily compromised as their technologies were not developed with security of communication as a top priority. Furthermore, these modes of communication could be easily disrupted in war, and with the threat of the nuclear bomb and intercontinental missiles shrinking the temporal horizon of destruction to mere seconds, the demand for secure and fast communication was of the highest urgency, lives were at stake.[11] The Defense Advanced Research Project Agency (DARPA)—formerly known as the Advanced Research Projects Agency (ARPA)—was created in 1958 when Eisenhower took steps in response to the Soviet's launch of Sputnik I to maintain US technological superiority. Backed by their funding, a group of scientists led by Lawrence G. Roberts set themselves the task of solving a problem Roberts' had laid out in 1964 which arose as a result of the limited capabilities of mainframe computers, namely that "the most important problem in the computer field before us at that time was computer networking; the ability to access one computer from another easily and economically to permit resource sharing" (Roberts, 1986, p. 51). Roberts' team successfully developed ARPANET, a network of interlinked computers, in 1969, and sensing the changed political climate that was becoming opposed to basic research, they framed it to Congress "with pragmatic economic or security reasons ... as an administrative tool for the military rather than as an experiment in computer science" (Abbate, 1999, p. 76). The British scientist, Donald Davies developed the concept of packet switching which ARPANET borrowed as the foundation

---

[11] "But let's go back to 1962, to the crucial events of the Cuban missile crisis. At that time, the two superpowers had *fifteen minutes*' warning time for war. The installation of Russian rockets on Castro's island threatened to reduce the Americans' warning to thirty seconds, which was unacceptable for President Kennedy, whatever the risks of categorical refusal. We all know what happened: the installation of a *direct line*—the "hot line"—and the interconnection of the two Heads of State! ... [T]he constant progress of rapidity threatens from one day to the next to reduce the warning time for nuclear war *to less than one fatal minute*—thus finally abolishing the Head of State's power of reflection and decision in favor of a pure and simple *automation* of defense systems" (Virilio, [1977] 2006, p. 155).

for this new technology. This technique decentralizes communication by making "it possible to route messages through many different computers and so the system does not depend on the survival of any one of its nodes" (Feenberg, 2012, p. 7). This framework represented a far more secure means of communicating which was a top military priority, so the project continued to receive government support.

Since it relied on wired communications, ARPANET still had limited applicability outside of institutional settings, so its uses had to be determined in that context. The unforeseen popularity of email was "crucial to creating and maintaining a sense of community among ARPANET users" and therefore pointed the way to its applications (Abbate, 1999, p. 110). In the mid-1970s it was recognized that this technology could be used as a form of mass communication, which could warn Americans of immediate threats without relying on the traditional and easily disrupted centralized communication networks. As ARPANET expanded, the potential for undesirable intrusions meant that there was a need to create a separate network for military applications than the one that would be used for mass communication. In the 1980s the ARPANET framework was translated into a standardized protocol language (TCP/IP) that could network with commercially available PCs across their different platforms. In the 1980s the networks were small and still largely contained in institutional settings, but by the 1990s "the Internet emerged as a public communications medium" (Abbate, 1999, p. 181). Like all technologies there was a lag in adaptation, but the internet experienced more rapid growth than many technologies due to how "it integrates both different modalities of communication (reciprocal interaction, broadcasting, individual reference-searching, group discussion, person/machine interaction) and different kinds of content (text, video, visual images, audio) all in a single medium" (DiMaggio, Hargittai, Neuman, & Robinson, 2001, p. 308). According to the Pew Research Center, when they began tracking American internet usage in 2000, 52% of adults were using the technology, by 2018 that number has climbed to 89% (2018a).

The ubiquitous rise of internet technology in postmodern societies was facilitated by the development of wireless technology. Wireless transmission has its roots in the development of radio technology, dating back to the late 19th and early 20th centuries, but these early forms could easily be interfered with and could not serve as the basis for the kinds of complex global systems and networks imagined by the late 20th century. Proof of concept of wireless transmission of data packets was accomplished in 1971 at the University of Hawaii (Binder, Abramson, Kuo, Okinaka, & Wax, 1975). Japan developed the first commercially viable wireless network in 1979, but it was based on analog technology that did not integrate with the digital processing done on computer networks (Seymour & Shaheen, 2011). Second-generation systems emerged in the 1990s and made the switch to digital which increased the network capacity and reliability of "voice communication" and added "text messaging and access to data networks" (Sarkar, Mailloux, Oliner, Salazar-Palma, & Sengupta, 2006, p. 160). In 1994, IBM released Simon, the first personal digital assistant (PDA) that combined the functions of a computer with those of a wireless telephone (Bidmead, 1994). The consumer market for these new 'smart phones' was solidified by the success of BlackBerry's PDAs in 1999 which allowed people to send and receive emails from their wireless device. The effect for business was that people were always "plugged in" and could carry the office with them in their pocket.[12] As the backbone networks were upgraded to handle larger data streams, Apple's release of the iPhone in 2007 again changed the game. Now people could carry around a device in their pocket that had more computing power than what had put humans on the moon, and it consolidated several technologies into one device that acted as a wireless extension of the human mind that was always within arm's reach. All one had to do to access the world's archive of information was pull a small device from their pocket and tap on the screen. By 2018 a full 95% of American adults carried cell phones with them and a whopping 77% of adults had smart phones (Pew Research Center, 2018b). Survey data collected in 2017 shows that on the basis of these technologies, the average American now spends 23.6 hours per week online (Center for the Digital Future, 2017), more than they spend eating and drinking, purchasing goods, caring for household and non-household members, engaging in organizational, civic, and religious activities, and general household activities (Bureau of Labor Statistics, 2017). As Jonathan Crary (2013) has suggested, the only time that hasn't been cannibalized by capital and technology is sleep.[13]

---

12 The effect was that the home was no longer a refuge from the office, alienation did not stop and start at the door, rather as Weber had identified, instrumental rationality demanded that official duties took priority in all spheres of modern life.

13 The television show, Futurama (Verrone, 1999), explored the idea of technology allowing advertisements to be directly transmitted into dreams thereby extending the reach of capital and technology to sleep. This fictionally counters Crary's (2013) laughable notion that "sleep can stand for the durability of the social" (p. 25), but even if we stay in the domain of material fact, there is no logical or scientific argument to be made for sleep as a means of defending the social from the totalizing logics of capital and information; the social has disintegrated in postmodern societies and his idea of reclaiming it in sleep is itself pure fiction. This is not to deny that "the boundary between science fiction and social reality is an optical illusion," after all "[t]he cyborg is a condensed image of both imagination and material reality," but at a minimum ideal theories of social change must account for how "the two joined centers structur[e] any possibility of historical transformation" if they

Mechanthropomorphism is, therefore, not merely a process of the human becoming machine and the machine becoming human, rather it splits itself along two axes. The first cyborg transformation is focused on the somatic realm and seeks to remediate psychical and biological attributes, the second focuses on the cognitive realm and seeks to remediate mental thought processes (More & Vita-More, 2013; Crombez, 2015). In postmodern societies humans have been transcribed within the discourse of the cyborg, in some ways biologically, and in every way mentally, making them something decidedly different than human. Failure to integrate one's self into this discourse is akin to welcoming social death as the consequences of resistance to this process places serious limitations and restrictions on the ability to function within the system of capital. The discourse of the cyborg is served up by mass society in exchange for accepting alienation. Unable to build ourselves as individuals in our alienated state, we are relieved of that burden in the cyborg existence. The cyborg is not "unalienated," rather alienation is its natural state and it is welcomed in exchange for the continued rewards of the informational stream which distract from the pains of alienated existence. There is no longer an economic dimension, a political dimension, a social dimension, and a cultural dimension, the whole of cyborg life is caught in the gravity of the mass and structured by it.

In the endless streams of information our brains are rewired away from reason into a script of unreason. Knowledge takes time, it involves solving riddles and struggling through uncertainty and anxiety on the journey to achieve the flash of ecstasy upon reaching the destination. But the destination reveals that the journey is always already only a beginning, the riddle dissolves and a new riddle appears, and it is within the process itself that we come to recognize long term changes in our "self." This pacing allows us to take a greater role in actively guiding the direction of the change we want to experience through continuous reflection. But information circulates at a speed which short circuits this process by playing on the reward centers of our brain. Each informational tidbit triggers a quick release of the feeling of mental reward, but it is passively achieved, thereby it recodes the self to align with the atemporal present by circumventing the struggle with anxiety and the time needed to reflect on what directionality we want to take in the shaping of the self. Planetary destruction becomes nothing but another science fiction, another competing

---

are to make any claims of being ground in the concrete material reality of the present (Haraway, 2004, p. 8). On this basis, Futurama's fiction more fully accounts for the totalizing logic of capital and information in its dystopian vision than Crary's utopian vision.

informational tidbit that cyborgs passively assimilate before moving on to the next sport score, television serial, news event, social media feed, meme, etc. It is comprehended as unreality because it is no more real than the experience of being plugged into a vast global network where we assume a swarm identity while maintaining the more powerful fiction that in doing so we somehow manage to retain individuality and access to the social. The mass conditions the swarm identity of cyborgs. Postmodern societies construct a posthuman reality, that appears as if it were nothing but a simulation as all differences start to align in a compendium of gradations.

The philosopher Nick Bostrom logically argues that one of three scenarios is true given this material reality: "(1) the human species is very likely to go extinct before reaching the "posthuman" stage; (2) any posthuman civilization is extremely unlikely to run a significant number of simulations of their evolutionary history (or variations thereof); (3) we are almost certainly living in a computer simulation" (2003, p. 243). The argument is based on the empirical fact that computing technologies point beyond themselves and as we run endless simulations of data in an attempt to see beyond the temporal curve into the future, it makes sense that as these technologies improve we will begin to run more complete civilizational simulations to account for a greater number of variables and extend information processing to all aspects of reality. Now there is a chance, given the current planetary crisis, that civilizations go extinct before they achieve the computing power needed to run these simulations. Therefore (1) is the least desirable outcome, as it implies our own immanent extinction. "In order for (2) to be true, there must be strong *convergence* among the courses of advanced civilizations ... [and] virtually no posthuman civilizations decide to use their resources to run large numbers of ancestor-simulations" (p. 252). This would imply that they have lost interest in history and have followed a totalizing logic that runs counter to that of capital and information, perhaps because the societies have achieved a level of post-scarcity that re-enables individual development by providing a social base; in this sense it would mean that as societies advanced their organizational modes they would revert to an early stage of cultural coding, which would be unprecedented, but not necessarily impossible. On the other hand, "if we do go on to create our own ancestor-simulations, then since this would be strong evidence against (1) and (2), we would therefore have to conclude that we live in a simulation" (p. 253). This is because ancestor-simulations would have to virtually stack simulations in order to complete the simulation of a reality which contains a simulation, and this stack would have an infinite regress, meaning that the total number of simulations would be so large that it would become an increasing statistical improbability that we live in the prime reality and not one of the simulations.[14] This argument builds a myth, a story that is based in reality, "but it is also the reason why it is useless to try and objectively verify these hypotheses through statistics" (Baudrillard, [1987] 2007, p. 22). And yet, it is a narrative that logically would explain the mass society of cyborgs, who through their differentiation are no longer individuals, but gradations of possibilities that are gravitationally attracted to the center of the structure but are fluid enough to not completely coincide with that center. Since the structure is dynamic ("at play"), it engages in destructuration and decentering at the same time that it reconstitutes and recenters itself in different structural forms (Derrida, [1967] 2002, pp. 351-370). Statistics can only be pulled along by the center of the structure; their weakness is in being unable to trace the decentering and predict where the next center will be reconstituted. This is something that can only be revealed in the unfolding of the simulation/history of the given society.

For Baudrillard, the simulation is not a metaphor of mass society, it is a material description of a society that engages in "the generation by models of a real without origin or reality: a hyperreal" (Baudrillard, [1981] 1994, p. 1). The endless stream of information and images of reality as they are broadcast on our technological screens, mediate the reality of the world and are in "essence" unreal, but they are processed in the mind as something more real than the real, the hyperreal. As such, it reveals the history of "the successive phases of the image: it is the reflection of a profound reality; it masks and denatures a profound reality; it masks the *absence* of a profound reality; it has no relation to any reality whatsoever: it is its own pure simulacrum" (p. 6). As the cyborg experiences more of its waking reality through the mediated technological image, the reality of the screen masks and replaces the reality of the material world. The technological image takes precedence and hides the disappearance of the human and its environment with the seductive procession of images that continuously trigger the mental reward mechanism. How many of us have witnessed the destruction of the planet on the screen? Asteroids, alien invasions, nuclear bombs, ice ages, tsunamis, earthquakes, world war, fiery apocalypse in all its forms march in the steady progression of simulacra. Often the world is saved by a small band of Americans haphazardly thrown together by coincidence in precisely the right configuration to overcome the mass event and save the species from complete annihilation. Our own material apocalypse plays out in the mind of the cyborg as nothing but another simulation of

---

14   For further clarification and responses to counter-arguments see Bostrom & Kulczucki, "A Patch for the Simulation Argument," 2011.

our collective end, and in the hero narrative the craving for individuality fuels the desire to witness the apocalypse as the culling of the mass and a reversion to an early form of our operational coding on the slight chance that, finally free of the material shackles of this world, our true selves will emerge and save the day. The unreality of this scenario is the hyperreality of the simulation of postmodern society. The myth of the cyborg inverts its profound unfreedom into a new enlightenment narrative of salvation only achievable in the complete destruction of the global order and, perhaps, nature itself.

This myth is what leads Donna Haraway to see a utopic side to the cyborg, and proclaim, "I would rather be a cyborg than a goddess" (Haraway, 2004, p. 39). Coming from a feminist perspective it is not hard to see the seeds of utopia in the possibility of the cyborg. Although, "[t]he main trouble with cyborgs ... is that they are the illegitimate offspring of militarism and patriarchal capitalism, not to mention state socialism," Haraway believes that "illegitimate offspring are often exceedingly unfaithful to their origins" (p. 10). Just as the feminist movement rose to demand equal legal rights within a patriarchal society, she sees a potential for the cyborg to be just as disobedient to its design. Here Haraway and Baudrillard part ways, as Baudrillard's thought aligns with the logic of the mass and Haraway's aligns with that of the cyborg. What Haraway holds on to is the idealism of the impulse behind the feminist movements, but in doing so she clouds the material reality which shows how far legal rights are from genuine cultural equality and thereby undermines her utopian cyborg impulses on materialist grounds. Baudrillard, engages in negative thinking by maintaining a focus on the material reality that emphasizes its immaterial forms, but in this, the ideal collapses in his embrace of nihilism, for in the face of the mass all actions are neutralized. In doing so he fails to place the totalizing logic of information in a dialectic, seeing its stream as too powerful a force on the mental structuring and therefore he refuses the utopian refuge of the mind as a safeguard for future hopes.

If Haraway is right, then her recommended mode of resisting the programed intentions of the cyborg is through writing. She grounds this in ideas taken from Audre Lorde's *Sister Outsider* ([1984] 2007), which highlights the ways that writing has aided the political struggles of oppressed peoples of color. But Haraway ignores a crucial part of Lorde's text, when she writes:

> Cyborg writing must not be about the Fall, the imagination of a once-upon-a-time wholeness before language, before writing, before Man. Cyborg writing is about the power to survive, not on the basis of original innocence, but on the basis of seizing the tools to mark the world that marked them as other. (p. 33)

This directly contradicts Lorde's warning that *"survival is not an academic skill ... For the master's tools will never dismantle the master's house.* They may allow us temporarily to beat him at his own game, but they will never enable us to bring about genuine change" ([1984] 2007, p. 91). Haraway's call to ignore the materiality of the Fall and focus on the ideality of the new, reverts to a mode of pure idealism that is completely disconnected from the concrete gravity of the material mass which negates attempts to actualize her ideal. The stronger lesson is the dialectical one offered by Adorno, one which recognizes that writing the future means writing the past, that liberation of the self without the liberation of society is impossible and, in that knowledge, new anxieties and pain await us, not a utopian revolution of reality. Recall the words of Fredrick Douglass when he discovered the power of reading and writing:

> The reading of these documents enabled me to utter my thoughts, and to meet the arguments brought forward to sustain slavery; but while they relieved me of one difficulty, they brought on another more painful than the one of which I was relieved. The more I read, the more I was led to abhor and detest my enslavers ... As I read and contemplated the subject, behold! that very discontentment which Master Hugh had predicted would follow my learning to read had already come, to torment and sting my soul to unutterable anguish. As I writhed under it, I would at times feel that learning to read had been a curse rather than a blessing. It had given me a view of my wretched condition, without the remedy. ([1845] 2009, p. 50)

Although the anxiety was enhanced by Douglass's ability to see and understand the horrors of slavery that were imposed upon him, reading and writing created the necessary preconditions for recognizing when there was a time to act, without which he may never have taken the chance to escape the bonds of his slavery. So, there is a redemptive value in these actions, even if it is important to counter Haraway's idealism by a continued accounting of the past and the given material conditions of the present. Although the current material conditions suggest nothing more than the amplification of our anxiety as we come to better understand our cyborg selves in a mass society, "the struggle for language and the struggle against perfect communication, against one code that translates meaning perfectly" (Haraway, 2004, p. 34), is precisely what is needed to understand exactly what we are dealing with at the level of the cyborg self.

On the other hand, if Baudrillard is right, and mass society is a simulation, then there is nothing that can hold the system in check except for the random interruptions of terrorism. For this reason, he concludes that "[t]heoretical violence, not truth, is the only resource left to us," but this "is utopian" because even terrorist actions have lost their meaning in mass society ([1981] 1994, p. 163). Since nothing can be produced that is not absorbed by the mass, his project is a nihilistic one because all it can do is trace out the disappearance of the forms that are neutralized by the mass. Just as the Frankfurt School traced out the disappearance of the individual, so too has Baudrillard traced out the disappearance of the social. Baudrillard notes the similarity:

> The trace of this radicality of the mode of disappearance is already found in Adorno and Benjamin, parallel to a nostalgic exercise of the dialectic. Because there is a nostalgia of the dialectic, and without a doubt the most subtle dialectic is nostalgic to begin with. But more deeply, there is in Benjamin and Adorno another tonality, that of a melancholy attached to the system itself, one that is incurable and beyond any dialectic. It is this melancholia of systems that today takes the upper hand through the ironically transparent forms that surround us. It is this melancholia that is becoming our fundamental passion ... Melancholia is the inherent quality of the mode of the disappearance of meaning, of the mode of volatilization of meaning in operational systems. And we are all melancholic. (p. 162)

With the failure of the 1960's revolutionary actions against the logic of capital, Baudrillard is somewhat representative of Jameson's claim that "postmodernism is the substitute for the sixties and the compensation for their political failure" (1991, p. xvi). This attitude presumes that anxiety has been dissipated in postmodern societies, because there are no material grounds to build an alternative future and anxiety is future-oriented, so it turns into a nostalgic mourning for the past, a melancholia.

Baudrillard is wrong to read this melancholia as the guiding factor in Adorno's negative dialectic, it is, in fact, precisely what Adorno warns against in the dangerous pull of pessimism. In tracing out the disappearance of these concepts, they are kept alive in the utopian refuge of the mind, as an ever-present reminder of what is lost. Again, the critical method maintains past hopes but does not mourn the past as an ideal, for this is false nostalgia. What needs to be done, what socioanalysis aims to do, is to place Haraway's concept of the cyborg self in a dialectic with Baudrillard's concept of the mass society, to aggravate our anxiety, to rouse it from complacency or melancholy, not by means of a terroristic interruption of the mass, but as a terroristic interruption at the level of the self, as a self-coded virus in the mind that waits in the hope that there will be a moment when it can infect the mass and use its own gravity against it, or escape its gravitational pull. In this way, socioanalysis in its attempt to alleviate anxiety must amplify it, because it is anxiety that must be kept alive to counter melancholia, on the chance that there is ever a moment to act, a moment to claim freedom, much as Fredrick Douglas accomplished all those years ago. It does not create an action plan for the external world, but guards knowledge in the inner world for safekeeping.

However, reading and writing are not sufficient to this task today, as they were for sparking movements in thought then. As Lyotard reminds us, under the totalizing logic of information these tasks are considered a waste of time, we cannot force people to engage in these practices and once they enter material reality they immediately succumb to commodification, but sociology persists in using these tools as the main way of disseminating its content to the public and as such, much of its content, becomes nothing more than information resting in some archive as part of the unused code of the simulation. Socioanalysis counters this by shifting the mode to one of conversation in the analytic setting. Conversation is a skill that is eroding in these technological times, not just conversation between people, but conversation with the self (Turkle, 2011; 2015). As Sherry Turkle says, now "is not a moment to reject technology but to find ourselves" (2015, p. 362). Socioanalysis provides the setting to do just that. It still requires commitment, but it plays on the fact that has made social media so compelling, people like to talk about themselves, only they often find that no one is really listening to them. Without that listener, the conversation with the "self" leans toward the narcissistic and anti-social, a screaming into the void. In socioanalysis, the role of the listener is reclaimed.

One final aspect of our scenic landscape must be included, before we turn our attention at last to the structure of socioanalysis, as these technological transformations are accelerating and what appears as immanent changes within our material reality deserve some initial reflections, if only to warn us of the coming tide.

3   The Coming Tide: Automation, Artificial Intelligence, and Space Colonization

Technology is such a dynamic construct that when looking at the present and the lag between release and adoption, it is difficult to separate existing technology, from technology in development and the roadmaps for future technologies. Since technology points beyond itself, the next generations are always biting at our heels, and technologists, recognizing their impact on the future, are constantly engage in forecasting that future (Barrat, 2013; Brynjolfsson & McAfee, 2012; [2014] 2016; Ford, M., 2016; Kaplan, 2015; McAfee & Brynjolfsson, 2017; Ross, 2017; Schwab, 2017; Susskind & Susskind, 2015). Much as a meteorologist might look at weather patterns and historical trends to give us a weather forecast, the social scientist can also engage in a practice of forecasting based on structural patterns and historical trends within a given society. There are a great many weaknesses to this practice. Forecasts are often wrong. We plan for a day at the beach, the forecast calls for sun, but when we arrive, a thunderstorm rolls in. Or, we trace the logic of capital to its internal contradictions which create barriers to its continuation, forecast that the logic will rupture and have to be replaced by something new, but then, capital finds a way to transcend its barriers. As often as weather forecasts are wrong, "social" forecasts are infinitely more complex and accordingly have a much more dismal rate of success. So why should we pay any attention to a mass forecasting of technology?

Because the mass is the black hole of the social, to track its gravitational pull we must look to the forces which cause cyborg collisions to release their energy and maintain the dynamism of the mass. These forces are, at present, the totalizing logics of capital and information. The best technological forecasts are those that are limited to critical evaluations of how these logics interpenetrate and shape the directionality of the mass into the future; in other words, these are negative forecasts.[15] The advantage of negative forecasts is that they recognize that the totalizing logics of capital and information do more to limit possibilities because they are forms of mass domination, than they do to bring about the conditions for freedom that would unleash mass possibilities. They track the narrowing of the possible in the material conditions, rather than its expansion, because despite the ways that technology opens new possibilities, so long as they are developed in societies that are guided by these logics, then they are most likely to only have possibilities that align with these forms of mass domination. As it stands, the vast surplus reserves of capital from our collective project of global wealth extraction have concentrated in the hands

---

[15] One positive technological forecast that has shown remarkable resiliency is Moore's Law, named after Gordon Moore, who in the late 1950s was one of the founders of what would become Silicon Valley. "In the microchip, Gordon Moore glimpsed an astonishing future. Trained as an experimental chemist, he observed and then, through his work, fulfilled his prophecy for silicon transistors within these microchips: that they would double and redouble relentlessly—with ever-increasing use in an ever-proliferating array of products—even as their cost tumbled across the decades. This repeated doubling with plummeting price is known as "Moore's Law"" (Thackray, Brock, & Jones, 2015, p. 18).

of a small number of people who dictate according to their own ideas and principles—which are themselves structured by these logics—how these logics will continue to shape the mass. According to Forbes' real time tracking of the world's billionaires, as of July 28, 2020, of the top ten richest people in the world, seven of their fortunes were primarily derived from the technology industry and seven of the seven are American.[16] As of that date, these seven people held a combined net worth of $659.6 billion, a number that is constantly in flux because it is in large part based on the speculative forecasting of exactly how they will continue to advance these logics within mass society. The political economy of capital has staked the largest part of its future on technological forecasting; therefore, the future of mass society is entangled with the decisions these people, and their ilk, will make. Meanwhile, "one in five U.S. households, over 19 percent [of the population], have zero or negative net worth" (Collins & Hoxie, 2017, p. 2); even among those "who *do* have some wealth [they] often don't have any liquid assets—cash or savings—at their disposal. Over 60 percent of Americans report not having enough savings to cover a $500 emergency" (p. 6). The mass unemployment caused by the COVID-19 pandemic has only accelerated this problem, with most news sources reporting that while American billionaires have gained over half a trillion dollars in wealth during this crisis between 20 and 30 million Americans face eviction.

Since mass society is still influenced first and foremost by the power of capital, if we want to gain some level of understanding as to the directionality of society, we must place this information within a dialectic to see how movements at the top will impact movements at the bottom. As was the case for Marx, it is far easier to forecast the actions of the elites in society because they often align in a shared set of goals and values that reflect their interests, which both produce and are reproduced by the totalizing logics of mass domination. Furthermore, they have the capital and control the avenues for putting it in motion. Those at the bottom, lacking the social as a cohesive regulating guide, are all too often forced into reactionary thought that does not align in any convenient class-based way precisely because they are caught in the current made by those at the top. Lacking capital, they are forced to react and respond to the movements of the top in whatever limited fashion they can muster.

An all too common trend in sociology is to study down as a way to empower the masses to gain control over their own destiny and create an ideal and socially responsible society. But this framework depends on the notion of the

---

16   Ranking: (1) Jeff Bezos, Amazon; (2) Bill Gates, Microsoft; (4) Mark Zuckerberg, Facebook; (7) Larry Elison, Oracle; (8) Steve Ballmer (9) Larry Page, Google; (10) Sergey Brin, Google (Forbes, 2018).

human, it targets humans as if they were rational actors with fully developed individuality which presumes a functional social basis, therefore it ignores the transformation of these subjects into their cyborg identities which in a mass society leads to struggles to separate the irrational from the rational across the scaled realities of our existence. If the goal of sociology is to understand modern/postmodern societies, then studying up must become a priority. I mean this in the sense suggested by the anthropologist Laura Nader (1972), that "[s]tudying "up" as well as "down" would lead us to ask many "common sense" questions in reverse" (p. 289). That is, instead of asking questions such as why do the poor support tax cuts on the wealthy, ask why do the wealthy insist that they need tax cuts? The answer to the first question is necessarily a reactionary one, whereas the answer to the second will tell us more about the directionality and composition of modern society that the masses are reacting to. Studying up may also lead to more novel policy implications, so instead of examining the consequences of a minimum wage hike (again, a reactionary tactic that simply extends the current system) this approach would examine the consequences of a maximum wage. It could not do this without also at the same time referring to the minimum wage and placing the concepts in a dialectic; so, this is not a call to avoid studying down, but a call to only do so in a dialectical way that involves studying up at the same time.

Since there has never been a mass social safety net that guaranteed the basic needs of those at the bottom, the erosion of the barebones welfare state of the mid-20th century has increased the precarity of the laboring classes. Lacking anything other than their labor power, they must continue to find a way to sell themselves as commodities in a market that increasingly has no use for their labor and cares little if at all about their health. "According to Marx, with the development of capitalist industrial production," and here we may now add postindustrial production of information, "the creation of material wealth becomes ever-less dependent on the expenditure of direct human labor in production" (Postone, 1993, p. 356). As a result, "[t]he emergence of the possibility of a future, in which surplus production no longer must be based on the labor of an oppressed class, is, at the same time, the emergence of the possibility of a disastrous development in which the growing superfluity of labor is expressed as the growing superfluity of people" (Postone, 2015, p. 21). In other words, the masses find themselves in a position in which technology is enabling a separation of the productive industries from their material basis as the source of labor-power and, lacking the capital reserves to maintain their livelihoods without the continued sale of their labor-power, the future of the masses is necessarily one of unbridled anxiety. Their futures hang on the decisions that this elite group of capitalists make; in other words, much of the current state of anxiety is based on the actions of a relatively small group of people. There is nothing democratic about their decision making, and even the so called "democratic" modes of state governance, that the masses could appeal to, to intervene on their behalf, appear to either be inept at controlling this process or, as the data suggests, are so tightly bound to furthering the logic of capital that they too are caught in the current of those at the top, so, they are first and foremost representatives of the elites' interests (Panageotou, 2017). This is not to suggest that those elites are not also caught up in the current of this anxiety of their own making. Many of their decisions are the direct result of their being unable to see a way to pull the mass away from its current course, which they too recognize is unsustainable, and so they undermine their own basis for survival at the same time that they are undermining the basis for the mass to continue.

One of the most prominent technological means of accomplishing this transformation has been the continued development of the technological basis of automated labor. On the one hand, by technologically automating the labor process, capitalists are responding to the increased pressures of a global economy in which competition demands cutting expenses to ensure the continuation of profitability. With the global sources of raw materials now in the hands of the elite class and the costs of raw materials largely dictated and stabilized by the speculation markets, labor is one of the only areas in the production process where they can still exercise a measure of managerial control over costs. On the other hand, by undermining the purchasing power of their labor/consumer base in denying them avenues to sell their labor-power, elites are aware that there is a diminishing rate of return on their modifications to the production process. No matter how cheap a commodity is, if consumers do not have money to spend, they cannot consume the products, and since they rely on these products for survival, cutting off their access is akin to denying them the continuation of their right to exist. But this later consideration is largely deemed of secondary concern to the present pressures of increased competition. A typical mindset is to focus only on solving the first problem, and then wait to deal with the repercussions once they materialize; after all, those repercussions might also require technological fixes which will demand new rounds of innovation and development thereby extending the reach of these industries and their potential capital earnings. For Paul Virilio, this is because there is a mass denial that we are dealing with a *"political economy of speed"* ([2009] 2010, p. 24) in which "we have long lost the depth of time of the past and of long durations, this 'post-historic' wreck actually not only invalidates the future ... it also invalidates the present ... [as it is] caught up and then outstripped by purely 'accidental' history whose tragic immanence no one wants to acknowledge" (p. 59). In other words, by following this model the mass is doomed to a history at the mercy of the accident, in which, failure to consider the effects on the mass can lead to mass ejections of the material that composes the mass (i.e. people) in such a way that elites are attempting to evade responsibility by keeping open their defense that there is no rational plan behind the mass ejection, it is, after all, merely an accident. To mitigate these "accidents" several have offered leftist based solutions that see in these automation technologies a path to a "post-work future" built on "commitments to (at the very least) open borders, the abolition of spatial mechanisms of control (like prisons and ghettos), the reduction/socialization of unwaged and waged work, the bolstering of the welfare state, and the provision of a global basic income" (Srnicek & Wiliams, 2015, p. 188). But such solutions require bottom up political interventions that have no basis in the material reality that the current configuration of "democratic" governance is signaling. Rather than build a political ideology, an examination of the material trends is a first step to painting a clearer picture of the immanent landscape of mass life and the complexity of addressing these problems.

In 2013, Carl Benedikt Frey and Michael A. Osborn of the University of Oxford, presented a paper on "The Future of Employment" (2017)[17] in which they asked the question: "How susceptible are jobs to computerisation?" Of the transformations in computing technology they considered, several fields were beginning to show signs of progress that had enormous synthetic potential. Of these were the development of 'big-data',[18] algorithms, sensing technology, user interfaces, and mobile robotics. The sensor devices are building big-data on the movements tracked in the production process in different environments and with different materials, algorithms process the data and determine how mobile robots need to move and interact in these environments, and the roles of multiple actors who previously used their labor power in the production process are consolidated into the user who monitors the robots

---

17   Published in 2017.
18   "Half a century after computers entered mainstream society, the data has begun to accumulate to the point where something new and special is taking place. Not only is the world awash with more information than ever before, but that information is growing faster. The change in scale has led to a change of state. The sciences like astronomy and genomics, which first experienced the explosion in the 2000s, coined the term "big data." … There is no rigorous definition of big data … One way to think about the issue today is this: big data refers to things one can do at a large scale that cannot be done at a smaller one, to extract new forms of value, in ways that change markets, organizations, the relationships between citizens and governments, and more" (Mayer-Schönberger & Cukier, 2013, p. 6).

through these interface devices. Although there are several limitations yet to be overcome, such as superior human perception and manipulation, creativity, and social intelligence tasks ("negotiation, persuasion, and care" p. 262), they estimate that the range of applicability of automation, as it stood in 2013, puts "47% of total US employment ... in the high risk category, meaning that associated occupations are potentially automatable" (p. 265). When expanded out to a sample of 32 OECD countries, researchers found that "the median job is estimated to have 48% probability of being automated" but there is significant variation across countries depending on how their economies are structured and how difficult it will be to integrate automation technologies into those industries (Nedelkoska & Quintini, 2018, p. 45). These numbers are obviously not fixed, as not only is the economy dynamic but so too is technological advancement which interrupts itself. Furthermore, the integration of automation into the production process is bound to reveal unforeseen sets of problems that delay implementation in some arenas while succeeding more quickly in others. However, with the current pandemic situation highlighting the dangers of human labor in the global supply chain, technological development along these lines is likely to accelerate, at least in advanced industrial and manufacturing heavy economies. The somewhat apocryphal quote, often attributed to the science fiction writer William Gibson, sums this up quite well: "The future is already here—it's just not evenly distributed."

For example, Elon Musk, CEO of Tesla, bought into an undeveloped political economy of speed, saying that speed "is the ultimate weapon when it comes to innovation or production" (Mitchell, R., 2018) but his lack of paying attention to the "accident" has proved costly. In an interview with CBS, Musk admitted that he ran into severe production delays with the Tesla Model 3 because of an over-reliance on automation. He was forced to agree that "using more humans than robots would speed up production" (Wong, 2018), but investors immediately responded, and Tesla's stock plunged (it has since rebounded and surged in valuation). But while automation within the factories has run into trouble, Tesla has, along with several other transportation companies, begun to perfect the automation of driving. With the "gig-economy" jobs already disrupting the consumer transportation industry (Stanford, 2017), Uber is attempting to unleash a fleet of self-driving cars to monopolize as much of that industry as they can, thereby eliminating most of the human element. Amazon, UPS, and Domino's are also investing heavily in automated drone delivery services to cut down on transportation and labor costs (Desjardins, 2018). This is just one of many signals that there is even more potential for economic disruption in the freight industry. According to the International Transport Forum (2017): "In Europe around 3.2 million were employed as heavy truck drivers in 2015, which represents 1.5% of the employed population ... In the U.S. around 2.4 million people or 1.7% of the employed population are estimated to drive heavy trucks" (p. 29). On the low-end, their conservative estimates show that by 2040 around 25% of this population will be disrupted by automated trucks, and on the high-end by 2020 40% will be disrupted and over 90% by 2040 (p. 28). For high-skilled factory work and transportation the technology still has a way to go, but the financial incentives are high for companies to continue this R&D so that they can implement these technologies as they become available and before their competitors do, to gain, if only for a moment, a competitive advantage and boost to their profit margin.

For the automation technology that is already robust, the lag of implementation has less to do with the perfecting of the technology, and more to do with the cost-benefit analysis of using disposable and cheap human labor over investing the necessary sums to integrate these technologies into the production process. Hideo Sawada, CEO of the Japanese tourism company H.I.S., in an interview with *The Atlantic* (Semuels, 2018), said that "It takes about a year or two to get your money back. But since you can work them 24 hours a day, and they don't need vacation, eventually it's more cost effective to use the robot." Low-wage, highly-repetitive, and long-hour jobs are some of the first targets. When the "Fight for $15" campaign took off in 2012 with fast-food workers demanding a pay rate that could afford them the basic necessities to reproduce their labor-power, Andy Puzder (2014), CEO of CKE Restaurants (the parent company of Carl's Jr. and Hardee's), came out ahead of the pack and blamed government for backing a minimum wage increase with a thinly veiled threat: "That sounds nice ... assuming they're still employed after the required raise." He immediately made good on his threat and began installing self-ordering kiosks to replace the front house staff in his restaurants and has plans to continue this automation trend in the back. Wendy's followed suit in 2017 by adding automation machines "to at least 1,000 restaurants, or about 15% of its stores" (Li, 2017). When McDonald's announced a similar initiative in 2017, their "shares hit an all-time high ... as Wall Street expects sales to increase from new digital ordering kiosks that will replace cashiers in 2,500 restaurants" (Kim, 2017). These transformations are not unique to fast-food industries, it is already common-place to see these kiosks replacing human labor in grocery stores and airports, and this technology is hardly more advanced than touch-screen computers.

There is a lot of truth to the arguments presented against saving these jobs, but it is contingent on several factors. One argument in favor of automating more of these jobs is that in eliminating the potential for human failure (especially dangerous in the freight industry) there are net health benefits and a decrease in the risk of death associated with the labor. Another argument is that these jobs are rarely rewarding and do not provide a high quality of life. While those are good reasons for people not to work those jobs, the alternative of not having a job at all is much worse. "Research suggests that displaced workers report higher levels of depressive symptoms, somatization, anxiety, and the loss of psychosocial assets. The increase in reported symptoms of depression and anxiety among displaced workers compared with nondisplaced workers is roughly 15 to 30%" (Brand, 2015, p. 365). With loss of employment also comes the loss of health benefits (if and when these jobs even provide them), so coupled with mental health concerns are those of physical health concerns and the inability to afford medication and doctor's visits which compounds the stress and anxiety of the unemployed. Although historically technological disruption in the labor market has simply shifted labor to new fields, with the reach of these new technologies it appears that this is simply a stopgap measure. The proposal for a basic income is one way that scholars and activists have proposed to alleviate this anxiety, however, if such proposals do not account for the fact that this must become a function of the logic of capital, then such proposals are "far more likely to increase further the ability of economic organizations and their leaders to impose their values on everyone else" (Dahms, 2006, p. 6).[19] It would be a mistake to think that these disruptions are only a blue-collar phenomenon that will not impact professional and white-collar jobs as well. It is not just physical labor that technology is automating, with the advances made in artificial intelligence (AI), intellectual and highly skilled/trained labor is also losing ground to technological advances. Anxiety is not just related to the potential of AI to displace labor, it also has implications that extend the reach of the logic of information across all spheres of the life experience, bringing a variety of concerns about uncertainty of the future into the present experience of anxiety.

Defining AI is a difficult task. Some of the founding fathers of the discipline defined it in 1955 as the problem of "making a machine behave in ways that would be called intelligent if a human were so behaving" (McCarthy, Minsky, Rochester, & Shannon, 1955). Jerry Kaplan has criticized this definition because it assumes that we can define and accurately measure human intelligence, but when you factor in "that *how* you approach a problem is as important as *whether* you solve it" the ways of thinking about intelligence begin to multiply and cannot be reduced to standardized models such as IQ

---

19 For more critiques on the relation of basic income to the logic of capital, see also Dahms, 2015; Harvey, 2018.

(2015, p. 2). Nil J. Nilsson, another pioneer in the field, defines AI as "that activity devoted to making machines intelligent, and intelligence is that quality that enables an entity to function appropriately and with foresight in its environment" (Nilsson, 2009, p. 13). Here he removes the human factor as the baseline for intelligence and offers up a way of conceptualizing intelligence, but this definition leads to "a repeating pattern known as the "AI effect" or the "odd paradox"—AI brings a new technology into the common fold, people become accustomed to this technology, it stops being considered AI, and new technology emerges" (One Hundred Year Study on Artificial Intelligence, 2016, p. 12).[20] One of the factors that causes this phenomenon is the disconnect between how the public has been conditioned to think of AI and what AI researchers are actually doing. Science fiction novels and films often depict AI as a machine that has gained self-awareness and consciousness, thereby allowing it to engage in self-reflection over the directionality of its coding and directives without being bound by its human-implanted programing. As such these narratives amplify the hopes and fears associated with "human" behaviors. Much in the same way that the ideals of humanity were projected on to divinity, AI is supposedly inclined toward extreme forms of rationality in science fiction with it either embracing the human ideal as a benevolent overlord, or in the more likely science fiction scenario, it sees the weakness and irrationality in humanity and turns against them as a superior being that can exploit their weaknesses. If AI were to achieve self-consciousness then the only sure thing is that it would be of a different order than human consciousness, but we are likely to fail in human ways when assessing and attempting to harness this potential (Crombez & Dahms, 2015). However, for most AI researchers' achieving machine consciousness is still a secondary objective in the field; science has still hardly scratched the surface of understanding what lies behind human consciousness let alone how to replicate it or create a novel form of consciousness in a strong, or general, AI.

A more compelling and materially grounded way of evaluating what AI is today is to look at "what AI researchers do" and while what they do draws on "several different fields of study, including psychology, economics, neuroscience, biology, engineering, statistics, and linguistics," it is still "primarily ... a branch of computer science that studies the properties of intelligence by synthesizing intelligence" (One Hundred Year Study on Artificial Intelligence, 2016, pp. 13-14). An obvious omission here is sociology, and with all the societal implications of AI, sociology must step up to the plate and demonstrate its

---

20   See also McCorduck, 2004.

relevancy in this field. Within that branch of computer science, most research is focused on weak, or narrow, AI, that is, the development of artificial intelligence that has specific applications within certain fields and environments so that its success can be measured in terms of its capacity to simulate the actions of other intelligences within the restricted confines of those fields and environments.

Susskind and Susskind (2015) have appropriately critiqued the use of the word "weak" as a potential misnomer to describe this kind of AI development, as several AI systems that would fall under this category "are becoming increasingly capable and can outperform human beings, even though they do not 'think' or operate in the same way as we think we do" (p. 275). For example, in 1994 the AI program CHINOOK successfully beat "the reigning human champion" at checkers, in 1997 Deep Blue beat the world chess champion, in 2010 WATSON beat the two all-time greatest players in Jeopardy (Bostrom, 2014, pp. 15-16), and in 2016 AlphaGo beat a 9 dan ranked Go master (DeepMind Technologies, 2018). Mastering these games has allowed researchers to develop more robust algorithms for machine learning, including self-taught AI that is not merely mimicking human action but developing its own problem-solving strategies (Silver, et al., 2017). Being able to develop strategies based on greater quantities of data than humans has led to some real benefits in AI deployment. For example, "the Department of Veteran affairs is using AI to better predict medical complications and improve treatment of severe combat wounds," likewise this approach "has also reduced hospital-acquired infections at Johns Hopkins University" (Executive Office of the President, 2016a, p. 13). IBM's WATSON is not just a Jeopardy champion, but has advanced the fields of oncology, genomics, drug discovery, and more (Watson Health, 2018). DeepMind, the division of Google responsible for developing AlphaGo, is also deploying their AI solutions in healthcare, across Google's consumer platforms, and toward gaining a deeper understanding of the ethical and social implications of AI; troublingly, again, sociologists are not making any major contributions to these discussions. While none of these applications represent general AI, there is a wide net cast by narrow AI applications.

"The current consensus of the private-sector expert community, with which the NSTC Committee on Technology concurs, is that General AI will not be achievable for at least decades" (Executive Office of the President, 2016a, p. 7). In its current form AI cannot be reduced to a singular technology, but since it is a multi-faceted array of technological solutions for enhancing automation, its current form still has the potential to disrupt or improve lives throughout the domains of our life experience. From a macro-economic perspective "AI has the potential to boost labor productivity by up to 40 percent in 2035" (Purdy & Daugherty, 2016, p. 17), but since the 1970s the majority of profits caused by increased productivity have funneled to the elites, so "[t]he economic pain this causes will fall more heavily on some than on others" (Executive Office of the President, 2016b, p. 7). What industries will be the hardest hit is difficult to predict, as such, "[r]esearcher's estimates on the scale of threatened jobs over the next decade or two range from 9 to 47 percent" (p. 2). According to Goldman Sachs (2015), AI applications are currently directed toward "data-heavy sectors," such as, content streaming platforms (e.g. Spotify, Netflix, Amazon), targeted advertising, digital personal assistants (e.g. Apple's Siri, Amazon's Alexa, Microsoft's Cortana, and Google Now), and industrial use in fields like healthcare and pharmaceuticals, media and financial services, other technological R&D, factory and warehouses, elder and patient care, and transportation. The market research company Deloitte (2016) agrees in their report with these findings and extends them to include the legal profession, where Ross Intelligence has built an AI legal assistant on IBM's Watson architecture and licensed it to law firms to answer legal questions in a fraction of the time that it would take human lawyers and legal assistants.

Since the combination of AI and automation are affecting so many spheres of the economy, there is no such thing as a single policy solution to fix these negative effects or alleviate the anxiety associated with the uncertainty of these coming changes. Several policy suggestions taken in tandem have been provided as ways to counter the negative effects experienced by those who will be displaced, such as, using state budgets to guide the development of AI toward socially beneficial uses, reshaping education to correspond more directly with the world of tomorrow than with the fleeting world of today, and aiding workers in transition by taking "steps to modernize the social safety net, including exploring strengthening critical supports such as unemployment insurance, Medicaid, Supplemental Nutrition Assistance Program (SNAP), and Temporary Assistance for Needy Families (TANF), and putting in place new programs such as wage insurance and emergency aid for families in crisis" (Executive Office of the President, 2016b, pp. 26-42). Bill Gates (2017) has suggested putting a tax on robot labor as one way that governments could pay for these necessary increases in social spending. However beneficial these policies could be in alleviating the pains of the coming disruptions in the labor market, rather than boost these programs the Trump administration and Republican leadership in Congress was determined to eliminate them and the Democratic party has more often than not shown no initiative to strengthen or maintain these programs either. Virilio ([2010] 2012) saw no real progress made in material attempts to veer us away from the 'accident' model of progress, they reside as solutions on paper but lack the means of being implemented, so he took a very critical evaluation of these facts, writing that they, point clearly to a future deindustrialization of nations, as a prelude to which the current delocalization of firms in favor of low-cost employees is nothing more than an early-warning sign of the spread of automation. Tomorrow, the robotics of artificial intelligence and its 'enhanced reality' will take on the bulk of productivity in a resolutely postmodern world. In the face of this assessment of affairs and for want of a political economy of speed and not just the wealth of nations, the 'speed box' (gearbox) of technical progress will go into automatic mode and the stock market crash brought on by speculation will end, sooner or later, in the crash of all job production. The futurism of the instant requires it and will force it tomorrow on the generations to come. (pp. 19-20)

If the state is going to play any positive role in protecting citizens from these mass challenges, then it will have to make a major move away from the political economic principles that it has had on display for the last 40 plus years and such a transformation appears unlikely in the current political climate as this would require enhancing the foundations of democracy, rather than the current trends of using these technologies to push the state in a more authoritarian direction.

It is not, therefore, just economic issues that are of concern with the rollout of AI applications. The totalizing logic of information is also structuring this process in ways that extend its mode of mass domination and undermine democratic freedoms. Under the National Security Agency's (NSA) data collection program, they collected details on 151,230,968 phone calls and text messages in 2016, this number jumped over 3 times in 2017 to a total of 534,396,285 phone calls and text messages (Office of the Director of National Intelligence, 2018). Using these mass surveillance tools and other cyber weapons the global powers have enhanced their "Cold" proxy wars, which they continue to fight in the semi-periphery countries, and now engage in direct and escalating cyberwar that undermines civil liberties and democratic action, "limiting liberty in the name of security" (Rodden, 2015, p. 407). Meanwhile, Facebook, the world's largest social networking site, and Google, the world's largest advertising company, are collecting data on all their users and even in many cases their nonusers to create targeted advertising campaigns. Facebook even admitted to experimenting with emotional manipulation of their users by controlling the content on their social media feeds (Calvo, Peters, & D'Mello, 2015). Several newspaper organizations are also turning to AI to write news articles (Zhao, 2017), and while they demonstrate remarkable abilities in summarizing information, they lack the ability to transmit knowledge and meaning in to the information; in other words, they lack theory, which is a problem when the consuming public has not developed their sociological imagination.

As recent U.S. presidential elections demonstrated, media manipulation can impact democratic processes and this information can all too easily be used to extend mass domination. Surprisingly though, "AI, in an admittedly backhanded way, actually reinforces the hypothesis of human distinctiveness by calling attention to the ambiguity-resolving, incomplete, and meaning dependent features of human minds" (Wolfe, 1991, p. 1093), but the cyborg self must work to maintain these abilities as they are eroded by the logic of information. At its most extreme form, failure to develop these abilities can buttress negative stereotypes and racial profiling as evidenced in how these new technologies are providing information to law enforcement and military officials that reinforce rather than weigh the consequences of existing patterns of structural discrimination and prejudice—particularly against Black Americans—since those are the patterns that shaped the information and now continue to shape it in a feedback loop. With surveillance technologies, including facial recognition software and traceable digital footprints, our actions are highly tracked and monitored in 'meatspace' as much as in the virtual worlds, and corporations and governments are gaining increasing levels of access to our most personal and private affairs with tools that increase our levels of unfreedom. In China, this technology is being used to enslave and control the Uighurs population (Byler and Sanchez Boe, 2020). Our cyborg selves are caught up in this logic just as tightly as we are with that of capital, and the consequences of the coming revolutions in AI and automation reveal how unprepared we are as a society for what lies ahead and how devastating the consequences can be for groups who have already suffered from marginalization under the logics of capital and information.

If this sounds like a doomsday scenario and one which people could not possibly allow to happen, then in conclusion it makes sense to turn our attention back on those actors who form the elite class that are funneling their capital into making these technologies. What are they doing with their capital other than investing in these technologies? In a telling example, as these corporate CEO's continue to lobby against taxes and social welfare programs that might sooth some of the sting from the coming disruptions, Jeff Bezos, the world's richest man and CEO of Amazon, Elon Musk, the 2nd richest person and CEO and founder of Tesla, SpaceX, Neuralink, and the Boring Company, and Dennis Muilenburg, former CEO of Boeing, are reigniting a space race to colonize the Moon and Mars. Bezos has said: "The only way I can see to deploy this much financial resource is by converting my Amazon winnings into space travel," to that end he has been funneling $1 billion a year into his company Blue Origin with aims of establishing a colony on the moon (Michaels, 2018). Part of his justification is that industrialism will be less environmentally damaging if it is done away from earth. Although there is no guarantee of success, this is deemed a good use of capital, but when Seattle attempted to tax corporations to the tune of $48 million annually "to combat Seattle's homelessness and affordable housing crises," Amazon and Starbucks successfully funded a campaign to repeal this tax; these problems are deemed a bad use of capital (Stein, 2018). Musk who has repeatedly warned against AI because of the dangers he believes it represents, is steering SpaceX toward building a colony on Mars, saying "if there's a third world war we want to make sure there's enough of a seed of human civilization somewhere else to bring it back and shorten the length of the dark ages. It's important to get a self-sustaining base on Mars because it's far enough away from earth that it's more likely to survive than a moon base" (Solon, 2018). Other than the challenges of getting to Mars and creating a sustainable living environment is the issue of cost. According to Musk (2017a), using his "architecture, assuming optimization over time, we are looking at a cost per ticket of <$200,000, maybe as little as $100,000 over time, depending on how much mass a person takes" (p. 56). With 60% of American's not being able to cover a $500 expense, it's clear that this is not a venture for the masses, despite his claim that the "key is making this affordable to almost anyone who wants to go" (p. 56). Dennis Muilenburg, while very well compensated as the former CEO of Boeing, is not one of the billionaire class and his moves are reactionary to their actions. In what he senses as a potential new market based on this renewed interest in private space colonization, Muilenburg sees a potential for Boeing to land more government and corporate contracts to develop the technology. At a recent space conference, Muilenburg said, "I certainly anticipate that we're going to put the first person on Mars during my lifetime, and I'm hopeful that we'll do it in the next decade. And I'm convinced that the first person that gets to Mars is going to get there on a Boeing rocket" (Boyle, 2018). Emphasizing the difference in motivation for the project, Musk responded to this news by saying "Do it" (2017b). The billionaire class appear to have taken Virilio's warning to heart: "Those who are looking, waiting for the revolution, have chosen the wrong planet" (2009, p. 43). Their project is fueled by the uncertainty they have over the future, which is great enough to self-fund this project, and they are looking out to new futures in space. They are pulling capital in their wake as others who are still focused on this planet in the present react and respond as they sense new arenas to continue the circulation of capital, not sensing just how critical the situation is and how in their acceleration of capital it has become a glutton of life and is extinguishing its energy supply. The lesson is that the billionaire class not only buy into the doomsday scenarios, but rather than attempt to prevent them they are finding avenues to escape, leaving the masses to deal with the collective mess we have made on planet earth.[21]

If this is the material condition that we face in our near future, then critical socioanalysis is needed more than ever, if not to dam the coming tide then at least to open our eyes to the reality of what is coming and how it will shape our lives. While everyone needs to feel the anxiety of these transformations and raise the tensions and contradictions of mass society to their conscious register, these transformations suggest two potential areas for the development of critical socioanalysis that must be considered. The first is that as AI continues to advance its capabilities, it is performing a kind of positivist socioanalysis on us—sorting through information about the world we have built and attempting to find a rational ordering of it—we must respond by performing a critical socioanalysis on AI, and if possible, developing ways to teach it to do the same on us, so that when it starts to take a bigger role in shaping policy it will account for the importance of our species to make new meaning out of our future lives. Second, if space colonization is in our immediate future then the cyborgs who live on those colonies will be in an entirely artificial environment that is decidedly different than that of mass society on earth. Their selves will be shaped by that environment just as they shape it. Critical socioanalysis will be a necessary practice if they are to understand the consequences of the co-construction of their self and society under those new conditions. Since we will be entering uncharted territory for the species, anxiety over these changes will demand a scientific accounting of this new kind of lifeworld. As such, those who are shaping the development of technology along these lines would be best suited to begin critical socioanalysis now to understand the difference between the world in which they thought of these ideas and the world that they are creating with them, if they are to understand what contradictions lie ahead and what levels of control they can exercise to shape the future in ways that more appropriately align with the reasonable demand of a better life for all.

21   It is worth pointing out that Musk has publicly voiced support for a tax-funded universal basic income as a way to mitigate some of these anxieties, but again, this is at best a stop-gap measure.

CHAPTER 6

Critical Methods 3

1   Critical Socioanalysis: Setting Up

The first step in becoming a critical socioanalyst is developing a deep and rigorous understanding of the material and historical co-construction of self and society. The way to accomplish this is through training in the critical method. This book serves only as a primer to this training. It is not and cannot be a replacement for the socioanalyst's own process of working through and appropriating the critical method from the early classics of modern theory (Marx, Durkheim, Weber, and Freud), the first generation of the Frankfurt School (Horkheimer, Marcuse, and Adorno), the French theorists of what I believe we are justified in calling postmodern society (Lyotard, Baudrillard, and Virilio), and what we will cover in this final section, the Neo-Freudian theories of Jacques Lacan that build on the critical method to revitalize the critical nature of psychoanalysis. Furthermore, this is not an exhaustive list. What it does provide is the basis for understanding the relation of anxiety to the symptoms critical socioanalysis targets: alienation, anomie, the Protestant ethic, repression, instrumental and technical rationality, one-dimensional thought, identity thinking, loss of meaning, the accident, and massification. As new symptoms are bound to emerge—as anxiety cannot be overcome but must be continuously worked through—the critical socioanalyst should never assume that they have a complete knowledge base. As different substantive areas of critical theory interrogate different aspects of existence, they should be incorporated according to the substantive interests of the analyst and the kinds of people who they wish to see as their analysands.

The "core" writers discussed here looked at the macro-trends in society, which form one of the hypotheses of socioanalysis, that mass society penetrates and shapes all cyborg selves with its totalizing logics. Sitting for analysis will expose the contradictions in those logics that produce the symptoms we suffer from. But how these symptoms manifest will vary with the life of the analysand as revealed in their analysis. The socioanalyst can prepare for these variations by supplementing their practice with critical theories of gender, sexuality, race, ethnicity, post-colonialism, religion, globalization, migration, crime and deviance, culture, media, and more, as well as new critical theories that will emerge to explain new features of our lived experience. The key to remember is that most of these theories are only supplementary, they are not replacements for understanding the totalizing logics because their object is not the whole but rather a part of that whole, so, if they are to be used to supplement critical socioanalysis then they must be understood in their relation to how those logics warp the reality that they are trying to illuminate.

The second step is to perform a critical socioanalysis of one's own self. The analyst must understand how their own constitution is shaped by the societies in which they developed their version of the cyborg self. Once critical socioanalysis has a more solid foundation as a practice in the material world, this should be performed under the analytic framework and guidance of someone who has already completed their own socioanalysis and has a full familiarity with the structural patterns that emerge in the discourse of narrating one's self and society. As this is a novel practice, those who adopt the practice at this early stage must struggle through the work of doing this on their own (see Crombez, *forthcoming*, as an example); it is an admittedly poor alternative, but every new practice needs those who take the first steps into the uncharted. Baudrillard's series of five books under the title *Cool Memories* ([1987] 1990; [1990] 1996; [1995] 1997; [2000] 2003; [2005] 2006), which take the form of a theoretical diary of sorts, is one of the best examples I have found of what this process might entail in a partial and fragmented form. It is only by performing this critical self-analysis, in which one performs a rigorous examination of the contradictory structures that shaped their identity, that the "truth" of the process of socioanalysis will reveal itself to the practitioner (as if in a flash).

As Marx's famous dictum stated, the point is not only to interpret but to change the world. The socioanalyst is an interpreter, but they perform this function on the grounds that once one has worked through interpreting critical theory and its relationship to their material existence, if it has been done successfully, then they cannot remain who they were prior to the process: it will change and transform them. In sticking with the hope of changing the world, critical socioanalysis keeps the hope alive in those who are changed by its effects, through a transference of the desire of the analyst to the analysand in the analytic session. There is no set amount of time that it takes to accomplish this task, therefore, the duration of analysis will vary with the complexity of the self and its integration in society with each analysand. In producing critical theory and in transferring the desire that it contains, these dual functions are the praxis of critical socioanalysis as it awaits the arrival of a new material reality.

After the critical socioanalyst has gathered their tools and gone through their own critical socioanalysis, then they are ready to perform the role of the analyst. The socio-therapeutic setting should be private, safe, quiet, and comfortable to put the analysand at ease to speak and begin their self-analysis.

Anonymity and confidentiality should be maintained to the extent that the identity of the analysand should be sufficiently masked when writing up case studies. Analysis can be a painful process, but for it to work the analysand must be confident that they are able to speak freely and free from the judgement of the socioanalyst; the setting of socioanalysis is therefore reserved as a special space and time for this activity and must be treated with the respect that such a personal conversation with the self deserves. The role of the analyst is to listen, and to encourage the analysand to explore and develop themes that arise in the analysis. When the analysand touches a thread in their narrative which the analyst recognizes as part of the structural logics, they should be encouraged to tug on and develop that thread to raise the object cause of their anxiety to the fore. Then by exploring their narrative around that object, the symptoms arising from the contradictions between the concept of the self and the concept of society will begin to manifest themselves to the analysand. Any set time limit for the analytic session is an arbitrary marker based on the logic of capital and its obsession with clock time, but for pragmatic reasons and for the necessary reflection post-analysis, it is necessary to set limitations on the length of sessions as appropriate to the place one is at in their analysis. Lacan introduced the variable length session in psychoanalysis as a way to provide punctuation on what the analysand says, so rather than letting them talk until the allotted time is up, it is better to let the session run until there is a point in their narrative that deserves punctuation and reflection. Whether that takes ten minutes or an hour, by punctuating their narrative at appropriate places it will help to focus the analysis by putting emphasis on what the analysand must think through and reflect on more fully.

Since analysis is based on a practice of conversation, the way that the analyst locates the appropriate threads is by listening for signal clues as to which discourse the analysand is speaking through at the given moment. There are two sets of overlapping discourses that the analyst must carefully listen for. The first is that of the psychological realm, the ways that the "self" responds to society to assert the primacy of the self over the society in which they are shaped. The second is that of the sociological realm, the ways that the "self" responds to society by giving it primacy over the self. In both sets of discourses, the dialectical relationship between the self and society is obscured by the contradiction of thinking that there is a harmonious self or a harmonious society, when the reality is that there are neither. Pinpointing where those contradictions emerge and punctuating those discourses when they appear, will help the analysand to place the two in tension and recognize where and when they must develop their critical desire to motivate them to go deeper by passing through their anxiety. Once they develop this desire to understand the contradictions, and can locate and deconstruct them, then we might say that we have achieved the goal of analysis. Only by achieving this alignment and the desire to engage with our own contradictory thoughts will the effects of living in a contradictory world be understood by the analysand thereby giving them some measure of control and acceptance of their anxiety, so that they can locate its object, and keep their anxiety at the ready for their inevitable confrontations with these contradictions in the future.

The first set of discourses are oriented to the psyche and were diagramed by the French psychoanalyst, Jacques Lacan. I have constructed the second set based on his model by diagraming the totalizing logics of capital and information and the cyborg self as the discourses of mass society. Since it is common for us to code switch (shifting between the language games we know) in our narratives and lose track of the fact that what is rational in one discourse is irrational in another, critical socioanalysis tracks movements in both sets of discourses, while ultimately being primarily concerned with the symptoms produced in the psyche by mass society, rather than those projected from the psyche onto society. If the socioanalyst determines that the analysand is suffering to a high degree from the latter to the extent that they cannot see the former, then they should recommend that the analysand engage in psychoanalysis first and only after working through their psyche should they then move on to socioanalysis.

2   Discourses of the Psyche and the Self: A Lacanian Framework

In the years following Freud's death, mainstream psychoanalysis began to stray from its critical roots and, in the attempt to justify its place among the positivist sciences, there was a move away from the unconscious. The unconscious, like the social or the mass, cannot be directly observed and studied in the same way that science treats most of its objects. They are the critical hypotheses upon which the theories and practices of psychoanalysis, sociology, and socioanalysis rest. Since they cannot be observed directly but only through the effects they produce, they do not align very well with the positivist sciences that insist the only good theory is falsifiable theory. The critical stance is that the persistence of the hypothesis depends on whether the theories derived from it can provide better explanatory power to observed phenomena than some alternative. The hypothesis is not to be reified but in its continuously being subjected to dialectical thought, the critical method reveals the movement of the concept, and if movement in the concept is no longer possible, but society's dynamism persists, then the concept no longer contains explanatory power and must be abandoned. In the case of sociology, the social has lost much of its explanatory power. Material reality is not moving in a way that there can be a corresponding movement in the concept. Recognizing this through their use of the critical method, the Frankfurt School tracked its diffusion into mass society and Baudrillard developed the hypothesis of the mass, as a new concept with a more robust explanatory power of the observable dynamism of the material world. In psychoanalysis, although the unconscious did not lose its explanatory power, the push toward positivism as the legitimate mode of science made psychoanalysts increasingly hesitant to base their practice on an object that was not empirically falsifiable. As previously mentioned, this led to a situation in which many psychoanalysts began to downplay the unconscious in favor of ego, individual, and object-relations models. Lacan ([1966] 2006), however, viewed this as the abandonment of "Freud's discovery [that] calls truth into question"; this discovery being the "unconscious" (pp. 337–338). As such, Lacan, in a similar fashion as the Frankfurt School, began to call into question the "truth" of the positivistic view of Science (with a capital S) that the mainstream sciences were claiming as the only legitimate mode of science.

By refusing to be party to this shift away from the critical side of psychoanalysis, by introducing the variable length session, and by writing in such a complex manner that it was condemned as either nonsensical or elitist, in 1953, Lacan's ([1973] 1998) "teaching ... [was] the object of censure" (p. 3). The perception was that his training of analysts was methodologically incompatible with the direction those in authority deemed worthy of the psychoanalytic title under the positivist conception of science. By refusing to conform to an authoritarian rule-based system of analytic training, Lacan's blasphemy ultimately led to his "major excommunication" (p. 3) from the International Psycho-analytic Association.[1] This excommunication, however, motivated Lacan to make a revolutionary return to Freud. He accomplished this, on the one hand, by a dialectical rereading of Freud which accounted for the changed socio-historical context of post-War France. And on the other hand, by synthesizing Freud's thought with new developments in the linguistic theories of Ferdinand de Saussure ([1972] 2009), the structural anthropology of Claude Lévi-Strauss (1963), and in some ways the economic structures of Marx (Tomšič, 2015). "Lacan inaugurates his return to Freud" (Gasperoni, 1996, p. 77) by demonstrating the synthetic potential of Freud's major discovery, the unconscious, and advancing it with his own major contribution: that "the unconscious, which tells the truth about truth, is structured like a language"

---

1 See also Miel, 1966.

(Lacan, [1966] 2006, p. 737). This discovery in turn justified the variable length session as a form of linguistic punctuation, or "scansion," as a way of parsing the unconscious and drawing attention to its "key"-words by stopping sessions at just the right moment. By developing this new theory of the unconscious to its full potential, Lacan's return to Freud reclaimed his first four-legged structure of psychoanalysis, "introduced by Freud as fundamental concepts, namely the unconscious, repetition, the transference and the drive" ([1973] 1998, p. 12).

In rejuvenating these Freudian concepts, Lacan places a question at the center of their development that concerns and preempts our turn to the discourses, "namely—what is the analyst's desire?" (p. 9). The analyst's desire assumes a central role within the development of the fundamental concepts because, for Lacan, "it is ultimately the analyst's desire that operates in psychoanalysis" ([1966] 2006, p. 724). "Desire springs from lack" (Fink, 1997, p. 44) making the two coextensive; in other words, if the subject lacks nothing then they can want for nothing, but it is the initial realization of the subject's incompleteness, translated as lack, that causes desire to spring forth and grow alongside the lack, thereby instituting their dialectical relationship. The lacking subject, or barred subject, is illustrated by the algebraic notation, or in Lacan's terms matheme, of an S (as a complete subject, but ultimately only an imaginary one) crossed by a bar / (which pertains to Saussure's placement of the bar between the signifier/signified, understood by Lacan as representative of the fundamental lack, which places the lacking subject in the real) giving us the symbolic notation $.[2] As alienated subjects we are lacking subjects, and this lack is what triggers our desire. When Lacan refers to the analyst's desire, though, he does not mean the desire that the analyst has as a lacking subject for their own subjective and personal gratifications. Rather this desire takes on a "social" dimension, in that it is projected outward and refers to the desire proper to the analyst as an analyst; that is, as the analyst who occupies one of the two needed subject positions (analyst and analysand) within the analytic process for it to occur. Since it is not only the subject who lacks, but it is also society that lacks, the analyst creates an artificial space in analysis in which they fill the role of the lacking society as if it was aware of its own desire. This is the desire that the analyst represents. This desire "is the desire of the Other" (Lacan, [2004] 2014, p. 22), or what Lacan calls the Big Other, the symbolic representation of what we may call the social (or the mass), that force beyond our self that we want recognition and acceptance from. In the analytic session

---

2   Due to font limitations I have substituted the barred S, which is the letter S crossed by /, with the $ as the closest approximation in this text.

the analyst serves as a representational stand-in of the Big Other, providing the framework in analysis for the analysand to confront their desire within this relation to the mass.

Lacan tells us that anxiety is what lies between our desire and our jouissance, or in other words, anxiety is the median point between what we desire and the pleasure of satiating that desire ([2004] 2014).[3] "[T]he real, an irreducible pattern by which this real presents itself as experience, is what anxiety signals" (p. 160). Since the real is marked by lack, it initiates desire, and desire is what fuels our quest for pleasure. However, we experience jouissance, which is not a pure form of pleasure. Our lack cannot be filled by what this society has to offer, any more than it can be filled by the analyst, so our desire can never be satiated, when this is realized the pleasure simultaneously produces pain, and in their union, we experience jouissance. As our desire circles our lack and gets ever closer to it, anxiety is aroused the closer our desire gets to the reality of our lack; the object of anxiety is a negative object, an objective lack which is always marked by uncertainty because no identifiable objects in the real align in a sufficient manner to fill the void caused by that lack, but desire is what forces us to keep trying to fill it. Since we are alienated subjects who in the current material circumstances cannot overcome our alienation, we cannot overcome our anxiety either, and the closer we get to identifying what it is that we lack—the object of our anxiety—the more we will be in the anxiety phase. Anxiety is therefore not only a warning of uncertainty, but an affect that signals that we are a lacking subject, and the more aware we are of that lack, the more anxiety we have. As capital attempts to fill in that lack with consumerist goods and other distractions, we get pleasure as they play on our desire, but the pleasure is fleeting because the lack cannot be filled and each time we attempt to fill it we pass through anxiety as the pleasure evaporates faster and faster with the increased speed of capital and the demands that logic places on our lives. Likewise, in the realm of information there is pleasure in its circulation, but as we engage in the endless circulation of information (on the internet and social media, or in gambling, for example) the pleasure is harder and harder to come by and eventually it turns into a painful addiction. The desire of the analyst is therefore the desire to recognize what this anxiety is and why, despite the pain it causes, we continue to pass through it, as if we truly enjoyed our symptom.

This specific kind of desire is necessary because the analytic process must be sustained against the resistance of the analysand. Although there is great pressure to conform to the ways that society constructs the self, and it is impossible

---

3   See also Harari, 2001.

to not conform as our survival depends on it to a large extent, this society also conditions that self to evade responsibility for its own actions, and for the self and society to be co-constructed, it means that the self is partially complicit in this process. Once this is raised to the conscious level, it is common for there to be mass repression and avoidance of confronting this truth about ourselves, so, the desire to continue must be maintained by the analyst when the analysand begins to shirk away from this pain. One way to cultivate this desire is, paradoxically, by revealing the lack in the subject. Lacan's difficult texts serve this precise function. In their complexity they reveal what the subject lacks is understanding, and if they want to get the pleasure of understanding, then they must cultivate their desire by working through that which they lack, only then does the riddle resolve itself in a flash, dissolve, and reconstitute itself in a new form. As the analysand confronts what is repressed and begins to pass through anxiety, the analyst must find ways to acknowledge and intervene as a way of dealing with "the unconscious [which] is the fact that being, by speaking, enjoys, and ... wants to know nothing more about it" (Lacan, [1975] 1999, pp. 104–5). In other words, the analysand is driven by an unconscious desire for nothing more than the supposed pleasure they assume awaits them at the end of the analytic process, but as the unconscious speaks in the sessions it does not primarily produce pleasure, pure and simple. Rather in speaking, the unconscious begins to reveal uncomfortable truths, which, due to the non-pleasure they produce, were repressed in the first place. By bringing them out into the open the analysand is in direct conflict with these repressive forces and without the analyst operating as the object cause of desire (object a, reduced in the discourses to the matheme, a)—that is, as the object that causes the analysand to desire, in spite of and against their resistance to continue the analysis—the analytic situation cannot sustain itself. This desire then, "is not a pure desire. It is a desire to obtain absolute difference" (Lacan, [1973] 1998, p. 276), a blasphemous desire to intervene in the analysand's unconscious, by becoming the cause of the analysand's desire for the analytic process to continue. When the transference of this desire is successful, then the analysand will want to confront their non-identity, that is, what is non-identical in their notion of their self and in the conceptual language they are structured into using as representative of who they are. The way to track these movements is through the use of a structural cartography of the possible discourses that the analysand will use as they speak.

Jacques-Alain Miller (1988) says of these algebraic symbols (see Figure 1), provided by Lacan for this purpose, that they are "proof of the way in which Lacan simplified theoretical questions," using them as short hand for complex concepts that are elaborated on in some places and condensed in others so as

S_1	the master signifier
S_2	knowledge; all-knowing; battery of signifiers
$	the barrred subject; lacking subject
a	object cause fo desire; surplus value/surplus jouissance

FIGURE 1   Key to mathemes (psychoanalysis)

(Discourse of the university) U $$\frac{S_2}{S_1} \rightarrow \frac{a}{$}$$	(Discourse of the master) M $$\frac{S_1}{$} \rightarrow \frac{S_2}{a}$$
(Discourse of the hysteric) H $$\frac{$}{a} \rightarrow \frac{S_1}{S_2}$$	(Discourse of the analyst) A $$\frac{a}{S_2} \rightarrow \frac{$}{S_1}$$

FIGURE 2   The four discourses (psychoanalysis)

to more easily understand their relations. In the following they are placed into schemas that trace four discourses that can be used to interpret analysis (see Figure 2). However, given that Lacan rarely left an idea alone after its initial development—revisiting concepts and expanding on them in his seminars and écrits—there are multiple ways of working-through each one. The route through this work, provided below, is only one such path and primarily follows Bruce Fink's (1995) outline, which is focused on clarity and brevity to help us think about what we are looking for in the socioanalytic sessions, rather than an exhaustive accounting of all possible reads. Each discourse speaks through the position of the agent, which is conceived of in relation to its other or what it sets to work, produces something, and ultimately has a hidden truth behind the discourse that must be revealed (see Figure 3).

Position 1: Uppper left **Agent**	*Position 2: Uppper right* **Work/Other**
Position 3: Bottom left **Truth**	*Position 4: Bottom right* **Production**

FIGURE 3   Schema for translating the discourses

In addition to the two mathemes introduced above—the barred subject ($) and the object cause of desire (a)—Lacan includes S_1 and S_2 to round out the structure of his four discourses. The S_1 represents the Master signifier: it is an imaginary position that inscribes and ushers the subject into the chain of signifiers and acts as "the signifier, the signifier function, that the essence of the master relies upon" (Lacan, [1991] 2007, p. 20). The master signifier, developed theoretically by Hegel, is our inner authoritarian voice and can take many forms—for example: the subject's name, race, gender or sexuality; or perhaps, profession, religious affiliation, disease, or hobby, as any primary referent that the subject attaches to and uses to anchor their sense of self, whether positively or negatively—but ultimately, whatever form the master signifier takes, it is only ever an artificially imposed regime that the subject attaches herself to as an anchor in the signifying chain. Frequently, it is only through analysis that the subject comes to the "truth" of the S_1, that its supposed anchor is not located in a concrete real that fixes its location but holds its position only on an imaginary foundation to which the subject prioritizes the investment of their mental energy and through which they structure the symbolic understanding of their self. The S_2, on the other hand, represents the "battery of signifiers" (p. 13) that is "not knowledge of everything … but all-knowing … as what is affirmed as being nothing other than knowledge, which in ordinary language is called bureaucracy" (p. 31). The invitation, then, is to consider S_2 as representative of knowledge proper to the system that lays claim to its production, pointing not to the hypothetical set of all Knowledge but to the set of known meanings in the system of knowledge that oversees its expansion. In other words, it represents the all-knowing, not of an external and cosmic pool of knowledge that enables universal omnipotence, but that which is internal to the modern system, making judgements as to inclusion and exclusion of its norms and values, as executed by its bureaucracies.

Lacan's discourses are historicized by their circulation around the domination of the discourse of the master, with its roots in Hegel's dialectic of lordship and bondage in the *Phenomenology* ([1807] 1977). Lacan's usage of the terms master and slave to describe this relationship is due to the vast influence of Alexandre Kojève's ([1947] 1980) famous lectures on Hegel that he delivered at the Sorbonne in Paris during the 1930's, which were attended by many influential French thinkers, including Lacan. In each of the discourses the dominant spot is in the upper left corner, the spot of the agent of the discourse, which in the master's discourse—historically the most dominant of the discourses—Lacan represents the master as S_1. The master, like the S_1, occupies a position of signification for no reason beyond its own authority as an agent; that is, because the master says something is so, and believes her power of self-identification, as the "I" that speaks, to be absolute, she acts as if her proclamations are representative of the real because she says they are. In the position of the work/other, in the master's discourse, is the slave, S_2, that representation of actual knowledge, as that which is the other to the master and the object of the master's work.

The master commands and the slave acts, but in acting it is the slave that possesses the knowledge as to how to act, a knowledge that eludes the master (who thinks they have command of absolute knowledge) and of which the master does not want to know (because it would rupture their claim to absolute knowledge). The master is content with the position of commanding knowledge, while masking ignorance, and "the a [in the bottom right, the position of what is produced in the discourse] is precisely identifiable with what the thought of a worker, Marx's, produced, namely what was, symbolically and really, the function of surplus value" (Lacan, [1991] 2007, p. 44), or surplus jouissance, as the desire that is retained by the master from the product of the slave's actions. Truth for the master is thus not in the realm of knowledge or as the earned fruit of one's own labor, rather the truth is that in remaining ignorant to the real, the master is, just as the slave, a lacking subject (in the bottom left corner, in the position of the truth of the discourse, represented as $). To speak then, from the position of the master, is to position oneself in

> the discourse of self-identity and the control of others, which institutes the dominance of the master signifier, S_1, thereby organizing the field of knowledge, S_2, into conformity with the values promoted by this master signifier. At the same time, mastery conceals subjective division [i.e. alienation], $, while generating a fantasmatic object, a, as its by-product. (Boucher, 2006, p. 275)

From this series of relations, we get the schema for understanding how to translate the remaining positions that the mathemes assume in each discourse as they make their quarter turn, shifting from one algorithmic configuration to the next.

In the discourse of the university, the dominant position is held by S_2, insofar as the university assumes the position of speaking as the agent of knowledge in modern/postmodern societies, as the rationalized justification of why things are the way they are, within the bureaucratic system. Lacan says that, "S_2 occupies the dominant place in that it is this place of the order, the command, the commandment, this place initially held by the master, that knowledge has come to occupy" ([1991] 2007, p. 104) in modernity. Because the university is bureaucratized by the master, it comes to speak from the position of bureaucratic knowledge in order to justify the actions taken by the master as a means of legitimation; in other words, it is not a discourse that de facto speaks facts, it is a discourse that rationalizes, by any means, through the claim that it speaks from the position of knowledge.

This is why the truth of the university discourse is the master signifier, S_1. Hiding behind the university's knowledge are the masters of the modern world as the drivers of the military-industrial-university complex that came to legitimate the modern system in the 20th century. This knowledge is addressed to and works on the other, as a, who as the mass of desiring subjects is reduced to the status of an object that must be dealt with and structured along the lines, not of individual desires, but of the modern system's bureaucratic desires that it constructs within the subjects of capital. The modern system comes to rely on the university discourse to confront the situation of a mass that, for a variety of reasons, is not simply satisfied with the master's proclamation that what she says is the case. However, by allowing the discourse of the university to work on the masses and structure their desire according to the master's ends, the product of this legitimating system can only be the barred, or alienated, subject, $. Max Weber (1978) came to the same conclusion when he explained that as a result of the bureaucratization of the university,

> through the concentration of such means in the hands of the privileged head [S_1] of the institute [S_2] the mass of researchers and instructors are separated from their "means of production [a],"—[that is, they are unable to produce pure knowledge as a product of their own desire, but the system works off of the fact of that desire, warping it to its own ends so as to produce that which rationalizes and legitimizes its own function]—in the same way that workers are separated from theirs by the capitalist enterprises [with $ as the result]. (p. 983)

It is important to note, however, that although the university is the primary location of scientific practice, science does not, by necessity, speak from the university's discourse. Critical scientific practices account for this contradiction and resist being inscribed into this discourse.

Philosophy, for Lacan, has always functioned in the service of the master, but it is not altogether clear that science, likewise, is condemned to the master's service. From within the university, at the level that it functions as the reproduction of the labor force, science frequently acts from the standpoint of the university's discourse, placing the student in the position of a, the mass desiring knowledge. By primarily fixing the financial burden of a university education on the student—inscribing them with a taste for a role that is potentially unstable within the economic structure—their desire as a subject is split by the knowledge of a science that sets them to work and disciplines their desires into alignment with market forces by instilling them into a system that is dictated by the demands of capital. Desires for fulfillment of the psychosocial self are thus replaced with structurally approved desires that are individualized around financial stability, consumerism, and the master's acknowledgement of the obedient servant in the form of good grades and praise, followed by, promotions and raises. This reproductive function of the university holds that "knowledge is not so much an end in itself as that which justifies the academic's very existence and activity" (Fink, 1995, p. 133). However, this functional task of science is not altogether the same as the scientific discourse that functions as the output of its research methods, indicating that science, too, can shift from speaking in one mode of discourse to another. So while Lacan ([1991] 2007) originally says that the discourse of the university is "what guarantees the discourse of science" (p. 104), which made sense given the student uprising in '68 and the troubled position that the university occupied in French society as it was becoming more bureaucratized, he eventually concluded, from the stand point of what is called "Scientific" research, "that scientific discourse and the hysteric's discourse have almost the same structure" ([1974] 1990, p. 19).

The hysteric's discourse begins, in the dominant position, with the split subject, $, as an agent who is acutely aware of their alienation. Although there are other psychodynamic positions in a psychoanalytic framework—i.e. the psychotic, the neurotic, and the pervert—it is the hysteric that is ascribed a special position in Lacan's discourses due to its relation to the production of knowledge. The hysteric works on the master signifier, S_1, that it has recognized as its other, calling that position, which commands knowledge for no other reason than its privileged position in the hierarchy of dominance, into question. By interrogating the master signifier and forcing it to do the work of situating itself in the chain of signifiers, the hysteric produces knowledge, S_2; as in the analytic session, where the hysteric challenges their ego-identity as a manifestation of purely conscious knowledge by producing unconscious knowledge, allowing them to demonstrate and work-through their self-contradictions; often by pushing them to their limits.

The hysteric is thus in a position that calls the master signifier out for being a false idol, an imaginary position of fictive power, that only claims a unified identity for the purpose of extracting pleasure from the slave's production. Lacan ([1991] 2007) says that "the hysteric's discourse reveals the master's discourse's relation to jouissance, in the sense that in it knowledge occupies the place of jouissance" (p. 94). In other words, the hysteric is the one who understands that the master's discourse is one that is attempting to hide the fact that they too are a lacking subject and alienated; meaning that the hysteric is not searching for a master signifier to cover up their alienation and provide them with that form of pleasure. Rather, the hysteric, unlike the master, is not afraid of being unmasked by knowledge: the "hysteric gets off on knowledge" (Fink, 1995, p. 133), accepting that knowledge brings both pleasure and pain. Here we see a parallel to the discourse of the scientist, who as a researcher, similarly is fueled by the pleasure of knowledge, even when that knowledge does not conform to their preconceived notions or hypotheses.

Since desire is related to lack, and since no natural or artificial interventions have yet demonstrated the ability to fill that lack, or dam it up completely, desire presents us with a real contradiction. Or perhaps it is a paradox, in that the desiring subject is always looking to fill the lack that bars them with objects that do anything to hide, but never manage to destroy, lack itself, thus doubling back in on and re-emphasizing the chasm made from lack and subsequently passing through anxiety time and again. By placing the object of desire, a, in the position of truth for the hysteric's discourse, Lacan is inscribing this position in the real. That is, unlike the master, who is anchored in an imaginary wholeness and whose truth is that they are incomplete and alienated, the hysteric accepts this position of the lacking subject as the real starting point; they know their lack, and the truth beneath that lack is that they desire something more than what the master can offer. In this way, the hysteric's discourse shows remarkable similarities to scientific discourse, in that:

> Hysterics, like good scientists, do not set out to desperately explain everything with the knowledge they already have ... [they accept that there is] something that it is impossible for us to know, a kind of conceptual anomaly ... [So science] does not set out to carefully cover over paradoxes and contradictions, in an attempt to prove that theory is nowhere lacking—that it works in every instance—but rather to take such paradoxes and contradictions as far as they can go.
> LACAN, [1991] 2007, pp. 134–5

To speak from the hysteric's discourse, then, is to speak as an alienated, or split, subject ($), who in accepting the reality of their condition, in truth desires (a) knowledge of that condition. Knowing that knowledge is necessarily incomplete, the hysteric interrogates any ideology that claims to have the answers (S_1), and by uncovering these false perspectives, they produce, unlike the other discourses, actual knowledge (S_2) limited and fragmented as it may be. This knowledge, or product, is jouissance for the hysteric, in that after the production, the knowledge is greater than before—causing pleasure—but complete knowledge is still lacking—causing pain—and yet, by remaining incomplete it wholly justifies the continuation of the discourse.

Of Lacan's four discourses, the final one, the analyst's discourse, is that which situates and executes the desire to intervene, the desire to set hysterics to work in challenging the master narrative supplied by society. It is in this discourse that Lacan provides his intervention in psychoanalysis, by providing the framework that is both descriptive and prescriptive for a mode of discourse proper to the psychoanalyst. The analyst's discourse challenges the mode of discourse that is attributed to the analyst by the analysand (the master's discourse), and also the one that many mainstream analysts mistakenly position themselves in (the university's discourse), which I will address in this order.

The agent in the analyst's discourse is the object cause of desire (a) which is a strange role to assume from a traditionally mainstream conception of psychoanalysis. The analysand seeks out the analyst, because "the analyst is the one who is given the function of the subject supposed to know" (Lacan, [1991] 2007, p. 38). That is, the analysand operates initially on the assumption that it is the psychoanalyst, who by means of the master signifier ("psychoanalyst"), occupies the position of commanding knowledge in the analytic session. This is why the master's discourse is the inverse of the analyst's discourse, the other side of psychoanalysis, because the analysand enters analysis by projecting the status of master on the analyst, believing the analyst to be in a position that will structure and organize the analysand's knowledge.

However, the truth of the analyst's discourse is a knowledge (S_2) that is split between the participants in the analytic process. On the one hand, the analyst only has the procedural knowledge (or bureaucratic knowledge) of psychoanalysis, of its theories and methods; in this sense, the analyst is the subject supposed to know. But this technical, bureaucratic knowledge, is not enough for a successful analysis. Analysis, on the other hand, can only function on the knowledge of the analysand's unconscious, which is only produced in the session through the gradual bubbling forth of those places where and when the unconscious speaks. In this way, the analyst can only learn the real nature of the analysand's lack, if the analysand can produce knowledge about it, which is why in both this model of psychoanalysis and in socioanalysis it is incumbent on the analysand to define the object of their anxiety as they work through it in analysis.

The analyst thereby sets the barred subject ($) to work, not by telling them the solution to their problems—which would only place the analyst in the position of the master and maintain the analysand in a position of dependence—but by temporarily standing in for desire by placing themselves in the analytic position of one who intervenes and sustains the analytic process. Analysts who speak from the master's discourse, do not "cure" the analysand, rather they only create a situation of dependency, but this confused position within the discourse is often a necessary starting point in the analysand's mind in order to get them to seek out psychoanalytic help in the first place. For Lacan, however, the analyst, in essence, becomes a placeholder for the desire of the analysand, who in speaking comes to realize that the analyst actually does not know them as a subject at all, but only has ideas about how the mass society has structured them as a subject. Despite all the informational content the analyst might have about the analysand, they come to recognize that at their core, they are non-identical to that informational concept of their self. It is only in working-through their own unconscious, as source material, that they come to the realization that they are actually the subject who knows (or does not know) and that they must take an active role in structuring their own self within the confines of mass society.

In this way the product of analysis is the revealing of new master signifiers (S_1) buried in the unconscious that the analysand was unaware of, and by bringing them to light it "grinds the patient's discourse to a halt" (Fink, 1995, p. 135) because it reveals the painful truth that the analysand's self is structured by things which they do not want to admit have power over them (i.e. mass society). The analyst fulfills their function when they help the analysand—with a word, a rephrasing, a shift in stress or emphasis, or some other form of verbal punctuation—to place these master signifiers back into the chain of signification, thus placing them in a dialectic so that the discourse may continue its movements. When the analyst intervenes and breaks up the discourse, it is to punctuate what the analysand has just said, to place the words in a different context by stressing how easily they slip from one to the other in the chain of signifiers, thus pulling out new meanings, new relations, and movements where formerly there was blockage. The analysand need not agree with how they are reframed, but even in disagreement they must work-through the context and speak it out, which has the effect of converting the analyst's desire into the analysand's desire to intervene in their own knowledge. In this way the analyst's goal in the analytic process is to become the object cause of the analysand's desire to intervene on their own behalf, not to become the master of the analysand.

The Lacanian psychoanalyst, Jacques Siboni (2014a; 2014b), has raised a different challenge to the psychoanalytic establishment from these discourses, namely the thesis that mainstream psychoanalysis runs the risk of primarily speaking the university's discourse. In fact, this raises a "social" problem in that modern society, like the barred subject, is also lacking. It is lacking in alternatives to the bureaucratic scheme, and while there are psychological challenges to structuring the barred subject's knowledge so as to facilitate pleasure, at the end of the day it is the external structure of society that delimits and orders the totality of desire by restricting the outlets of pleasure available to the barred subject. Recalling Adorno's ([1951] 2005) famous quote that "wrong life cannot be lived rightly" (p. 39), those who do not conform to the bureaucratic ideals of society are often denied the sources of pleasure modern society has to offer, forcing them to conform to a system that denies them the very authenticity they seek. This highlights the temptation that one might have to assume the discourse of the university so as to place the analysand in sync with societal norms as dictated by the modern system so that they can feel some of the pleasure offered by this life.

Speaking from the university's discourse in the analytic setting elevates the pleasure of the system, not of the analysand, by aligning the barred subject ($) with the supposed aims of the barred Other (A).[4] That is, it hides the master signifier and speaks from the position of a knowledge that is part and parcel of the bureaucratic system's desire, exchanging the quest for a self-satisfied jouissance for the momentary pleasures of a system that glosses over and tries to bury the pain with an ever-ready supply of consumerist pleasure in the now. Mainstream psychoanalysis, then, by speaking the university's discourse instead of the analyst's discourse, suffers from the same problem as traditional theory: it affirms rather than challenges the status quo of the modern system of capital by inscribing the analysand into the structure of bureaucratic knowledge, trading the quest for the authentic self for a series of momentary satisfactions. Since we must see movements in both self and society, and this framework is one for getting dialectical movement at the level of the lacking self ($), we must now examine the discourses of mass society in an examination of how to track movements in discourse at the level of lacking society (A).

The figures on the following page will be a useful reference for the next section. Figure 4 introduces Lacan's "fifth" discourse, which serves as a gateway to the discourses of socioanalysis. Figure 5 shows the mathemes used in those discourses and Figure 6 shows their structural relations.

---

4 In French the word for other is *autre*, so the Other is symbolized by the matheme, A, and as it too is lacking, the lacking Other is, A.

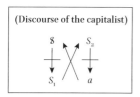

FIGURE 4   The "fifth" discourse

C	totalizing logic of capital
I_∞	totalizing logic of information
A	the barred other, the lacking society
a	object cause of desire; surplus value/surplus jouissance

FIGURE 5   Key to mathemes (socioanalysis)

(Discourse of the archive) A $$\frac{I_\infty}{C} \rightarrow \frac{a}{A}$$	(Discourse of capital) $C_{\$}$ $$\frac{C}{A} \rightarrow \frac{I_\infty}{a}$$
(Discourse of the cyborg) C $$\frac{A}{a} \rightarrow \frac{C}{I_\infty}$$	(Discourse of the socioanalyst) S $$\frac{a}{I_\infty} \rightarrow \frac{A}{C}$$

FIGURE 6   The four discourses (critical socioanalysis)

3   The Other Side of Socioanalysis: A Guide for Talk Therapy

In 1972, at a lecture delivered in Milan, Lacan ([1972] 1978) violated the rules on which he based the structure of the four discourses presented above by rearranging the order of these mathemes to highlight one additional configuration (see Figure 4). Recognizing the sociohistorical predominance of the capitalist in modern societies as a variant of the Master, Lacan provides, what Tomšič (2015) refers to as his "fake 'fifth discourse'" (p. 221): the capitalist's discourse. Lacan only ever presents it in an incomplete fashion, providing us with the structural algebra but never with a fully developed argument of the relations involved, leading it to be largely underutilized in psychoanalysis. However, by following the formula for translating these structures, it provides a useful launching point for shifting the scale from the self to society and the analytic framework from psychoanalysis to socioanalysis.

In his Milan lecture, Lacan ([1972] 1978) points to "the crisis, not of the master discourse, but of the capitalist discourse, which is its substitute" (p. 10), in that it follows a logic that is "wildly clever, but headed for a blowout" (p. 11). The problem, for Lacan, is related to the fact that the capitalist's is a discourse that forms a feedback loop where "it inverts the position of truth and agent, which makes the subject appear as autonomous agent and the initiator of an infinite circulation, from which there is no breakout" (Tomšič 2015: 220). In other words, it represents a totalizing logic as a feedback loop. By inverting the S_1 and the $, from the master's discourse (see Figure 4), and shifting the movement from an open system of difference to a closed and self-referential system, we first see that the truth of the lacking subject, as agent, is the master signifier as capital. They are then presented as a dynamic inversion where we read capital as agent and alienated subject as capital's truth—which works on bureaucratic knowledge (S_2) to justify the production of objects of desire as a fetishized essence of surplus value and surplus jouissance (a) the consumption of which reconstitutes the barred subject by increasing not their pleasure but their truth: an inscription in the discourse of capital. In other words, once the subject is inscribed in this discourse, they become the active agents of their own alienation, abdicating the subject position of history to capital and working to create a system of knowledge that justifies this state. When they are finally convinced by that knowledge that this is the legitimate way of life, they produce their own temptations, and, by succumbing to the fruits of this alienating process, the alienated masses come to insist on seeing capital, as the agent, as the real "truth" of this world. This process happens ad infinitum, or at least it would if it did not rely on material objects in the form of finite resources. Going beyond the dimension of the subject, by putting the subject in direct relation to the material object, this feedback loop is unsustainable in that it consumes itself because it consumes the real, leading to the destruction of the system as the extracted excess collapses under its own vacuous weight (a paradoxical situation that must reach its limit, unless capital can transcend the barriers of the real).

Even though this algorithm lays the ground for working-through the discourse of the capitalist, it still resides primarily at the level of the self, only it now adds to the complexity by incorporating compounded layers of the mass structure. Although the psychoanalytic approach is appropriate for diagnosing the function of these discourses at the level of the self, moving to the level of society—to understand "mass" discourses—requires a different analytic approach. In his programmatic essay on critical theory, Horkheimer ([1968]

1972) noted that "the distinction within this complex totality between what belongs to unconscious nature and what to the action of man in society cannot be drawn in concrete detail" (p. 201) meaning that the links between the fractal layers of the totality, as we move from the self to society, are bound to remain murky. This is evident in both Lacan's discourses and those that follow in that there are no distinct boundaries to delimit precisely between the domains of self and society within the discourses because of the complex and contradictory symbiosis of self and society. However, this does not negate the fact that the structure produced at each level of the system can attain some measure of analytic clarity when it is placed in focus. Because socioanalysis is a branch of the critical method, it "starts out from the contention that modern society is inherently contradictory" (Dahms, H. F., 2008b, p. 36) and it provides the appropriate framework for working-through the false totality of mass society. Furthermore, because critical theory is what first "recognized the responsibility of economic conditions for the totality of the established world" and it "is essentially linked with materialism" (Marcuse, [1968] 2009, p. 99) socioanalysis is needed as a vehicle for diagnosing the contradictory discourse of capital, which in modern societies is exactly what we must confront as the other side of our socioanalytic desire, that which causes societal lack and initiates our desire to interrogate our condition.

Unlike the first four discourses, the capitalist's does not operate in a system of temporal distinction; in other words, it is atemporal in that it knows no time horizon beyond its own theoretical limits, imagining them to be inexistent in the ecstasy of an ever-changing but eternal present of circulation and exchange. For example, while each subject can from one moment to the next speak from the four Lacanian discourses, producing speech that is proper to the function of each, the capitalist's discourse does not, and cannot, temporally break its rhythm without imploding. Rather this discourse structures the self by executing its code first at the external level of sociodynamics, and then by triggering an internal subroutine at the level of psychodynamics, operating behind the scenes of all other discourses. In other words, the discourse of the capitalist is totalizing because it is always running, regardless of the compounding discourses, no matter their distinction, which concurrently run at the level of the self.

By pushing it to the level of "mass" discourse, capital is symbolically illustrated in the following cartography as the matheme, C, to see how it influences other discourses and the structural relations it forms, including its own. The key difference that emerges by shifting levels is that, at the mass level, there is no longer the insistence of the master that grants power, rather it sees that

> capital is ... the *power to command* labor and its products. The capitalist [as master signifier] possesses this power not on account of his personal or human properties but in so far as he is an *owner* of capital.
> MARX [1844b] 1992, p. 295

By defining capital as the power to command, we recognize that it not only grows from its roots in the production/consumption model of economic capital to new branches, like finance capital, but creates new root systems bestowed with the power of capital in cultural, social, and symbolic forms as it totalizes reality under its logic (Bourdieu, 1986; 1989). Capital accomplishes this feat of transcending its roots in the material economy by allowing the totalizing logic of information to emerge as a historic process that attempts to solve two problems that arise as a result of capital's discursive model of self-replication.

These problems are related to the way that the logic of capital advances its control over the human and technical elements in its system. The first problem is that capital is the chosen organizational structure for modern life, therefore it must find a way to inscribe people in that structure to prevent alternative organizational structures from competing with its power. The second problem is that due to its atemporal, but dynamic, nature, capital cannot solve the human problem once and for all, but rather is constantly confronted with crises because of its technical progress. As Marx ([1939] 1973) pointed out, "capital itself is the moving contradiction" (p. 706) of modernity, so it creates nonsensical configurations as it attempts to solve these problems. He ([1867] 1990) outlined this process beginning from the fact that

> Modern Industry never views or treats the existing form of a production process as the definitive one. Its technical basis is therefore revolutionary ... it is continually transforming not only the technical basis of production but also the functions of the worker and the social combinations of the labour process. At the same time, it thereby also revolutionizes the division of labour within society, and incessantly throws masses of capital and of workers from one branch of production to another ... We have seen how this absolute contradiction [between the technical necessities of Modern Industry and the social character inherent in its capitalistic form] does away with all repose, all fixity and all security as far as the worker's life-situation is concerned: how it constantly threatens, by taking away the instruments of labor, to snatch from his hands the means of subsistence, and, by suppressing his specialized function, to make him superfluous. (pp. 617-618)

As a result of this ongoing revolution in the means of technology, human productive labor is needed less and less to sustain the process while the technological base assumes this role more and more. On the surface it seems like capital could free life from labor, but as "it presses to reduce labour time to a minimum ... it posits labour time, on the other side, as sole measure and source of wealth" ([1939] 1973, p. 706), meaning that it cannot complete the decoupling of life from labor without threatening the maintenance of wealth, so the human problem is reproduced alongside the technical revolution in the means of production. "Hence it diminishes labour time in the necessary form so as to increase it in the superfluous form" (p. 706), so if Marx's labor theory of value is an apt description of how capital determines value, in order to sustain the consumptive side of the equation and the continuation of wealth extraction, humans must be put to work regardless of the "social" necessity of their production.

To maintain this inscription of laborers in the mass discourse of capital, it has to manufacture desire in its subjects, and as the labor becomes more tedious, repetitive, and machinic, it has to spend more of its resources on maintaining the workers cooperation. It does this by alternating between disciplinary tactics—used when too much stress is placed on the system for alternative organizational structures by modern subjects who either object to the nonsensical nature of superfluous labor or are excluded from it—and seductive techniques—used on subjects who are tempted by the alluring jouissance of consumerism and spectacle. In both cases we are dealing with the political economy of the sign, as the subject is convinced or coerced to labor because the organizational structure of capital maintains a society that controls the flow of information as signs that emphasize societal lack as individual lack (Baudrillard [1968] 2005; [1972] 1981). For advanced modern/postmodern societies, this lack is translated into both the presentist atemporalization of capital—with financialization pulling the future into the present with markets dedicated to credit and speculation (we lack the future, so let's bring it to us!)—and "the annihilation of space through time" (Harvey, 1990, p. 205) by means of virtualization—as physical space is morphed into its digital other (we lack time, so let's eliminate distance!). This enables capital to increase the speed of information flows, which are its new life blood, and mobilize these flows for the production of capital and the maintenance of an obedient labor force. Capital, C, comes to stand in for all of these processes and forms, and it replaces the S_1 of the master signifier because there is no master signifier that can compete with the power of capital in structuring the material and ideological conditions of the mass in modern/postmodern societies. Capital, as the symbolic representation of this power to command, is treated as a material base for the free-floating master signifier, tolerating different capitalist identities, as signs, so long as they continue to draw their power from the ultimate sign of capital.

Under the totalizing logic of capital, as Lyotard ([1979] 1984) rightly noted, knowledge is instrumentalized by the system to coincide with the revolution of the technical base of industry. This instrumentalization occurs, not only in the bureaucratic mode that codes subjects to operate as functions of the organizational structure of capital, as is demonstrated in the university discourse as the function of S_2, but through the systematic realignment and supersession of the use and exchange values of knowledge to those of information and data. Knowledge is embodied in the subject, which serves as storage receptacle and transmitter of that knowledge to other subjects (Liew, 2007; Stenmark, 2001), and therein lies its weakness for capital. Knowledge transmission that relies on a human subject is constrained by the biological speed limit of the species. With the prioritization of speed in advanced modern/postmodern societies, as evidenced by the socio-technical move toward increased space-time compression (Harvey, 1990), information and data are disembodied to the extent that they are technically constrained only by the barrier of physics, the speed of light. As such, information and data are transmitted between objects, rather than subjects. The totalizing logic of information works to convert the whole of reality into its objects, to prioritize its own function in much the same way as capital. The problem with this model, of course, is that information and data are symbolic and do not provide the same level of understanding as knowledge, even though they are treated in capital as representative of the real and are used as justifications for intervening in the real.

Mirroring the feedback loop system of capital, "information is operational in a way that knowledge is not to the extent that it alters the system and selects states that are open and mutable in its structure" (Malik, 2005, p. 47). Like capital, information acts on itself for the purpose of its own expansion by feeding on external sources while reconfiguring its inner system. However, capital continues to have a material anchor and thus a limit that it cannot overcome in the form of raw material resources. Since information imagines itself as infinite, that is, as the perfect mirror of reality with the power to convert all materiality into its symbolic representation, capital aligns itself with information imagining that it too can be relieved of this finite material anchor without itself being subsumed by the totalizing logic of information. In this sense, the digitization of the real coincides with the aim of capital to eliminate its weaknesses by transcending them, which it attempts to do in its finance and symbolic forms that play off of the fact that "the movement [or circulation] of capital is ... limitless" (Marx, [1867] 1990, p. 253). According to Scott Lash (2002), this "informationalization opens up a new paradigm of power and inequality," to such an extent that "critique, and the texts of critical theory, must be part and parcel of this general informationalization" (p. 10). Just as they must assume the commodity form, while at the same time resisting commodification and informationalization with their knowledge. Continuing then in accepting the contradictory nature of modernity, critical theory must critique both capital and information, while also recognizing that it is reliant on the very forms that are the subject of its critique. As capital circulates around all discourses in modern society, so too does this self-replicating mode of information—which replaces the S_2, with the matheme, $I\infty$, as information and data—with its utopic view of an infinite horizon—circulates as the primary language of capital with aims beyond the scope of a finite capitalist materialism.

Due to this reign of capital in modern history, which initiates the transformation and separation from nature by inscribing its agents into a techno-scientific culture that alternates between the rational and irrational, the truth of the Other is also revealed. The big Other—which has no concrete reality, but is rather a "symbolic order, society's unwritten constitution ... the second nature of every speaking being" (Žižek, 2006: 8)—is unmasked in modernity, by the successive failures of its ultimate referents: God, the social, and finally, the least convincing, the ideal Human. Despite each of these stand-ins failing to act when the code of this constitution is violated, they persist by entering the realm of simulation, allowing subjects to continue to act *as if* they served the function, in spite of their ultimately being ineffectual guarantors of the real and its structure. Since each is found to be lacking the completeness that would grant the power to enforce these rules, by only existing as simulation, the Other, as modern/postmodern society, is also marked by a lack, giving us the Lacanian concept of the barred Other, (from the French, Autre, marked by the bar for lack, and represented by the matheme: $\cancel{A}$). This lacking Other mirrors the lacking subject, $, at the level claimed by the mass, and seeing as modern/postmodern society cannot fill the lack experienced by the subject by virtue of its own lacking nature, the $\cancel{A}$ serves the function of the lacking society in these discourses.

The final matheme of the following set of discourses, remains, *a*, as the object cause of desire. "The minimal common ground," Tomšič (2015) uses to link "the critique of political economy and that of psychoanalysis" is that "Marx and Freud both insisted that the symbolic networks operate beyond consciousness and are endowed with causality, the power to work back on conscious subjects" (p. 200). In other words, they operated according to the same structural principles in building their critical theories, the process of which in Lacanian terms is fueled by a desire to know and to work through anxiety. Remembering, as outlined above, that it is not a subjective desire for pleasure, such as the boost of one's own ego or the projection of an ideal future that aligns with one's Weltanschauung, but a desire proper to the role of the socioanalyst, as an analyst who is diagnosing the contradictions between the cyborg self and the mass society. By putting these algebraic symbols in a cartography of their structural relations, we end up with: the discourse of capital, the discourse of the archive, the discourse of the cyborg, and the discourse of the socioanalyst.

Consider first the discourse of capital. When Marx turned his attention to the critique of the commodity form, he revealed a very peculiar set of features, namely, that the commodity is "abounding in metaphysical subtleties and theological niceties" (Marx, [1867] 1990, p. 163). The peculiarity of the commodity form is that it has features that go beyond its use-value, as something embedded within the labor process itself, to mask the "definite social relation between men themselves which assumes here, for them, the fantastic form of a relation between things" (p. 165), which Marx points to as the fetish characteristic of the commodity. The fetish with its history in religious discourses is appropriate here because, like the fetish object of religious origins, the commodity fetish is imbued with forces that are materially absent but are attached to it nonetheless *as if* it had real spiritual properties. These properties directly relate to the sign that each commodity assumes as the symbolic representation of the commodity exudes an ethereal presence.

"[E]very product of labour" is transformed "into a social hieroglyphic" when they are presented to consumers as commodities, and it is the task of socioanalysis "to decipher the hieroglyphic, to get behind the secret of [our] own social product" (p. 167). Value is transfigured by the hieroglyphic sign that the commodity assumes. Rather than being able to see the commodity as a series of social relations, in which all commodities have the same history of a capitalist setting labor to the task of their production, the commodity appears as a thing without a history, as something divorced from the production process. It appears to us on the market as an always finished product that, due to the power of its sign, allows us to see it as something more than the sameness of the labor process that went into its production. It is the *magic* of capital that turns the commodity into a fetish object that actively hides its material roots in the labor process. Opting instead for the rule of the transcendent sign as the artificial byproduct of capital production, in modernity we reached the stage of the enchantment of the commodity.

A hypothesis emerges here by way of explaining this commodity feature in synthesis with the Lacanian discourse model. Specifically, since C, capital, in the position of the agent, is a self-referential system that assumes the role of the totality: it has no lack and thus no desire in and of itself. Lack is not a feature internal to capital itself, but rather it emerges from the way that capital sets itself to work on artificially reconfiguring reality and structuring its subjects. These subjects of capital, by working in an organizational process that abdicates to capital the role of the subject and agent of history, are displaced from the role they historically assigned themselves. As a result, they become more sensitized to the fact that modern society, which is geared toward the needs and ends of capital, lacks what they need as subjects in and for themselves; thus their desire is not triggered by capital itself, but by the lacking society made for capital's ends.

By means of science and rationalization, the capital form merely amplified "the disenchantment of the world" in modernity as the continuation of Western thought (Weber, 2004, p. 30). The primary explanatory models and narratives, that legitimated a conviction in the success of society by grounding it in the transcendental (i.e. God, the social, the Human), could not stand-up to rational scrutiny and can no longer provide satisfactory levels of understanding and meaning for our modern condition. The point, however, is that this lack predates capital since it was the disenchantment process that first revealed society's lack by exposing its foundational referents as failures. The religious fetish object, for example, used to stand in as the embodiment of completeness, the spiritual whole denied to material beings, but as nature raged against humanity these objects were exposed for their ineptitude and came to symbolize lack rather than completeness. Without abandoning the model, however, in capital this object was simply transformed and replaced with the psychologically more powerful commodity fetish object that often brings with it at least some level of necessary use-value making it appear more powerful than the lifeless religious fetish. As a result, capital set itself the task of managing this lack by trying to fill it with the commodity fetish so as to avoid the consequences of being held responsible for the real lack in and of society.

This is what is revealed by the mass discourse of capital. C, acting as the agent, sets all of the available information and data, $I\infty$, at its disposal to work on reality. Capital does not, and cannot, have knowledge of its subjects since it is a subject of a different order and communicates as an object. What capital does have at its disposal is information and data about the subjects internal to its system, in the form of social media posts, market research, demographics, statistical surveys, legal codes, engineering and architectural plans, etc. By treating this information as representative of reality, the subjects of capital hold out on the promise that capital can, in fact, by controlling information, control and manipulate their reality freely, and that it will, eventually, do so on a rational basis for their own betterment.

The result, however, of $I\infty$ being put to work is that out of it comes products that are based on a refracted idea of the real, as the embodiment of the crystallization of surplus value in commodity fetish objects: symbolized as a. Each commodity produced is imbued with a spiritual essence that works to either align desire with the goals of capital, or to mask over the current truth of capital on the auspices that it will one day act as an extension of the desire of its willing subjects. The contradiction being, of course, that capital artificially produces the desire (a) for its commodities by basing their production on mere information gleaned from social symbols that it created in the first place to structure society and not from any real knowledge of us as subjects! These commodities are only ever representative of an artificially manufactured desire for capital's products. Even when they meet real needs with their use-value, they do not and cannot fill the lack of a society made for capital by unknowable subjects working in concert with it, they can only mask it. Commodities work then in the service of capital to hide the lacking society, $Ⱥ$, which is the truth of the real that capital masks. This truth exposes capital as a sham covering over the real, but we cannot discount the vast power of this discourse at dissimulating this truth, since capital is largely able to maintain order in modern and postmodern societies through this manufacture of desire, as the power of seduction trumps the power of discipline in consumer society.

To speak then from the discourse of capital is to speak on behalf of capital, in alignment with its needs. The speaker functions as capital's agent, C, by treating information and data, $I\infty$, produced by the system, on the system, as the reality of the system. In other words, by taking the simulation of the real, in the form of opinion polls, surveys, and statistical reductions as accurate representations of reality for the purpose of intervening in and acting on reality. As this information is put to work, the product, or speech/text/thought, appears in the form of a commodity fetish, a: an object geared toward the alignment of people with the will of capital. By restructuring individual desires, this product serves its function when people act in concert with the promotion of the atemporal extension of capital in an eternal present. By speaking in this discourse, then, one denies that the truth of modern society is that it is marked by a lack, $Ⱥ$, and that capital fails its subjects by ignoring the fact that this society fails to meet the genuine needs of its subjects, ignorant as it is of its own role in the expansion of this lacking society. For socioanalysis this reality is the other side of its discourse, the side which must always be the subject of critique so long as capital lays claim to the totality, if we are to avoid affirming life in the service of capital by merely reproducing in our thoughts the desire capital has instilled in us all to be its willing servants.

The next discourse is that of the archive. While many in the social sciences are today aware of the thin line they must traverse between affirming and critiquing capital in their research, often the more immediate problem faced is that most research and thought on reality, rather than enlightening the masses to our shared condition, is destined to a muted life as a text in the archive. When Freud ([1925] 2006) wrote about the mystic writing-pad, he highlighted the positive feature of storing information in an archival format when he mused that it "is as it were a materialized portion of my mnemic apparatus" (p. 20). In other words, the written notation extends the body as it serves the purpose of preserving a thought from one moment to the next without requiring a human, as agent, to retain the thought in their internal memory. This bodily extension is made all the more material in devices, like the smartphone, which allow users to record their own memories externally, quickly and easily, with text, photo, and video applications, and by plugging into the internet they link their users to the largest archive ever to exist on planet Earth.

All of these contributions to the archive assume the form of information, but "information devours its own content; it devours communication and the social, and for two reasons:

1. Instead of causing communication, *it exhausts itself in the act* of staging the communication; instead of producing meaning, it exhausts itself in the staging of meaning ... [That is, information and information technologies are focused more on how the information is stored and presented, than on the content of that information.] Thus communication as well as *the social* functions as a closed circuit, as a lure—to which is attached the force of a myth.
2. Behind this exacerbated staging of communication, the mass media, with its pressure of information, carries out an irresistible destructuration of the social. Thus information dissolves all meaning and the social into a sort of nebulous state leading not at all to a surfeit of innovations but to the very contrary, to total entropy. (Baudrillard, [1978] 2007, pp. 101–102)

Since $I\infty$ serves the function of the agent in the discourse of the archive, we must keep in mind that the goals of communication between subjects and the supposedly resultant social, are not in alignment with information acting as the agent. This is why C sits in the position of truth in this discourse, since information is the form of capital's communication, which is between objects, and it is held in the archive primarily for the purposes of extending capital's goals, not Human (i.e. social) goals. As the structure of capital is more deeply embedded in modern society, this continued destructuration of the social is the result.

$I\infty$ sets itself to work on a, as information works on the structure of artificial desires in modern society which it treats as real social desires, it puts desiring subjects to work to produce more information about themselves as if it were for themselves, but it masks the truth that this is in the service of capital. Thinking of social media for instance, the service is free to consumers who think that they are using it to express themselves, when the reality is that they are the product and they are creating information for capital, which hides behind the surface. Remembering that information is distinct from knowledge because it shifts the mode of communication from subjects to objects, what is frequently lost in the transfer of information is the meaning and understanding (as *Verstehen*, in Weberian terms) about that information. Here we have the paradox that "information, all information, act in two directions: outwardly they produce more of the social, inwardly they neutralize social relations and the social itself" (Baudrillard, [1978] 2007, p. 79–80). In other words, the vast amounts of informational content available to us widens our reach across the globe and expands our horizon to levels never before seen in human history which, according to the hypothesis that the more information provided in the model the better the outcome, should have increased and expanded the communication between subjects on a global scale, thus elevating the social to new heights. Rather, all historical indications are that this information flow is primarily a function of capital, and in setting the desires produced by the system of capital to work, information structures that desire in a manner that negates the social while affirming its subjects as a mass, which is the result of actively producing A: the lacking society.

Thinking back to the example of the smartphone, with it we postmoderns have more information at our fingertips than any premodern or modern human, and yet, since the archive does not discriminate the content of information but only its form, users of the archive are able to curate their own experience by only looking at information that already conforms to their worldview, leaving the many critiques of those worldviews buried on the digital shelf where they collect dust on the second, third, fourth, etc., pages of Google search results (if they are produced and archived at all).[5] Furthermore, as a communication medium, the smartphone allows near instantaneous communication with anyone anywhere, but rather than increase the quality of communication between subjects, it is largely shortened to acronyms and emoticons in text messages so as to communicate the least possible content in the quickest

---

5 Based on 2013 data, items "listed on the first Google search results page generate 92% of all traffic from an average search," with just the first listing pulling 32.5% of viewers (Chitika, 2013).

possible manner, as is the goal of communication between objects. What this society lacks, and what information destroys, according to Baudrillard, is an effective means of communicating between subjects, as this lack is amplified by our increasing reliance on information and its *Verstehen*-free form of communication between objects.

Although this discourse of the archive produces a result that is obviously worthy of critique, academics largely assign it a positive function, since it provides the framework for a historical catalogue of information for those who wish to commune with historical discourses, seeking to further the discourse on the topic of their preference across space and time with their own contribution to its vaults in the form of endless social taxonomies. Lacan's ([1991] 2007) example of the citation serves as a useful illustration of this process:

> What does a citation consist in? In the course of a text where you are making more or less good progress, if you happen to be in the right places of the class struggle, all of a sudden you will cite Marx, and you will add, "Marx said." If you are an analyst you will cite Freud and you will add, "Freud said." This is fundamental ... A citation is like this. I make a statement, and for the remainder, there is the solid support you will find in the author's name for which I hand responsibility back to you ... When one cites Marx or Freud ... one does so as a function of the part the supposed reader takes in a discourse. The citation is in its own way also a half-said. It is a statement about which someone is indicating to you that it is admissible only insofar as you already participate in a certain structured discourse ... (p. 37)

Lacan's explanation of citations points to a key way that academics who speak the discourse of the archive come to understand their contributions: as plugging into a historical discourse by furthering its chain of signification in the hope that there is "a supposed reader" to fill that side of the function. Particularly among those academics who ascribe to a humanist ideology, "the writer ... sends his work out into the world without knowing the recipient ... writing not only creates a telecommunicative bridge ... but it sets in motion an unpredictable process" (Sloterdijk, 2009, p. 13). At least this is purported to be the desire of the subject who speaks in the discourse of the archive: that their contribution to the archive, which often has no immediately visible impact, will in some unknown future become the cause of some unknowable, yet presumably "socially" desirable result when the *right* reader stumbles upon it. However, we must ask ourselves, if the archive is ever expanding, at what point in the future will it suddenly reclaim the capacity to transmit meaning and understanding for each item residing in the depths of its vaults, if they are only composed of information? Failing to address this question is what leads information into a feedback loop situation, where, from an academic standpoint, the primary output of the discourse of science becomes the mere continuation of the chain of citations, the justification of which can only be the advancement of information qua information in the service of capital as the only artificial intelligence for whom this information could have any value.

When the analysand is speaking through the discourse of the archive, it is worth prodding them to see how their sense of self aligns with this discourse, by challenging them to find out what is non-identical in their self and the informational representation of their self that they broadcast. By locating what is non-identical, not only will their sense of self move, but so to there should be movement in their assimilation to the discourse of the archive.

Since these discourses are, however, aimed at revealing the structuration of the mass, this subjective justification for speaking in the discourse of the archive must be seen for what it really is. Here we enter "a universe where there is more and more information, and less and less meaning" (Baudrillard, [1978] 2007, p. 99). The scientist, speaking on behalf of $I\infty$ works on the presumed desires of society (a) failing to account for the fact that these desires are structured as a part of the totality that is C, capital: information's truth. In speaking this discourse, then, scientists merely reproduce the simulation of the real by increasing the mythology of the informational version of the real as an extension of its actual product: A. This lacking society is what the archivist gets off on since it serves a dual function in that it is both the product of this informationalization process and the justification of the very attempt to fill that lack with more information. The irony being that as the information grows, the product is a society whose lack becomes more apparent as the infinite horizon of information becomes the infinite horizon of societal lack. "Disregarding the obligation to help people in the shaping of their most important concerns, the accumulation of knowledge [(here we would read this as information)] has degenerated into an end in itself, a fetish" (The Frankfurt Institute for Social Research, [1956] 1972, p. 8). As a result of this discourse the lacking society is thus affirmed, rather than critiqued, and meaning is sacrificed for a slot on the shelf of the archive as a gateway to a presumed future.

The discourse of capital and the discourse of the archive dominate modern and postmodern societies today with two very distinct but interrelated impacts on the lives of their subjects. The first is under the domination of the discourse of capital: as the function of human labor is gradually replaced by the continuous revolution of the technical means of production, the human assumes roles that gradually appear more and more repetitive, information driven, and superfluous to human necessity, in a word: machinic. The second is under the sub-domination of the discourse of the archive: because information is more focused on the form than the content of its messages, the predominant mode of communication in modern society is more suited to objects, i.e. machines, than subjects, i.e. humans, and the social withers in this mode while capital blooms. In sum, while there remains something that we persist in labeling as "human" and as "social", the signifier/signified relationship of these words to their historical meaning has gone the way of the religious fetish object in our mechanthropomorphism. As the power invested in the categories of the human and the social lost their force, something with the characteristics of the machine came to take over materially, while these anachronisms persist as simulation. The discourse to which we now turn our attention requires a bit more historical background than was provided in the previous chapter, for while it too dominates modern/postmodern societies, it is ideologically denied by many who cling to the human and the social as the only guideposts lighting the path to freedom.

The result of the aforementioned discourses is the prevalence of a military-industrial-university complex that functions as the primary system for managing the totality of capital on behalf of capital, which triggers at the level of the subject a 'border-war between organism and machine' (Haraway, 2004, p. 8). This war materialized with the introduction of cybernetics in 1947 by Norbert Wiener at the Macy conferences, which he then expounded on in his 1948 book *Cybernetics; or Control and Communication in the Animal and the Machine* ([1948] 1965). In that book he outlined cybernetics as a novel field of study that combined "digital electronic computing, information theory, early work on neural networks, the theory of servomechanisms and feedback systems, and work in psychology, psychiatry, decision theory, and the social sciences" (Pickering, 2010, p. 3). Despite its overt implications for political economy and military applications, Wiener's (1954) research did not simply focus on the technical aspects of cybernetics but paid enormous attention to the sociological consequences of these modern developments:

> When human atoms are knit into an organization in which they are used, not in their full right as responsible human beings, but as cogs and levers and rods, it matters little that their raw material is flesh and blood. *What is used as an element in a machine, is in fact an element in the machine.* (p. 185)

What Wiener understood, which many today either refuse to acknowledge or remain ignorant of, is that the human transformation into a machinic subject is the result of the organizational model of modern society; that is, under the dominance of capital and information, whether or not the "human" subject persists as a meat-suit of flesh and blood, by functioning as a machine in our everyday tasks, we *are* in fact machines. These new machinic subjects that act like objects, are best described by Clynes and Kline's (1960) term, cyborg. In other words, as capital began to move toward self-regulation and organized society in a neoliberal manner to enable this process, its subjects unconsciously began to mimic the drive to self-regulate as a means of adapting to their new environment. What began as a human, under modern capital became something else, something other, something posthuman, in a word, a cyborg.

In light of this history, in which the transformation of the human into the cyborg materialized, the irony is that the cyborg is without a true origin point. In fact, the cyborg has abandoned origins stories because it recognizes that it has moved beyond "the myth of original unity ... and the twin potent myths inscribed most powerfully for us in psychoanalysis and Marxism," in the form of a spiritual reality guaranteed by God, a complete Human in the sense of "pre-Oedipal" symbiosis, or a fully developed social sphere that enjoys "unalienated labor" (Haraway, 2004, p. 9).[6] Whether fully realized or not, by having some sense that modern society does not provide everything that its subjects need, in the discourse of the cyborg this lacking society, A̶, is in the position of the agent. The cyborg is a direct consequence of those who recognize that modern society is organized according to a principle that produces conditions that are primarily suitable for self-replicating, self-referencing, and self-regulating systems, i.e. capital. Since this society does not produce the conditions needed to sustain human and social elements—that is to say, it is marked by a lack—it is by way of coping and survival that this new life-form emerges and demands its own means for self-regulation as a form of protest against their status as the abandoned and illegitimate offspring of the world

---

6 This position calls out both mainstream psychoanalysis and mainstream Marxism, for ignoring material conditions and falling into ideological patterns that make them anything but critical. It is, however, wholly compatible with certain critical Marxian interpretations, such as that offered by Moishe Postone (1993). Similarly, Lacanian psychoanalysis provides a critical interpretation of Freud that does not portend to the complete subject, seeing as it is the barred subject that is inscribed in the real. Rather than promising a cure as a form of completeness, Lacanian psychoanalysis offers a cure only insofar as the subject comes to understand that they cannot nail down their drives once and for all, and in accepting this they are relieved of the burden of thinking otherwise. Since "the drive achieves satisfaction taking no heed of repression; it mocks repression ... [a successful analysis moves the subject to have] love for something unknown or radically other," a love that cannot exist in a model of completeness and sameness (Dunand 1995: 256).

capital made. As Baudrillard ([1970] 1998) framed it, "because the system produces only for its needs, it is all the readier to systematically hide behind the alibi of individual needs" (p. 65); in other words, because society has become a function of capital, the lack in society that its subjects experience is sold to them as an individual need and not a societal one. The irony, of course, is that it is the very attempt that individuals make to fill the lack of society, by means of embodying that society's self-regulatory tools, that the process compounds the lack of society and emphasizes how we have demanded a transition from the human to the cyborg as a means of surviving the symptoms produced by these societies.

But this knowledge does not stop the A from putting C to work. Since the lack in modern society is exacerbated by capital, cyborg subjects come to demand that capital work on their behalf to make up for some of this lack, not in an unequal exchange that places the entirety of the blame on capital for the lack, but as partners in the process, with the cyborg demanding new tools from capital so as to improve their symbiotic relationship to capital in a contradictory system that is both inter-dependent and self-regulatory. As a result, capital does not produce things that are geared toward the liberation and furthering of the human and the social, but, by accepting the condition of society, the cyborg demands that capital produce more in the form of information, $I\infty$, so as to capture and duplicate this reality in a more manipulable form. This is a key difference between the discourse of the hysteric operating at the level of the self to produce knowledge, and the discourse of the cyborg, which operating at the level of the "mass" produces mere information. Perhaps the justification, if one were needed at the subjective level, would go something like this: "The world does not provide me what I want or need, but at least, through the magic of capital, I can manipulate the world in its virtual, informational, form to simulate the fulfilment of my wants and needs." Cyborgs take this information, that they use capital to produce, believing it helps them: on the one hand, as they adapt to the dominating mode of communication between objects, and on the other hand, as they fill roles in society that demand more and more machinic tasks. By prioritizing the form it takes, information assumes a seductive allure, hinting just enough to convince some cyborgs that it could be the last hope of freedom. If information "longs to be free" (Hughes, [1993] 2001, p. 82) it tantalizingly suggests possible futures that it might enable, such as, freedom from the system as cyborgs become fully self-reliant symbionts with information, or freedom from biological limits, that is freedom from the body, by becoming wholly machine; or at least such are the hopes of the growing numbers of techno-libertarians and transhumanists.

The truth of the cyborg, however, is a, which like the hysteric, inscribes the cyborg in the domain of the real. That is to say that the cyborg is not a metaphor for how humans live in modern societies, rather the reality of subjects in modern societies is that they already *are* cyborgs. What is hidden is that the truth of a for the cyborg is not altogether different from the desires of supposedly human agents for a social basis on which to base a society that will more easily meet their needs. Rather, in a frequently unconscious manner the cyborg, believing that it is acting on subjective desires for need fulfillment, adapts their desire to the dictates of capital and channels it as a reaction to the organizational structure of modern society by demanding new ways of covering over this lack in society as a means of completing their adaptation to the modern system.

Speaking from the discourse of the cyborg then is a coping mechanism for those who understand the rigidity of the modern totality of capital. The academic, working in this mode, demands that capital work on behalf of the lacking society to produce more information about this reality and fill the gaps of its informational double, to the end of harnessing that information to produce conditions that re-inscribe their function and their use-value in a system, whose model of self-reliance, is beginning to exclude any need for these subjects. If capital comes to not need cyborg subjects, through processes of automation and artificial intelligence, the loss of their place, their deletion from the system of capital, is too horrifying for them to confront. So knowledge produced under this mode begins from the real standpoint of a lacking society, A, which sets the tools of capital, C, to work, to produce more information, $I\infty$, about our reality, in the hopes that it will eventually respond to the reality of a hidden desire, a, to maintain and justify a space and a role for the cyborg in a society that increasingly appears to no longer need them.

Recall that critical psychoanalysis must stay vigilant so as not to fall into the discourse of the master or the discourse of the university, as its mainstream variants are wont to do. If it begins to speak from those discourses, then it risks either placing the analyst in the position of the master who simply commands the analysand and creates a situation of dependency or the analyst assumes the role of systemic knowledge and merely inscribes the analysand into the bureaucratic scheme of modern society making them complacent. Speaking from the discourse of analysis, on the contrary, the analyst puts the analysand to work and instills in them a desire to interrogate the way things are, in short, the analysand obtains a critical desire to understand this life by challenging it.

Likewise, critical socioanalysis must avoid speaking in the discourse of capital or that of the archive, as is the tendency of traditional theory. "The traditional idea of theory," according to Max Horkheimer ([1937] 2002), "corresponds to the activity of the scholar which takes place alongside all the other activities of a society but in no immediately clear connection with them" (p. 197). In other words, traditional theory is just a job, like any other, that the theorist imagines that they perform from a position in which they have somehow escaped the compounded levels of discourse, structure, and alienation in their self and society. Producing theory in this way, traditional theorists assume "an affirmative perspective on ... societal reality" (Dahms, H. F., 2008, p. 33) that ignores the "characteristic uniqueness of the reality in which we move" (Weber, 1949, p. 72). What differentiates critical socioanalysis, from the traditional variant, is that critical socioanalysts take seriously the characteristic uniqueness of engaging in this practice in modern and postmodern societies, given that theory is by necessity produced from within the very framework of the reality that is the subject of critique.

Starting with the critical concept of the totality—which, "formulated provocatively, totality is society as a thing-in-itself, with all the guilt of reification" (Adorno, T. W., et al., [1969] 1976, p. 12)—critical socioanalysts are able to confront the contradiction of capital, as a self-referencing system made in its own image, only by treating it on its own terms. That is to say that critical socioanalysts must treat capital as a thing-in-itself, since capital is a system that thingifies social relations and doubles the process back in on itself, "reification [becomes] ... the principle of intelligibility of capitalism" (Feenberg, 2015, p. 122). This does not mean that critical socioanalysis escapes the domination of capital, but that by applying the concepts of capital, on capital, critical socioanalysis can aim at a "dialectical critique [that] seeks to salvage or help establish what does not obey totality" (Adorno, T. W., et al., [1969] 1976, p. 12). As a result of applying the concepts of capital to capital, however, critical socioanalysis must perform a double function. First it must critique capital using these concepts, but then it must critique the critique of capital in an attempt to neutralize the absorption of the critique back into the system of capital as it is prone to do as a strengthening measure that operationalizes critique for its own ends. Due to the fact that the discourse of capital is the other side of the discourse of the socioanalyst (the mathemes are inverted), capital's feedback loop attempts to reduce all knowledge—especially critiques that threaten capital by exposing that which disobeys it—to mere information, so that it can redouble this information into capital and short-circuit the intended intervention of the critique before it can take root in the analysand.

Seeing as capital operates at the level of structuring the totality and critical socioanalysis operates from within the totality, with critical socioanalysis "the interpretation of facts is directed towards totality, without the interpretation itself being fact" (Adorno, T. W., et al., [1969] 1976, p. 12). "The commandment to remain within a framework of the given reality thus begins to change into its opposite ... [a] confrontation of the object [capital] with one's own concept" (The Frankfurt Institute for Social Research, [1956] 1972, p. 11). What this means then is that in order to critique capital, critical socioanalysis must act in a fashion that is as contradictory as capital itself, in a manner that befuddles or at the very least delays capital's attempts to reduce critique to its informational content. On the one hand this means that critical socioanalysts must produce knowledge about a scale of reality that cannot be directly observed but is experienced in everyday life, and on the other hand, they must also be aware that such knowledge will always be fragmented, partial, and potentially always-already thwarted by the system the moment it enters the information banks.

Recalling Scott Lash's (2002) point, that critical theory and critique must become a part of this informationalization process, the critique of capital must be followed up with a critique of information. If critical socioanalysis speaks from the discourse of the archive, then all it does is reinforce the system of capital by contributing to the production of a refracted reality that masks its lack by claiming an existence in the supposedly infinite space of information. "But only a critical spirit can make science more than a mere duplication of reality by means of thought, and to explain reality means, at all times, to break the spell of this duplication" (The Frankfurt Institute for Social Research, [1956] 1972, p. 11). In other words, as a result of acknowledging the discourse of the archive, critical theory must deal with two layers of the totality: the first, as capital in a material sense; the second, as information, in a virtual sense. Each of these discourses acts as if they are in full control of the totality of social relations by reducing everything they confront to their overwhelming logic. The only way for critical socioanalysis to respond, is with a logic of its own.

The discourse of the socioanalyst explains that logic by placing *a* as its agent position, as the object cause of desire. Just like the psychoanalyst who must become the cause of desire for the barred subject, the critical theorist must act as a stand-in of the cause of desire for the lacking society, Ⱥ, in the socioanalytic session. This lacking society sits in the position of the socioanalyst's other, what it puts to work, because without a lacking society, there would be no reason for socioanalysis. By pulling from the pool of information, $I\infty$, that is attributed to the socioanalyst as the one in the position of the subject supposed to know about society, the socioanalyst challenges that information by critiquing its position as the truth of socioanalysis. While socioanalysis must engage with the information that is produced by the system as representative of it, as this is what the analysand will channel as they first begin to self-analyze the contradictions between their self and society, in challenging this position the critical socioanalyst aims to trigger a desire in the analysand to recognize the lacking society and its use of information to cloud out knowledge. By becoming the object cause of desire in the lacking society (within the confines of analysis), the critical socioanalyst thus has the goal of making the analysand (who is a reflection of the lacking society) question the information of the world, so as to produce C, not as capital in the traditional sense, but as knowledge as capital. That is, the goal is to create a reservoir of knowledge that has the function of capital, in that it can be invested and pay dividends by self-valorizing. The desire of the analyst is reinforced by the analysand's production of this knowledge-based capital, so the analyst consumes the surplus enjoyment of the analysand's jouissance as they come to recognize the lacking society, pass through anxiety, and arrive at knowledge of capital's structuring functions on their self and their society. Since the world does not change with these movements, but only their concepts move, the pain of the world remains, but since there is a real pleasure in revealing this knowledge, the analysand will come to embody the desire of the analyst as they continue the process of interrogating the contradictions and tracking the movements of the concepts they use to make meaning of their lives on the hope that when material conditions change they will be able to invest this knowledge materially, meaning that so too will the configuration of what they lack change.

In other words, critical socioanalysis must trigger a desire in the analysand that the lacking society has blocked, to expose the fictional representations of reality as the simulation of modern/postmodern society's efforts to control its subjects. While critical socioanalysis can simply state the critique of capital, and indeed this is a component of it, by being the only voice in society that issues this challenge to the totality, critical theory is doomed to a life on the shelves of the archive. Sitting in this archive, the likelihood is that it will only be other critical theorists who will stumble upon these text as they enter the chain of citations. While this preserves critical theory by reproducing its function in the bureaucratic system and meets the productive requirements of that system, by staying at this level critical theory fails to take seriously the concrete gravity of the sociohistorical transformations of the late 20th and early 21st centuries and slides back into traditional theory. It is only by recognizing that the totality is specifically designed in a such a way as to castrate critical theory of its most radical potential that critical socioanalysis can befuddle the process by making its transmission in analysis a part of its praxis.

Speaking the discourse of the socioanalyst, then, the critical socioanalyst must first learn how to become the object cause of desire. Lacan ([2004] 2014) tells us that "anxiety appears prior to desire" (p. 280) and the lacking society is in no short supply of anxiety. By diagnosing the compounding layers of structure, of discourse, and of alienation that orients the self in modern/

postmodern societies, critical socioanalysis exposes the organizational principles of that society and the contradictions that they imply by working through anxiety. It is only by pushing these contradictions with a negative dialectic to the point that they reach their limit as paradoxes that the location of anxiety can be revealed as arising from the lacking society. Anxiety is not primarily a negative affect when it comes to critical socioanalysis, rather it is the affect that is revealed by the contradictory nature of capital and information in modern societies and has the potential to serve as a trigger to action, but only if and when material circumstances create an opening for this action. This action is not, as some subjectively desire, the revolution that will usher in the reign of the social and the realization of the ideal Human. The material transformations that capital has wrought on modern society have already transformed us into the cyborg subject, which is a different kind of subject whose future is not destined for the alternate history of humanity that was opened in the 19th century and closed in the 20th. This action, on the contrary, is a desire from within the lacking society for absolute difference which can only be obtained once the lacking society takes up the role of critiquing capital as a system of sameness by producing challenges to that system that cannot be easily assimilated into it.

The goal of outlining these discourses is to provide a framework for the socioanalyst to recognize what is structuring the thought of the analysand as they speak in analysis and move them through the discourses to the point that they will come to embody the desire of the analyst, a desire to continuously flush out the contradictions of life in which they are lacking subjects in a lacking society. The goal is to work toward meeting the precondition of thought so that if and when material conditions change, the totalizing logics that control the directionality of mass society can be deconstructed from within and replaced with something that more closely aligns with the desire for a more complete and harmonious life with and for all.

Conclusion to Part 3

It is evident by the material changes in the conditions of society during the late 20th and early 21st century that we are dealing with a new set of circumstances that necessitate an evolution in the critical method if we are to continue to expose the contradictions that shape our lives. By performing a recombinant innovation of the critical elements of psychoanalysis and the critical method of sociology into the structure of a critical socioanalysis, what these eight discourses of the self and society provide is a diagnostic model for understanding the murky relationships between the two in modern/postmodern societies within the analytic session. They underscore the critical need to expose, not just what the self and society portend to be, but the hidden truth that operates behind the scenes in their structural formation. But this is clearly not enough. If it were, then simply exposing the lacking subject and the lacking society would be sufficient to fulfilling the function of psychoanalysis and critical theory, and we could count their histories by the success of empirical transformations that align with their logics. Instead we must confront the fact that the problems they exposed in their origination have continued to compound, meaning that our method must be just as dynamic as the world in which we practice it if we are to continue to track these problems and their effect on the constitution of our self and our society.

By placing Lacan's discourses of psychoanalysis in a dialectic with the discourses of socioanalysis we can bolster the critical method by engaging in this new methodological practice so as to better account for the ways that it must deal with our socio-historical context, one in which the self is a cyborg and the society is a mass, and go beyond the level of merely exposing the realities of modern life for those whose subjective curiosity makes them want to better understand this condition. The necessity of confronting these transformations cannot be overstated, as Dahms (2008b) highlighted when he wrote that

> *to the degree that we refuse to address in a systematic manner, as an integral component of our work, the link between our practices as social scientists and the contradictions of modern society, with regard to concrete and specific consequences of our research, we may not only betray the claim to be social scientists, we actively—albeit unintentionally—may sabotage the possibility of social science.* (p. 17)

With mainstream theory and sociological practice dominating the scene as primarily a function of the discourse of capital and of the archive, critical theory must actively respond by targeting the lacking society as the object of its anxiety. Not merely to reproduce the function of critical theory in the archive, but as a means of triggering the desire for critical thought in society itself. This requires that we ask the right questions in socioanalysis by radically realigning theory so that the questions are based on the knowledge that

> if the world is hardly compatible with the concept of the real which we impose upon it, the function of theory is certainly not to reconcile it, but on the contrary, to seduce, to wrest things from their condition, to force them into an over-existence which is incompatible with the real.
> BAUDRILLARD [1987] 2012: 79

In other words, by setting the lacking society to work in our minds and uncovering the nonidentical features of our concepts in their contradiction to material reality, critical socioanalysis must overexpose the anxiety that this configuration produces. Furthermore, it must be more seductive than the commodity that capital uses, and the boundless promise of information, to cover up this lack so as to trigger a desire in the lacking subject, not for the production of capital or the escape into a virtual otherness, but for the production of knowledge as the critique of capital and information. Only in this manner can the lacking subject come to the realization that its existence is wholly incompatible with the world that capital is building for itself.

The discourse of socioanalysis provided above is only a first step in this direction by demonstrating how the model of critical socioanalysis can work within the analytic setting and how it functions in opposition to the discourses it is structurally linked to. The next step is to establish this as a necessary practice in the consciousness of critical social scientists while also continuing to reexamine the motives and impulses of critical theory. In that way we will make progress by critiquing how critical theory is practiced in modern societies today and how it must change to face these new realities. Like Lacan's return to Freud, there is a need for sociology to make a return to the critical method and to dialectically reread critical theories against our socio-historical context. This is perhaps the only way that we will be able to push anxiety to its limit and transfer the desire of critical socioanalysis to the lacking subjects who make up this lacking society, so that it can confront the unprecedented and myriad challenges that face the world today. What we face today is the amplification of the consequences of what Norbert Wiener (1954) summed up in his sociological diagnosis of this world that cybernetics remade and in it lies our challenge for the future:

> Whether we entrust our decisions to machines of metal, or to those machines of flesh and blood which are bureaus and vast laboratories and armies and corporations, we shall never receive the right answers to our questions unless we ask the right questions ... The hour is very late, and the choice of good and evil knocks at our door. (p. 185-186)

Conclusion

This project has traced the history of modern society from its troubling beginnings in traditional life, through the contradictions of modern life, to the absurd and chaotic nature of postmodern life. Each period related to the totalizing logics of religion, capital, and information in unique ways as the type of social domination that each form produced within material reality shaped the mind in ways that translated into segments of society thinking along particular avenues, as these logics push specific epistemologies and ontologies. Anxiety, with its future-orientation, is necessarily present in all societies that undergo social change, and as technology became embedded in everyday life, change largely began to be experienced through technology as it is guided by capital. With the rate of change linked to the generational advance of technology, our perception of reality has accelerated and so too has the prevalence of anxiety. But the object of our anxiety is difficult to locate because it is obscured by society and, given the diffusion of the means to cultivate the individual and social—which reached a saturation point under capital that rather than strengthening the forms made their essence disappear in the proliferation of their simulated and commodified versions—the ways of cultivating minds that can locate and develop these objects on their own became increasingly foreclosed upon in modern and postmodern societies. Furthermore, those who are most sensitive to the contradictory effects of living in these societies and who are, therefore, most likely to interrogate their reality, are also the most likely to suffer from anxiety and lose their sense of self as they become paralyzed by that anxiety; especially if they lack the tools to do so. But as critical socioanalysis suggests this is not a necessary occurrence, rather by working through the anxiety as a component of the development of the self in its relation to society, anxiety can be kept in check and reclaimed as a positive affect that warns us when there are happenings in our reality that require our attention and intervention, to the extent that the latter is even possible. Since society entered its mass stage, and the "self" assumed cyborg properties, locating our anxiety requires placing the concepts that we have historically used to think of our self and our society in a dialectical relationship with the material reality that we face to expose the contradictions inherent in them. To do this, we must pass through our anxiety, time and again, as we draw closer to the sober truth of our reality. It can be a painful process to reshape our thought patterns, especially when living in a dynamic society that requires our constant reshaping, and although there are some avenues in society that attempt this process, especially in the arts and in a variety of therapeutic approaches, if they do not incorporate a dialectical critique of how their approach interacts with and is in turn shaped by the structure of mass society, then there is a high likelihood of their being unintentionally coopted into the service of the totalizing logics. This is why the critical socioanalyst must create an artificial environment that is continuously subjected to a critique of its own foundations and which stands in as the object of this desire to know for those who wish to gain a better sense of how they fit in to this world, why their "self" is at odds with society, where their anxiety is coming from, and how to avoid the paralysis it causes.

Given the problems that we face as a species and as a part of the planetary totality, the only way that we will open the possibility of confronting these problems in a rational manner is by embracing a radical means of revolutionizing the way that we conceive of ourselves and our societies. The cost of our failure to accomplish this increasingly appears to be devastating to the future of life itself. The elites in society, those who have access and control over the dwindling supply of material resources and have a disproportionate share of the available means to mobilize the populace to use labor to build paths to different futures now, are increasingly deciding that such qualitative changes are impossible, and as such, they are making plans to escape and protect their own interests while the masses are abandoned and left behind to sort through the wreckage of earth. This does not mean that the elites have the power to radically alter the mass of society in a rational manner, only that given that the primary driver of the mass is capital and information, and that the elites control most of these resources, they have the greatest ability within the current framework of pushing for alternative modes of societal organization by deploying those resources. Paradoxically, however, this also would mean that they would have to abandon the societal configurations that grant them power by using their resources in ways that would undermine those sources of power in modern and postmodern societies.

Existential threats abound, not only environmentally, but also politically, economically, and technologically, as the global order is being reshaped into a new divide between those who are at best instituting reactionary policies that attempt to hold onto the status quo for as long as possible by delaying the inevitability of the coming tide, and those who are encouraging people to ignore what is coming by enhancing and building new types of social domination in an attempt to control the increasing anxiety of the masses,[1] of which they

---

1 Such as, but not limited to, enhanced physical and virtual surveillance—with the most overt version of this being China's social credit system which tracks and rates individuals and assigns them a score based on that tracking—, the continued dismantling of the welfare state—which compels people by limiting their access to basic necessities thereby

too are a part, and which threatens the structure of the power that they wield in mass society because of its destabilizing effects. Neither side is, however, seriously considering ways to reconfigure society in such a manner that it will meet the needs of all, assuming that such an outcome is itself possible and not merely an ideal that lingers from the domination of religiously inspired fantasy. As the history reconstructed in this text and as the do-nothing political response to the COVID-19 pandemic in the United States illustrates, there is no limit to the horrors that the system of capital is willing to inflict upon the masses to maintain its circulation and growth imperative. But as the generations who experienced the World Wars firsthand are passing away, so too, it seems, the social lessons that our species should have learned from these mass events are disappearing from the collective conscious. Rather than funnel our resources and social power into strategies that could improve the lives of peoples around the world, nation states and those who control the mass of capital have reinvested in the technologies of war and social control as they fear the loss of their power over mass society; which as noted above, is largely an illusory power and one which as it is exposed is leading to more overt forms of control in an attempt to hang on to that power for as long as possible. Recognizing that the elites are also a part of the masses, they too are subject to the anxiety that emanates from this configuration of society, and they are increasingly aware that their interventions do not immunize them from this anxiety and may, in fact, only serve to amplify it. There is every indication that unless new models of addressing these problems are developed, then the types of social domination we experience today will continue to intensify, at least until the elites are convinced that their escape plans are secure, and they abandon the maintenance of these types of social domination and leave the masses to deal with the burning planet they abandoned.

To address this scale of problems, our collective sense of self needs to be reprogramed, but without turning to authoritarian means this can only be accomplished in a dialectic of self and society that is undertaken through an internal reprograming of the self by the self, and not in any top down manner that imposes a new code from without which would only recreate another type of social domination and individual control. Rather this must be a reprograming of the self that originates in the self. This is what critical socioanalysis has as its goal, but given the vast control that elites have over institutional

---

compelling them to work in worse conditions for lower pay—, and the increased reliance on AI algorithms to predict future individual behaviors—for example, in the criminal justice system with these algorithms being used to calculate the sentencing of repeat criminal offenders.

avenues which could support the widespread implementation of such a practice, its success is bound to be limited unless they are convinced to support such endeavors. To demonstrate the value of this practice, elites should be one of the first and primary targets of critical socioanalysis because to convince them that it is of mutual benefit, and not just of their own personal benefit, they must revolutionize their sense of self and come to see that the endless growth of capital will never fill their lack, just as it will never fill that of society, and failures to address these problems now will only mean that they will be recreated in whatever utopian escape they dream up. Working from the critical perspective of the mass, which includes the totality of the planet and its subjects, critical socioanalysis takes a step in the direction of aiding those who are concerned with the future to better locate this lack and in doing so recognize what possibilities for the future remain available to us.

Whereas psychoanalysis often targets specific issues that the analysand is confronting in their personal life, and therefore, once the desired effect is achieved the analysand can claim to have been "cured" by the process, since the subject of socioanalysis is mass society there is no end to the necessity of confronting this reality and its effects on us unless the mass disappears. Participating in critical socioanalysis should therefore become a permanent feature in the lives of modern and postmodern subjects so long as they desire to construct their own rich life histories and biographies rather than simply fill the prescribed roles dictated to them by the logics of capital and information. To the extent that we are concerned with the future it is imperative that we gain a critical understanding of the past and the ways that the human species constructed the posthuman reality we now inhabit. As new challenges arise that exist at the level of the mass, the necessity of working together through the division of labor to shape the directionality of the future whether on planet earth, in virtual worlds, or through space colonization, working through the relationship of the self and society and the impact anxiety has on our psychosocial makeup is of paramount importance. The earlier that we begin this process, the better prepared people will be to face the coming tide as new generations will have to face the intensification of the system that we live in now.

One way to address this would be to integrate critical socioanalysis in the educational system and begin to train children not just in the bureaucratic knowledge that prepares them for jobs that demand the application of technical rationality, but to help them learn how to make meaningful lives that are fulfilling and that will open new avenues for that fulfillment in the future. By implementing critical socioanalysis in the educational system, it would also provide scientific researchers with a more in depth and personal look at these interconnections as they evolve generationally so as to better recognize when and how material transformations in society recode the minds of the species. By tracking these changes throughout the life course, not only will we have a better understanding of how new technologies alter the experience of youth, but it will better help the older generations to understand how their actions shape the future that the youth will inherit, rather than continue to operate under the atemporal assumption that the future is simply a continuation of their experience.

For adults the process will in all likelihood be more painful, than it would be with children who early on make it a part of their routine, as it requires going back further and deeper to access the damage caused by living in the "wrong" society. While this practice promises to be of value for all, it will be of special importance for those who are directly influencing technology and capital, so that they can come to a better understanding of the effects their actions have on the mass. Given the division of labor, it is not practical to assume that all politicians, engineers, computer scientists, finance capitalists, and many others, will have or take the time to seek out for themselves the historical narratives in which they are participating to fully understand the impact of their actions on the totality. However, by collaborating with critical socioanalysts who have performed that labor, they can integrate these critiques into their designs by working through the issues that they hope to address within the analytic session and develop for themselves a more critical manner of thought that is not guided by instrumental and technical rationality. In this way sociological knowledge can be used at a personal level in a manner that it has largely never seriously confronted by working in a new model of collaborative critical thought with the public. If the species is to take seriously the need to alter and change our sense of identity in this new reality, then beginning this process is of urgent necessity. Critical socioanalysis does not promise a return to our humanity, rather it explores what and how facets of humanity that are worth preserving can be integrated with our technological reality and examine if there is a possibility of building a reality that excludes the problematic features of both while it incorporates the desirable aspects of each without sacrificing that which allows us to make meaning and find fulfillment in life.

By beginning the process of confronting the anxiety of modern and postmodern societies, it allows us to get a better grasp on how our symptoms are produced by the material conditions that shape our lives. As we track alienation, anomie, the Protestant ethic, repression, instrumental and technological rationality, one-dimensional thought, identity thinking, the accident, loss of meaning, and other symptoms of mass society, we will begin to recognize precisely what we are up against in this life, and although it will agitate our anxiety, it will also have a therapeutic effect as we begin to see the truth of this world and our roles in it. Only then will we have any hope of applying reason and the tools available to us in ways that are conducive to the future of life itself on planet earth in a manner that reduces the tensions of our collective existential threats.

This text is only a primer to critical socioanalysis. It traces certain aspects of our history and the evolution of the critical method, but it has several limitations. I have only briefly evaluated the coming technologies, but artificial intelligence, automation, and the renewed interest in space colonization with its billionaire backers, must be more fully theorized and tracked than they are here. It will be especially important to see how these technologies integrate with world militaries and government, as well as what steps, if any, government will take to intervene in their corporate use. Sociology needs to insert itself in these processes if it wishes to remain a relevant science to present concerns and those that are likely to arise in the near future. Lacking the social as a clear object, this means that sociology must renew its attempts to study the mass no matter how elusive and contradictory such an object is, even if this means the development of a new science that is more appropriate to this task. The only way that this can be accomplished is with a renewed commitment to the critical method by tracking the history of the totalizing logics. These technologies are not the true object of anxiety, although they are often labeled as the cause of anxiety about the future as they threaten the labor force and the autonomy of the species. But this is not entirely due to the technologies themselves, rather, this is a result of their being coded according to the totalizing logics of capital and information, which they amplify as a form of rationality that is not aligned to the interests of life itself, but rather to those of capital as an artificial intelligence that has become the subject of history. Those logics are the source of why these technologies appear in an uncertain and potentially hazardous light, but that uncertainty has more to do with the ways that they will amplify the totalizing logics and further enhance social domination than with the forms and projects themselves. The environmental catastrophes that science is predicting are likewise of this order, it is not the environment and climate change that are the real object of anxiety, the uncertainty of the climate is not due to the manifestly obvious increase in ecological and natural disasters. Rather those manifest changes only remind us of the hidden reality of our world, that which seems to persist beneath our simulation, and by passing through anxiety its negative object is revealed. The cause of these problems is the blind reliance on the totalizing logics as if they represented the rational interests of life itself, rather than the interests of capital and a select few of its representatives who benefit from the destruction and exploitation of the masses and our environment in the present at the cost of the future. However, lacking the social as a means to challenge this state of affairs means that our anxiety cannot be discharged on the object that produces it through a bottom up approach. It is the lack of a visible object, our lack and the lack of our societies, that we must come to understand in the socioanalytic session and how working through and from within those logics we can develop viral recoding strategies to change their directionality and thrust. For the socioanalyst, we need to build a better understanding of where these logics are heading if we are to adequately explore them in analysis with our analysands as a part of the scenic landscape of our lives, and to do this, we must produce more critical knowledge on these coming technological changes.

Furthermore, as artificial intelligence, automation, and space colonization progress, they open new avenues for the possible development of socioanalysis. By following the methods laid out in this text, socioanalysts could contribute to these spheres. If automation leads to mass unemployment, then assuming that the needs of the species are either met or ignored by the system, the reconfiguration of the self in that society will require serious efforts to delink in thought the commodification of labor-power from the self as it will be an absolute necessity to reconceive of the purpose of life and find new avenues of meaning and fulfillment that are not guided by the imperatives of capital. Socioanalysis can also help AI researchers to distinguish how they are imprinting the totalizing logics in their coding of AI. Socioanalysis can serve as a useful setting for exploring how we may think of alternative guiding logics that do not merely reproduce the social domination of modern society in a new technical form, and therefore can help us to think of novel uses for AI that benefit the mass rather than using it to control and manipulate the mass. Finally, with space colonization our cyborg selves will be entering a fully artificial world and the effects it has on the self must be tracked unless we are willing to leave the quality of life in these new worlds to chance. By integrating critical socioanalysis into the life of those who live in these worlds, not only will it help them to better understand and confront the anxieties that are bound to arise from their new conditions, but it will also help us produce a comparative case to life on earth; the value for the social sciences of this comparative case cannot be overstated.

The other limitation to this text is that it offers a theory of the discourses and a heuristic model for tracking them in socioanalysis, but it does not go through and illustrate how these discourses will emerge with an empirical example. What this text does is empirically demonstrate the need for socioanalysis, but it does not empirically demonstrate the effect of socioanalysis. As Freud said of psychoanalysis, the proof of the effect can only be experienced by going through the analytic process. If we want to explore what the incorporation of critical socioanalysis can offer in our lives, then we will only discover this within the practice as different life histories and desires meet in the analytic session. And since we are dealing with an accelerating society, projects such as socioanalysis cannot be the result of one person's efforts alone. It will take the efforts of several like-minded scientists, driven by the desire of the critical method, to build the full foundation for this new practice. Therefore, this text represents my desire as an analyst for critical thought, and it is my hope that this desire will be sparked in my readers. For those who recognize the seed of the desire to know growing within them, this is an invitation to help cultivate that desire in others by working together to establish critical socioanalysis as a new practice for the 21st century as we brace for the coming tide.

Bibliography

Abbate, J. (1999). *Inventing the Internet.* Cambridge, MA: The MIT Press.

Aberbach, D. (2003). *Major Turning Points in Jewish Intellectual History.* New York: Palgrave Macmillan.

Abromeit, J. (2011). *Max Horkheimer and the Foundations of the Frankfurt School.* Cambridge, UK: Cambridge University Press.

Adorno, T. ([1951] 2005). *Minima Moralia: Reflections from Damaged Life.* (E. F. Jephcott, Trans.) London: Verso.

Adorno, T. (1967). Sociology and Psychology (Part 1). *New Left Review I, 46.*

Adorno, T. (1968). Sociology and Psychology (Part II). *New Left Review I, 47.*

Adorno, T. W. ([1931] 2000). The Actuality of Philosophy. In B. O'Connor (Ed.), *The Adorno Reader* (pp. 23–39). Oxford, UK: Blackwell Publishers.

Adorno, T. W. ([1958] 1991). *Notes to Literature, Volume 1.* (R. Tiedemann, Ed., & S. W. Nicholsen, Trans.) New York: Columbia University Press.

Adorno, T. W. ([1962] 1989). *Kierkegaard: Construction of the Aesthetic.* (R. Hullot-Kentor, Ed., & R. Hullot-Kentor, Trans.) Minneapolis, MN: University of Minnesota Press.

Adorno, T. W. ([1963] 1993). *Hegel: Three Studies.* (S. W. Nicholsen, Trans.) Cambridge, MA: The MIT Press.

Adorno, T. W. ([1966] 2007). *Negative Dialectics.* (E. B. Ashton, Trans.) New York: Continuum.

Adorno, T. W. ([2003] 2008). *Lectures on Negative Dialectics.* (R. Tiedemann, Ed., & R. Livingstone, Trans.) Cambridge, UK: Polity Press.

Adorno, T. W. (2005). Education After Auschwitz. In T. W. Adorno, *Critical Models: Interventions and Catchwords* (H. W. Pickford, Trans., pp. 191–204). New York: Columbia University Press.

Adorno, T. W. ([2010] 2017). *An Introduction to Dialectics.* (C. Ziermann, Ed., & N. Walker, Trans.) Cambridge, UK: Polity Press.

Adorno, T. W. (2018a). Social Sciences and Sociological Tendencies in Psychoanalysis. In W. Bock, *Dialektische Psychologie: Adornos Rezeption der Psychoanalyse* (pp. 623–642). Weisbaden, Germany: Springer vs.

Adorno, T. W. (2018b). Theodor W. Adorno on 'Marx and the Basic Concepts of Sociological Theory'—From a Seminar Transcript in the Summer Semester of 1962. *Historical Materialism*, 1–11. doi:10.1163/1569206X-00001619.

Adorno, T. W., Albert, H., Dahrendorf, R., Habermas, J., Pilot, H., & Popper, K. R. ([1969] 1976). *The Positivist Dispute in German Sociology.* (G. Adey, & D. Frisby, Trans.) New York: Harper & Row.

Alexander, J. C. (1986). Rethinking Durkheim's Intellectual Development I: On 'Marxism' and the Anxiety of Being Understood. *International Sociology, 1*(1), 91–107.

Alexander, J. C. (2013). *The Dark Side of Modernity.* Cambridge, UK: Polity.

American Diabetes Association. (2015, March 16). *American Diabetes Association.* Retrieved from Insulin Pumps Need Greater Safety Review: American Diabetes Association Issues Joint Statement with European Association for the Study of Diabetes: http://www.diabetes.org/newsroom/press-releases/2015/insulin-pumps.html.

American Psychiatric Association. (1952). *Diagnostic and Statistical Manual of Mental Disorders (DSM-I).* Washington, D.C.: American Psychiatric Association.

American Psychiatric Association. (1968). *Diagnostic and Statistical Manual of Mental Disorders (DSM-II).* Washington, D.C.: American Psychiatric Association.

American Psychiatric Association. (1980). *Diagnostic and Statistical Manual of Mental Disorders (DSM-III).* Washington, D.C.: American Psychiatric Association.

American Psychiatric Association. (1994). *Diagnostic and Statistical Manual of Mental Disorders (DSM-IV).* Washington, D.C.: American Psychiatric Association.

American Psychiatric Association. (2013). *Diagnostic and Statistical Manual of Mental Disorders (DSM-5).* Washington, D.C.: American Psychiatric Association.

Antonio, R. J. (1981). Immanent Critique as the Core of Critical Theory: Its Origins and Developments in Hegel, Marx and Contemporary Thought. *The British Journal of Sociology, 32*(2), 330–345.

Antonio, R. J., & Bonanno, A. (2000). A New Global Capitalism? From "Americanism to Fordism" to "Americanization-Globalization". *American Studies, 33*, 77.

Antonio, R. J., & Bonanno, A. (2012). Fordism. In H. K. Anheier, & M. Juergensmeyer, *Encyclopedia of Global Studies* (pp. 582–583). Thousand Oaks, CA: SAGE.

Anxiety and Depression Association of America. (2018). *Facts & Statistics.* Retrieved from: https://adaa.org/about-adaa/press-room/facts-statistics.

Anzaldua, G. ([1987] 1999). *Borderlands/La Frontera* (Second ed.). San Fransisco, CA: Aunt Lute Books.

Arrighi, G. ([1994] 2010). *The Long Twentieth Century: Money, Power, and the Origins of our Times.* London: Verso.

Bain, A. (1999). On Socio-Analysis. *Socio-Analysis, 1*(1), 1–12.

Bainbridge, W. S. (1977). *The Space Flight Revolution: A Sociological Study.* Hoboken, NJ: John Wiley & Sons.

Bainbridge, W. S. (2015). *The Meaning and Value of Space Flight: Public Perceptions.* London: Springer.

Baker, W. M. (1970). A Case Study of Anti-Americanism in English Speaking Canada: The Election Campaign of 1911. *The Canadian Historical Review, 51*(4), 426–449.

Barrat, J. (2013). *Our Final Invention: Artificial Intelligence and the End of the Human Era.* New York: St. Martin's Press.

Bataille, G. ([1949] 1991). *The Accursed Share: An Essay on General Economy.* (R. Hurley, Trans.) New York: Zone Books.

Bataille, G. (1985). *Visions of Excess: Selected Writings, 1927–1939.* (A. Stoekl, Ed., A. Stoekl, C. R. Lovitt, & D. M. Leslie Jr., Trans.) Minneapolis, MN: University of Minnesota.
Baudrillard, J. ([1968] 2005). *The System of Objects.* (J. Benedict, Trans.) London: Verso.
Baudrillard, J. ([1970] 1998). *The Consumer Society.* (C. Turner, Trans.) London: Sage.
Baudrillard, J. ([1972] 1981). *For a Critique of the Political Economy of the Sign.* (C. Levin, Trans.) St. Louis, MO: Telos Press.
Baudrillard, J. ([1978] 2007). *In the Shadow of the Silent Majorities or the End of the Social.* (P. Foss, J. Johnston, P. Patton, & A. Berardini, Trans.) Los Angeles: Semiotext(e).
Baudrillard, J. ([1981] 1994). *Simulacra and Simulation.* (S. F. Glaser, Trans.) Ann Arbor, MI: The University of Michigan Press.
Baudrillard, J. ([1986] 2010). *America.* (C. Turner, Trans.) London: Verso.
Baudrillard, J. ([1987] 1990). *Cool Memories.* (C. Turner, Trans.) London: Verso.
Baudrillard, J. ([1987] 2012). *The Ecstasy of Communication.* (B. Schütze, & C. Schütze, Trans.) Los Angeles: Semiotext(e).
Baudrillard, J. ([1990] 1996). *Cool Memories II.* (C. Turner, Trans.) Durham, NC: Duke University Press.
Baudrillard, J. ([1990] 2008). *Fatal Strategies.* (P. Beitchman, & W. G. Niesluchowski, Trans.) Los Angeles: Semiotext(e).
Baudrillard, J. ([1992] 1994). *The Illusion of the End.* (C. Turner, Trans.) Stanford, CA: Stanford University Press.
Baudrillard, J. ([1995] 1997). *Fragments: Cool Memories III.* (E. Agar, Trans.) London: Verso.
Baudrillard, J. ([2000] 2003). *Cool Memories IV.* (C. Turner, Trans.) London: Verso.
Baudrillard, J. ([2005] 2006). *Cool Memories V.* (C. Turner, Trans.) Cambridge, UK: Polity.
Bauman, Z. (1993). *Postmodern Ethics.* Oxford, UK: Blackwell.
Bauman, Z. (1995). *Life in Fragments: Essays on Postmodern Morality.* Oxford, UK: Blackwell.
Bauman, Z. (1997). *Postmodernity and its Discontents.* Cambridge, UK: Polity Press.
Bauman, Z. (2014). *What Use is Sociology?: Conversations with Michael-Hviid Jacobsen and Keith Tester.* Cambridge, UK: Polity Press.
Beck, U. (1992). *Risk Society: Towards a New Modernity.* London: Sage Publications.
Belich, J. (2016). The Black Death and the Spread of Europe. In J. Belich, J. Darwin, M. Frenz, & C. Wickman, *The Prospect of Global History* (pp. 93–107). Oxford, UK: Oxford University Press.
Bell, D. ([1960] 2000). *The End of Ideology: On the Exhaustion of Political Ideas in the Fifties.* Harvard University Press.
Bell, D. ([1973] 1999). *The Coming of Post-Industrial Society.* New York: Basic Books.
Benjamin, W. (2008). *The Work of Art in the Age of its Technological Reproducibility and Other Writings on Media.* (M. W. Jennings, B. Doherty, T. Y. Levin, Eds., E. Jephcott, R. Livingstone, & H. Eiland, Trans.) Cambridge, MA: The Belknap Press.

Bernard, G. W. (2005). *The King's Reformation: Henry VIII and the remaking of the English church.* New Haven, CT: Yale University Press.

Bidmead, C. (1994). IBM's Simon. *Which Computer?, 17*(1), 17.

Bijker, W. E. (1995). *Of Bicycles, Bakelites, and Bulbs.* Cambridge, MA: The MIT Press.

Binder, R., Abramson, N., Kuo, F., Okinaka, A., & Wax, D. (1975). ALOHA Packet Broadcasting—A Retrospect. *AFIPS '75 Proceedings of the May 19-22, 1975, National Computer Conderence and Exposition* (pp. 203-215). Anaheim, CA: Association for Computing Machinery.

Bion, W. R. ([1961] 2004). *Experiences in Groups and Other Papers.* New York: Brunner-Routledge.

Blackbourn, D. (1991). The Catholic Church since the French Revolution. *Comparative Studies in Society and History*, 778-790.

Bloch, M. ([1939] 2014). *Feudal Society.* London: Routledge.

Bose, C. (1979). Technology and Changes in the Division of Labor in the American Home. *Women's Studies International Quarterly, 2*(3), 295-304.

Bostrom, N. (2002). Existential Risks: Analyzing Human Extinction Scenerios and Related Hazards. *Journal of Evolution and Technology, 9*(1).

Bostrom, N. (2003). Are We Living in a Computer Simulation. *The Philosophical Quarterly, 53*(211), 243-255.

Bostrom, N. (2013). Existential Risk Prevention as Global Priority. *Global Policy, 4*(1), 15-31.

Bostrom, N. (2014). *Superintelligence: Paths, Dangers, Strategies.* Oxford, UK: Oxford University Press.

Bostrom, N., & Kulczucki, M. (2011). A Patch for the Simulation Argument. *Analysis, 71*(1), 54-61.

Boucher, G. (2006). Bureaucratic Speech Acts and the University Discourse: Lacan's Theory of Modernity. In J. Clemens, & R. Grigg (Eds.), *Jacques Lacan and the Other Side of Psychoanalysis: Reflections on Seminar XVII* (pp. 274-291). Durham, NC: Duke University Press.

Bourdieu, P. (1986). The Forms of Capital. In J. Richardson (Ed.), *Handbook of Theory and Research for the Sociology of Education* (pp. 241-258). New York: Greenwood.

Bourdieu, P. (1989). Social Space and Symbolic Power. *Sociological Theory, 7*(1), 14-25.

Bow, B., & Chapnick, A. (2016). Teaching Canada-US relations: Three great debates. *International Journal, 71*(2), 291-312.

Bowen, W. H. (2006). *Spain During World War II.* Columbia, MO: University of Missouri Press.

Boyle, A. (2018, April 12). *Boeing CEO Denis Muilenburg Says that Humans Could Get to Mars Within a Decade.* Retrieved from GeekWire: https://www.geekwire.com/2018/boeing-ceo-dennis-muilenburg-mars-decade/.

Brand, J. E. (2015). The Far-Reaching Impact of Job Loss and Unemployment. *Annual Review of Sociology, 41*, 359–375.

Braudel, F. (1958). Histoire at Sciences sociales: La longue durée. *Annales. Économies, Sociétés, Civilisations, 13*(4), 725–753.

Braverman, H. ([1974] 1998). *Labor and Monopoly Capital: The Degradation of Work in the Twentieth Century.* New York: Monthly Review Press.

Bridgstock, M., Burch, D., Forge, J., Laurent, J., & Lowe, I. (1998). *Science, Technology and Society: An Introduction.* Cambridge, UK: Cambridge University Press.

Broadberry, S., Federico, G., & Klien, A. (2010). Sectoral developments, 1870–1914. In S. Broadberry, & K. O'Rourke (Eds.), *The Cambridge Economic History of Modern Europe, Volume 2: 1870 to the Present* (pp. 59–83). Cambridge, UK: Cambridge University Press.

Brownlee, W. E. (2004). *Federal Taxation in America: A Short History* (Second ed.). Cambridge, UK: Cambridge University Press.

Brynjolfsson, E., & McAfee, A. (2012). *Race Against the Machine: How the Digital Revolution is Accelerating Innovation, Driving Productivity, and Irreversibly Transforming Employment and the Economy.* Digital Frontier Press.

Brynjolfsson, E., & McAfee, A. ([2014] 2016). *The Second Machine Age: Work, Progress, and Prosperity in a Time of Brilliant Technologies.* New York: W. W. Norton & Company.

Buck-Morss, S. (1977). *The Origin of Negative Dialectics: Theodor W. Adorno, Walter Benjamin, and the Frankfurt Institute.* New York: The Free Press.

Bueno de Mesquita, B. (1990). Pride of Place: The Origins of German Hegemony. *World Politics, 43*(1), 28–52.

Bureau International des Expositions. (2018, February 2). Retrieved from Bureau International des Expositions: http://www.bie-paris.org/site/en.

Bureau of Labor Statistics. (2017, June 27). *American Time Use Survey.* Retrieved from: https://www.bls.gov/charts/american-time-use/activity-by-sex.htm.

Bush, V. (1945). As We May Think. *The Atlantic.* Retrieved from https://www.theatlantic.com/magazine/archive/1945/07/as-we-may-think/303881/.

Byler, D. and C. Sanchez Boe. (2020). "Tech-enabled 'terror capitalism' is spreading worldwide. The surveillance regimes must be stopped." *The Guardian*, July 24. Retrieved from: https://www.theguardian.com/world/2020/jul/24/surveillance-tech-facial-recognition-terror-capitalism.

Calhoun, C. (1989). Classical Social Theory and the French Revolution of 1848. *Sociological Theory, 7*(2), 210–225.

Calvo, R. A., Peters, D., & D'Mello, S. (2015). When Technologies Manipulate Our Emotions. *Communications of the ACM, 58*(11), 41–42.

Campbell-Kelly, M., & Aspray, W. (2013). *Computer: A history of the information machine.* New York: Westview Press.

Cardwell, D. S. (1994). *The Fontana History of Technology*. London: Harper.
Carlson, W. B. (2013). *Tesla: Inventor of the Electrical Age*. Princeton, NJ: Princeton University Press.
Carreras, A., & Josephson, C. (2010). Aggregte growth, 1870–1914: growing at the production frontier. In S. Broadberry, & K. H. O'Rourke (Eds.), *The Cambridge Economic History of Modern Europe: Volume 2, 1870 to the Present* (pp. 30–58). Cambridge, UK: Cambridge University Press.
Casson, H. ([1910] 1922). *The History of the Telephone* (Tenth ed.). Chicago, IL: A. C. McClurg & Co.
Caudwell, C. ([1939, 1949] 2009). *Studies and Further Studies in a Dying Culture*. New York: Monthly Review Press.
Cavalletto, G. (2016). *Crossing the Psycho-Social Divide*. Abingdon, UK: Routledge.
Center for the Digital Future. (2017). *Surveying the Digital Future*. Los Angeles, CA: Center for the Digital Future at USC Annenberg.
Chadeau, E. (1987). *De Blériot à Dassault: Histoire de l'industrie aéronautique en France, 1900–1950*. Paris: Fayard.
Chancer, L., & Andrews, J. (2014). *The Unhappy Divorce of Sociology and Psychoanalysis*. (L. Chancer, & J. Andrews, Eds.) New Yorrk: Palgrave Macmillan.
Chandler Jr., A. D. (1977). *The Visible Hand: The Managerial Revolution in American Business*. Cambridge, MA: The Belknap Press.
Cheney, M. (1981). *Tesla: Man Out of Time*. New York: Simon & Schuster.
Chitika. (2013, June 12). *The Value of Google Result Positioning*. Retrieved from Chitika: https://chitika.com/2013/06/07/the-value-of-google-result-positioning-2/.
Churchill, W. (1946). *Winston Churchill's Iron Curtain Speech*. Retrieved from The History Guide: Lectures on Twentieth Century Europe: http://www.historyguide.org/europe/churchill.html.
Clancy, B. (2017). Rebel or Rioter? Luddites then and now. *Society, 54*, 392–398.
Clark, C. (1996). Confessional Policy and the Limits of State Action: Frederick William III and the Prussian Church Union 1817–1840. *The Historical Journal*, 985–1004.
Clark, C. (2006). *Iron Kingdom: The Rise and Downfall of Prussia, 1600–1947*. New York: Penguin.
Claussen, D. ([2003] 2008). *Theodor W. Adorno: One Last Genius*. (R. Livingstone, Trans.) Cambridge, MA: The Belknap Press.
Clynes, M. E., & Kline, N. S. (1960, September). Cyborgs in Space. *Astronautics*, 26–27, 74–76.
Coleman, F. M. (1977). *Hobbes and America: Exploring the Constitutional Foundations*. Toronto, Canada: University of Toronto Press.
Collins, C., & Hoxie, J. (2017). *Billionaire Bonanza: The Forbes 400 and the Rest of Us*. Washington, DC: Institute for Policy Studies. Retrieved from https://inequality.org/wp-content/uploads/2017/11/BILLIONAIRE-BONANZA-2017-Embargoed.pdf.

Collins, R., & Makowski, M. (1993). *The Discovery of Society* (5th ed.). New York: McGraw-Hill.

Collopy, P. D. (2004). Military Technology Pull and the Structure of the Commercial Aircraft Industry. *Journal of Aircraft, 41*(1), 85–94.

Comin, D. A., & Mestieri, M. (2010). The Intensive Margin of Technology Adoption. *Harvard Business School BGIE Unit Working Paper No. 11–026*, 1–45.

Comin, D., Hobijn, B., & Rovito, E. (2008). Technology Usage Lags. *Journal of Economic Growth, 13*(4), 237–256.

Comte, A. (1858). *The Catechism of Positive Religion.* London: John Chapman.

Comte, A. ([1896] 2000). *The Positive Philosophy of Auguste Comte, Vol. 1.* (H. Martineau, Trans.) Kitchener, Ontario: Batoche Books.

Cotesta, V. (2017). Classical Sociology and the First World War: Weber, Durkheim, Simmel and Scheler in the Trenches. *History: The Journal of the Historical Association, 102*(351), 432–449.

Cowan, R. S. (1976). The "Industrial Revolution" in the Home: Household Technology and Social Change in the 20th Century. *Technology and Culture, 17*(1), 1–23.

Cowan, R. S. (1997). *A Social History of American Technology.* New York: Oxford University Press.

Crary, J. (2013). *24/7: Late Capitalism and the Ends of Sleep.* London: Verso.

Creary, S. J., & Gordon, J. R. (2006). Role Conflict, Role Overload, and Role Strain. In C. L. Shehan, *The Wiley Blackwell Encyclopedia of Family Studies.* Hoboken, NJ: Wiley-Blackwell.

Crocq, M.-A. (2015). A history of anxiety: from Hippocrates to DSM. *Dialogues in Clinical Neuroscience, 17*(3), 319–325.

Crombez, J. (2015, May). *After the Human: Theory and Sociology in the Age of Fractal Ambiguity, Dromology, and Emergent Epi-spaces.* Master's Thesis, University of Tennessee, Knoxville. Retrieved from http://trace.tennessee.edu/utk_gradthes/3356.

Crombez, J., & Dahms, H. F. (2015). Artificial Intelligence and the Problem of Digital Ontotheology: Toward a Critical Rethinking of Science Fiction as Theory. *Bulletin of Science, Technology and Society, 35*(3–4), 104–113.

Crombez, J. *(forthcoming).* Critical Socioanalysis and the Critique of Religion, or, Why I Read Theory: Gloria Anzaldúa, Jacques Lacan, and Memories of Latin America. In *Planetary Sociology: Linking Identity and Social Structure*, edited by H. F. Dahms.

Cummins, L. (1976). *Internal Fire: The Internal Combustion Engine 1673–1900.* Lake Oswego, OR: Carnot Press.

Curmer, H. L. (1843–44). *L'Industrie Exposition des Produits de l'Industrie Française en 1844.* Paris: L. Curmer. Retrieved from http://tinyurl.galegroup.com/tinyurl/5sxdP2.

Curry, A. (2003). *Essential Histories: The Hundred Years' War 1337–1453.* New York: Routledge.

Daggett, S. (2010). *Costs of Major U.S. Wars (CRS Report for Congress)*. Washington, D.C.: Congressional Research Service.

Dahlman, L. (2017, September 11). *Climate Change: Global Temperature*. Retrieved from National Oceanic and Atmospheric Administration: https://www.climate.gov/news-features/understanding-climate/climate-change-global-temperature.

Dahmer, H. (2012). Adorno's View of Psychoanalysis. *Thesis Eleven, 111*(1), 97–109.

Dahms, H. F. (2006). Capitalism Unbound? Peril and Promise of Basic Income. *Basic Income Studies: An International Journal of Basic Income Research, 1*(1), 1–6.

Dahms, H. F. (2007). Confronting the Dynamic Nature of Modern Social Life. *Soundings: An Interdisciplinary Journal, 90*(3/4), 191–205.

Dahms, H. F. (2008a). Alienation. In V. N. Parillo (Ed.), *Encyclopedia of Social Problems* (pp. 40–42). Thousand Oaks, CA: SAGE.

Dahms, H. F. (2008b). How Social Science is Impossible Without Critical Theory: The Immersion of Mainstream Approaches in Time and Space. (H. F. Dahms, Ed.) *Current Perspectives in Social Theory, 25*, 3–61.

Dahms, H. F. (2011). *The Vitality of Critical Theory*. Bingley, UK: Emerald.

Dahms, H. F. (2015). Which Capital, Which Marx? Basic Income between Mainstream Economics, Critical Theory, and the Logic of Capital. *Basic Income Studies, 10*(1), 115–140.

Dahms, H. F. (2017a). Critical Theory in the Twenty-First Century: The Logic of Capital Between Classical Social Theory, the Early Frankfurt School Critique of Politcal Economy and the Prospect of Artifice. In D. Krier, & M. P. Worrell (Eds.), *The Social Ontology of Capitalism* (pp. 47–74). London: Palgrave Macmillan.

Dahms, H. F. (2017b). Critical Theory as Radical Comparative-Historical Research. In M. J. Thompson (Ed.), *The Palgrave Handbook of Critical Theory* (pp. 165–184). New York: Palgrave Macmillan.

Dahms, H. F. (2018). Critical Theory, Radical Reform, and Planetary Sociology: Between Impossibilitiy and Inevitability. In L. Langman, & D. A. Smith (Eds.), *Twenty-First Century Inequality & Capitalism: Piketty, Marx and Beyond* (pp. 152–168). Leiden, Netherlands: Brill.

Dahms, H. F. (2019). Critical Theory Derailed: Paradigm Fetishism and Critical Liberalism in Honneth (and Habermas). In V. Schmitz (Ed.), *Axel Honneth and the Critical Theory of Recognition*. Houndmills, UK: Palgrave.

Dahms, H. F. (Forthcoming). *Modern Society as Artifice: Critical Theory and the Logic of Capital*. Abingdon, UK: Routledge.

Dalton, D. (2011). The Object of Anxiety: Heidegger and Levinas and the Phenomenology of the Dead. *Janus Head, 12*(2), 67–83.

Daniell, C. (2013). *From Norman Conquest to Magna Carta*. London: Taylor and Francis.

Davies, R. O. (2012). *Sports in American Life: A History*. Malden, MA: Wiley-Blackwell.

de Chardin, P. T. ([1959] 1964). *The Future of Man*. (N. Denny, Trans.) New York: Image Books.

De Grande, A. (1982). *Italian Fascism: Its Origins & Developments*. Lincoln, NE: University of Nebraska Press.

de Grazia, V. (2005). *America's Advance through 20th Century Europe*. Cambridge, MA: The Belknap Press.

de Saussure, F. ([1972] 2009). *Course in General Linguistics*. (C. Bally, A. Sechehaye, Eds., & R. Harris, Trans.) Chicago, IL: Open Court.

de Zeeuw, Daniël. (2014). The Dissappearance of the Masses: The Future of a True Illusion. Krisis (2), 56–61.

DeepMind Technologies. (2018). *AlphaGo*. Retrieved from DeepMind: https://deepmind.com/research/alphago/.

DeFleur, L. B. (1982). Technology, Social Change, and the Future of Sociology. *Pacific Sociological Review, 25*(4), 403–417.

Deleuze, G., & Guattari, F. ([1987] 2007). *A Thousand Plateaus: Capitalism and Schizophrenia*. Minneapolis, MN: University of Minnesota Press.

Deloitte. (2016). *Artificial Intelligence Innovation Report*. London: Springwise Intelligence Ltd.

Department of State. (2018a). *Lend-Lease and Military Aid to the Allies in the Early Years of World War II*. Retrieved from Office of the Historian: https://history.state.gov/milestones/1937–1945/lend-lease.

Department of State. (2018b). *U.S.-Soviet Alliance, 1941–1945*. Retrieved from Office of the Historian: https://history.state.gov/milestones/1937–1945/us-soviet.

Deranty, J.-P. (2014). Feuerbach and the Philosophy of Critical Theory. *British Journal for the History of Philosophy, 22*(6), 1208–1233.

Derrida, J. ([1967] 2002). *Writing and Difference*. (A. Bass, Trans.) London: Routledge.

Descartes, R. (1985). *The Philosophical Writings of Descartes* (Vol. 1). (J. Cottingham, R. Stoothoff, & D. Murdoch, Trans.) Cambridge, UK: Cambridge University Press.

Descartes, R. (2008). *Meditations on First Philosophy: With Selectons for Objections and Replies*. (M. Moriarty, Trans.) Oxford, UK: Oxford University Press.

Desjardins, J. (2018, March 11). *Amazon and UPS are Betting Big on Drone Delivery*. Retrieved from Business Insider: http://www.businessinsider.com/amazon-and-ups-are-betting-big-on-drone-delivery-2018-3.

Dienstag, J. F. (1996). Serving God and mammon: the Lockean sympathy in early American political thought. *American Political Science Review*, 497–511.

Diggins, J. P. (1972). *Mussolini and Fascism: The View From America*. Princeton, NJ: Princeton University Press.

DiMaggio, P., Hargittai, E., Neuman, W. R., & Robinson, J. P. (2001). Social Implications of the Internet. *Annual Review of Sociology, 27*, 307–336.

Dittmar, J. E. (2011). Information Technology and Economic Change: The Impact of the Printing Press. *The Quarterly Journal of Economics, 126*, 1133–1172.

Dobin, F. R. (1993). The Social Construction of the Great Depression: Industrial Policy during the 1930s in the United States, Britain, and France. *Theory and Society, 22*(1), 1–56.

Donly, A. W. (1920). The Railroad Situation in Mexico. *The Journal of International Relations, 11*(2), 234–251.

Doray, B. (1988). *From Taylorism to Fordism: A Rational Madness*. London: Free Association.

Douglas, F. ([1845] 2009). *Narrative of the Life of Fredrick Douglass, An American Slave, Written by Himself*. Cambridge, MA: The Belknap Press of Harvard University Press.

Drake, H. A. (1995). Constantine and Consensus: An analysis of political motives and sincerity of spiritual conversion in early Chrisitianity. *Church History, 64*(1), 1–15.

Dreijmanis, J. (2008). Introduction. In M. Weber, & J. Dreijmanis (Ed.), *Max Weber's Complete Writings on Academic and Political Vocations* (G. C. Wells, Trans., pp. 1–24). New York: Algora Publishing.

Dreyer, W. A. (2012). The amazing growth of the early church. *HTS Teologiese Studies/Theological Studies, 68*(1), 1–7.

Du Bois, W. E. (1915). The African Roots of War. *The Atlantic*. Retrieved from https://www.theatlantic.com/magazine/archive/2014/08/the-african-roots-of-war/373403/.

Du Bois, W. E. ([1956] 2002). *I Won't Vote*. Retrieved from The Nation: https://www.thenation.com/article/i-wont-vote/.

Dunand, A. (1995). The End of Analysis (II). In R. Feldstein, B. Fink, & M. Jaanus (Eds.), *Reading Seminar XI: Lacan's Four Fundamental Concepts of Psychoanalysis* (pp. 251–256). Albany, NY: State University of New York Press.

Dunham, A. L. (1955). *The Industrial Revolution in France, 1815–1848*. New York: Exposition Press.

Dupré, S. (2002). *Galileo, the Telescope, and the Science of Optics in the Sixteenth Century*. Ghent, Belgium: Universiteit Gent.

Durkheim, E. ([1893] 2013). *The Division of Labor in Society* (2nd ed.). (S. Lukes, Ed., & W. D. Halls, Trans.) London: Palgrave Macmillan.

Durkheim, E. ([1895] 1982). *The Rules of the Sociological Method*. (S. Lukes, Ed., & W. D. Halls, Trans.) New York: The Free Press.

Durkheim, E. ([1897] 2002). *Suicide: A Study in Sociology*. (G. Simpson, Ed., J. A. Spaulding, & G. Simpson, Trans.) London: Routledge.

Durkheim, E. ([1912] 1995). *The Elementary Forms of Religious Life*. (K. E. Fields, Trans.) New York: The Free Press.

Durkheim, E. (2005). The Dualism of Human Nature and its Social Conditions. *Durkheimian Studies, 11*, 35–45.

Durkheim, E., & Karsenti, B. ([1915] 2017). *L'Allemagne as-dessus de tout.* Paris, France: Audiographe.

Eckes Jr., A. E., & Zeiler, T. (2003). *Globalization and the American Century.* Cambridge, UK: Cambridge University Press.

Eisenstein, E. (1979). *The Printing Press as an Agent of Change.* Cambridge, UK: Cambridge University Press.

Elias, A. J. (2018, April 22). *The Voices of Hayden White.* Retrieved from Los Angeles Review of Books: https://lareviewofbooks.org/article/the-voices-of-hayden-white/#!.

Ellenberger, H. F. (1970). *The Discovery of the Unconscious: The History and Evolution of Dynamic Psychiatry.* New York: Basic Books.

Elliot, T. G. (1992). Constantine's Preparations for the Council of Nicaea. *The Journal of Religious History, 17*(2), 127–137.

Ellman, M., & Maksudov, S. (1994). Soviet Deaths in the Great Patriotic War: A Note. *Europe-Asia Studies, 46*(4), 671–680.

Ellul, J. ([1954] 1964). *The Technological Society.* (J. Wilkinson, Trans.) New York: Vintage Books.

Enriquez, J., & Gullans, S. (2016). *Evolving Ourselves: Redesigning the Future of Humanity—One Gene at a Time.* New York: Penguin.

Evans, A. B. (2013). Jules Verne's Dream Machines. *Extrapolation, 54*(2), 129–146.

Executive Office of the President. (2016a). *Preparing for the Future of Artificial Intelligence.* Washington, DC: Executive Office of the President. Retrieved from https://obamawhitehouse.archives.gov/sites/default/files/whitehouse_files/microsites/ostp/NSTC/preparing_for_the_future_of_ai.pdf.

Executive Office of the President. (2016b). *Artificial Intelligence, Automation, and the Economy.* Washington, DC: Executive Office of the President. Retrieved from https://obamawhitehouse.archives.gov/sites/whitehouse.gov/files/documents/Artificial-Intelligence-Automation-Economy.PDF.

Falk, R., & Krieger, D. (2012). *The Path to Zero.* Boulder, CO: Paradigm Publishers.

Featherstone, M., & Burrows, R. (1995). *Cyberspace/Cyberbodies/Cyberpunk: Cultures of Technological Embodiment.* (M. Featherstone, & R. Burrows, Eds.) London: SAGE.

Federal Communications Commission. (2003–2004). *A Short History of Radio: With an Inside Focus on Mobile Radio.* Washington, DC: Federal Communications Commission. Retrieved May 7, 2018, from https://transition.fcc.gov/omd/history/radio/documents/short_history.pdf.

Feenberg, A. (2012). Introduction: Toward a Critical Theory of the Internet. In A. Feenberg, & N. Friesen (Eds.), *(Re)Inventing the Internet* (pp. 3–17). Rotterdam, The Netherlands: Sense Publishers.

Feuerbach, L. ([1842] 2012). Preliminary Theses on the Reform of Philosophy. In L. Feuerbach, *The Fiery Brook: Selected Writings* (pp. 153–173). London: Verso.

Fink, B. (1995). *The Lacanian Subject.* Princeton, NJ: Princeton University Press.

Fink, B. (1997). *A Clinical Introduction to Lacanian Psychoanalysis.* Cambridge, MA: Harvard University Press.

Fink, B. (2007). *Fundamentals of Psychoanalytic Technique: A Lacanian Approach for Practitioners.* New York: W. W. Norton & Company.

Fischbach, F. (2008). Transformations du concept d'aliénation. Hegel, Feuerbach, Marx. *Revue germanique internationale, 8,* 93–112.

Fisk, H. E. (1924). *The Inter-Ally Debts: An Analysis of War and Post-War Finance, 1914–1923.* New York: Bankers Trust Company.

Fitzpatrick, K. (2006). *The Anxiety of Obsolescence: the American Novel in the Age of Television.* Nashville, TN: Vanderbilt University Press.

Fletcher III, W. M. (1982). *Search for a New Order: Intellectuals and Fascism in Prewar Japan.* Chapel Hill, NC: The University of North Carolina Press.

Forbes. (2020, July 28). *Forbes.* Retrieved from *Forbes*: The World's Billionaires (Real Time Ranking): https://www.forbes.com/billionaires/list/#version:realtime.

Ford Motor Company. (2018). *Company Timeline.* Retrieved from Ford: Our History: https://corporate.ford.com/history.html.

Ford, H. (1926 [1988]). *Today and Tomorrow.* Cambridge, MA: Productivity Press.

Ford, H., & Crowther, S. (1922). *My Life and Work.* Garden City, NY: Garden City Publishing.

Ford, M. (2016). *Rise of the Robots: Technology and the Treat of a Jobless Future.* New York: Basic Books.

Fournier, M. ([2007] 2013). *Émile Durkheim: A Biography.* (D. Macey, Trans.) Cambridge, UK: Polity.

Fowler, K. A. (1973). The Hundred Years War. In A. Marwick, *The Study of War and Society: Thucydides to the Eighteenth Century* (pp. 171–210). Bletchley, UK: Open University Press.

Freeman, J. B. (2018). *Behemoth: A History of the Factory and the Making of the Modern World.* New York: W. W. Norton & Company.

Freud, M. (1957). *Glory Reflected: Sigmund Freud, Man and Father.* London: Angus and Robertson.

Freud, S. ([1899] 2010). *The Interpretation of Dreams: The Complete and Definitive Text.* (J. Strachey, Ed., & J. Strachey, Trans.) New York: Basic Books.

Freud, S. ([1907] 1959). *Jensen's 'Gradiva' and Other Works.* (J. Strachey, Ed., & J. Strachey, Trans.) London: Hogarth Press.

Freud, S. (1910). The Origin and Development of Psychoanalysis. *The American Jounral of Psychoanalysis,* 181–218.

Freud, S. ([1910] 1961). *Five Lectures on Psycho-Analysis.* (J. Strachey, Ed., & J. Strachey, Trans.) New York: W. W. Norton & Company.

Freud, S. (1912). Concerning "Wild" Psychoanalysis. In S. Freud, *Selected Papers on Hysteria and Other Psychoneuroses* (A. A. Brill, Trans.). New York: The Journal of Nervous and Mental Disease Publishing Company.

Freud, S. (1915). *Thoughts for the Times on War and Death.* Retrieved from Panarchy: https://www.panarchy.org/freud/war.1915.html.

Freud, S. ([1915–17] 2012). *General Introduction to Psychoanalysis.* Renaissance Classics.

Freud, S. ([1920] 1990). *Beyond the Pleasure Principle.* (J. Strachey, Ed., & J. Strachey, Trans.) New York: W. W. Norton & Company.

Freud, S. ([1923] 1990). *The Ego and the Id.* (J. Strachey, Ed., & J. Strachey, Trans.) New York: W. W. Norton & Company.

Freud, S. ([1925] 2006). A Note Upon the Mystic Writing-Pad. In C. Merewether (Ed.), *The Archive* (pp. 20–24). London: Whitechapel.

Freud, S. ([1926] 1959). *Inhibitions, Symptoms and Anxiety.* (J. Strachey, Ed., & A. Strachey, Trans.) New York: W. W. Norton & Company.

Freud, S. ([1926] 1978). *The Question of Lay Analysis.* (J. Strachey, Ed., & J. Strachey, Trans.) New York: W. W. Norton & Company.

Freud, S. ([1927] 1961). *The Future of an Illusion.* (J. Strachey, Ed., & J. Strachey, Trans.) New York: W. W. Norton & Company.

Freud, S. ([1930] 2010). *Civilization and Its Discontents.* (J. Strachey, Ed., & J. Strachey, Trans.) New York: W. W. Norton & Company.

Freud, S. ([1933] 1989). *New Introductory Lectures on Psycho-Analysis.* (J. Strachey, Ed., & J. Strachey, Trans.) New York: W. W. Norton & Company.

Freud, S. ([1935] 1952). *An Autobiographical Study.* (J. Strachey, Ed., & J. Strachey, Trans.) New York: W. W. Norton & Company.

Freud, S. ([1939] 1967). *Moses and Monotheism.* (K. Jones, Trans.) New York: Vintage Books.

Freud, S. ([1940] 1989). *An Outline of Psycho-Analysis.* (J. Strachey, Ed., & J. Strachey, Trans.) New York: W. W. Norton & Company.

Freud, S. ([1958] 2001). *The Case of Schreber: Papers on Technique and Other Works.* (J. Strachey, Trans.) London: Vintage.

Freud, S. (1954). *The Origins of Psycho-Analysis: Letters to Wilhelm Fliess, Drafts and Notes: 1887–1902.* (M. Bonaparte, A. Freud, E. Kris, Eds., E. Mosbacher, & J. Strachey, Trans.) New York: Basic Books.

Freud, S. (1963). *Psychoanalysis and Faith: The Letters of Sigmund Freud & Oscar Pfister.* (H. Meng, E. L. Freud, Eds., & E. Mosbacher, Trans.) New York: Basic Books.

Freud, S. (2003). *The "Wolfman" and Other Cases.* (L. A. Huish, Trans.) New York: Penguin.

Freud, S., & Einstein, A. (1931–1932). *The Einstein-Freud Correspondence.* Retrieved from Arizona State University: http://www.public.asu.edu/~jmlynch/273/documents/FreudEinstein.pdf.

Frey, C. B., & Osborn, M. A. (2017). The Future of Employment: How Susceptible are Jobs to Computerization? *Technological Forecasting and Social Change, 114,* 254–280.

Friedan, B. ([1963] 2001). *The Feminine Mystique.* New York: W. W. Norton & Company.

Friedel, R., Isreal, P., & Finn, B. S. (2010). *Edison's Electric Light: The Art of Invention.* Baltimore, MD: The Johns Hopkins University Press.

Frieser, K.-H., & Greenwood, J. (2005). *The Blitzkreig Legend: The 1940 Campaign in the West.* Annapolis, MD: Naval Institute Press.

Fromm, E. ([1941] 1969). *Escape from Freedom.* New York: Henry Holt and Company.

Fromm, E. ([1955] 1990). *The Sane Society.* New York: Henry Holt and Company.

Fry, H. (1896). *The History of North Atlantic Steam Navigation with Some Account of Early Ships and Shipowners.* London: Sampson Low, Marston and Company Limited.

Fuchs, C. (2017). Donald Trump: A Critical Theory-Perspective on Authoritarian Capitalism. *TripleC, 15*(1), 1–72.

Fuks, B. B. (2008). *Freud and the Invention of Jewishness.* (P. H. Britto, Trans.) New York: Agincourt Press.

Fukuyama, F. (2002). *Our Posthuman Future: Consequences of the Biotechnology Revolution.* New York: Picador.

Fuller, S. (2011). *Humanity 2.0: What it Means to be Human Past, Present and Future.* New York: Palgrave Macmillan.

Gabriel, M. (2012). *Love and Capital: Karl and Jenny Marx and the Birth of a Revolution.* New York: Back Bay Books.

Gadamer, H.-G. ([1975] 2006). *Truth and Method* (Second ed.). (J. Weinsheimer, & D. G. Marshall, Trans.) London: Continuum.

Garfinkel, H. ([2008] 2016). *Toward a Sociological Theory of Information.* (A. W. Rawls, Ed.) New York: Routledge.

Garner, J., & Benclowicz, J. (2018). The only solution is revolution: the Spanish Confederación Nacional de Trabajo and the problem of unemployment in Republican Spain, 1931–1932. *Labor History*, 1–23.

Gaskell, P. (1833). *The Manufacturing Population of England.* London: Baldwin and Cradock.

Gasperoni, J. (1996). The Unconscious is Structured like a Language. *Qui Parle, 9*(2), 77–104.

Gates, B. (2017, February 17). Bill Gates: We Should Tax the Robot that Takes Your Job. (K. Delaney, Interviewer) Retrieved from https://www.youtube.com/watch?v=nccryZOcrUg.

Gay, P. ([1950] 1989). Sigmund Freud: A Brief Life. In S. Freud, *The Question of Lay Analysis* (pp. ix–xxvi). New York: W.W. Norton & Company.

Gay, P. (1998). *Freud: A Life for Our Time.* New York: W.W. Norton & Company.

Gerth, H. (1940). The Nazi Party: Its Leadership and Composition. *The American Journal of Sociology, 45*(4): 517–541.

Gibbs, J. P., & Martin, W. T. (1958). Urbanization and Natural Resources. *American Sociological Review, 23*(3), 266–277.

Gibbs, J. P., & Martin, W. T. (1962). Urbanization, Technology, and the Division of Labor: International Patterns. *American Sociological Review, 27*(5), 667–677.

Gibson, R. (1989). *Social History of French Catholicism 1789–1914*. London: Routledge.

Giddens, A. (1977). *Studies in Social and Political Theory*. New York: Basic Books.

Giddens, A. (1991). *Modernity and Self-Identity: Self and Society in the Late Modern Age*. Cambridge, UK: Polity Press.

Gidron, N., & Hall, P. A. (2017). The Politics of Social Status: Economic and Cultural Roots of the Populist Right. *The British Journal of Sociology, 68*(S1), S57–S84.

Gilbert, A. (1978). Marx on Internationalism and War. *Philosophy and Public Affairs, 7*(4), 346–369.

Gilmore, M. P. (1952). *The World of Humanism: 1453–1517*. New York: Harper & Brothers.

Giménez, E. L., & Montero, M. (2015). The Great Depression in Spain. *Economic Modelling, 44*, 200–214.

Goffman, E. (1959). *The Presentation of Self in Everyday Life*. New York: Anchor.

Goffman, E. (1966). *Behavior in Public Places: Notes on Social Organization of Gatherings*. New York: Free Press.

Goffman, E. (1981). *Forms of Talk*. Philadelphia, PA: University of Pennsylvania Press.

Goffman, E. (1982). *Interaction Ritual: Essays on Face-to-Face Behavior*. New York: Pantheon.

Goldman Sachs. (2015). *Fortnightly Thoughts (Issue 85)*. New York: Goldman Sachs. Retrieved from http://live-cognitive-scale.pantheonsite.io/wp-content/uploads/2015/03/FT-Artificial-Intelligence.pdf.

Graeber, D. (2018). *Bullshit Jobs: A Theory*. New York: Simon and Schuster.

Gramsci, A. (1971). *Selections from the Prison Notebooks*. (Q. Hoare, G. N. Smith, Eds., Q. Hoare, & G. N. Smith, Trans.) New York: International Publishers.

Green, D. (2014). *The Hundred Years War: A Peoples History*. New Haven, CT: Yale University Press.

Greenberg, G. (2011). *Manufacturing Depression: The Secret History of a Modern Disease*. New York: Simon & Schuster.

Greenberg, G. (2014). *The Book of Woe: The DSM and the Unmaking of Psychiatry*. New York: Blue Rider Press.

Greenspon, A. J., Patel, J. D., Lau, E., Ochoa, J. A., Frish, D. R., Ho, R. T., ... Kurtz, S. M. (2012). Trends in Permanent Pacemaker Implantation in the United States From 1993–2009: Increasing Complexity of Patients and Procedures. *Journal o the American College of Cardiology, 60*(16), 1540–1545.

Gregor, A. J. (1979). *Young Mussolini and the Intellectual Origins of Fascism*. Berkeley, CA: University of California Press.

Griffin, R. (2007). *Modernism and Fascism: The Sense of a Beginning under Mussolini and Hitler*. New York: Palgrave Macmillan.

Gutmann, M. (2015). The Nature of Total War: Grasping the Global Environmental Dimensions of World War II. *History Compass, 13*(5), 251–261.

Habermas, J. ([1981] 1984). *The Theory of Communicative Action: Reason and the Rationalization of Society* (Vol. One). (T. McCarthy, Trans.) Boston, MA: Beacon Press.

Habermas, J. ([1981] 1989). *The Theory of Communicative Action: Lifeworld and System: A Critique of Functionalist Reason* (Vol. Two). (T. McCarthy, Trans.) Boston, MA: Beacon Press.

Habermas, J. (1983). Modernity—An Incomplete Project. In H. Foster, *The Anti-Aesthetic* (S. Ben-Habib, Trans., pp. 3–15). New York: The New Press.

Habermas, J. ([1985] 1987). *The Philosophical Discourses of Modernity*. (F. Lawrence, Trans.) Cambridge, MA: The MIT Press.

Hacker, B. C. (2005). The Machines of War: Western Military Technology, 1850–2000. *History and Technology, 21*(3), 255–300.

Haining, P. (2002). *The Flying Bomb War: Contemporary Eyewitness Accounts of the German V1 and V2 Raids on Britain 1942-1845*. London: Robson Books.

Hall, P., & Preston, P. (1988). *The Carrier Wave: New Information Technology and the Geography of Innovation, 1846–2003*. Boston, MA: Unwin Hyman.

Hamerow, T. S. (2008). *Why We Watched: Europe, America, and the Holocaust*. New York: W. W. Norton & Company.

Hamilton, R. F., & Herwig, H. H. (2003). World Wars: Definition and Causes. In R. F. Hamilton, & H. H. Herwig (Eds.), *The Origins of World War I* (pp. 1–44). New York: Cambridge University Press.

Hanson, S. E. (2010). The Founding of the French Third Republic. *Comparative Political Studies, 43*(8/9), 1023–1058.

Harari, R. (2001). *Lacan's Seminar on "Anxiety": An Introduction*. (R. Franses, Ed., & J. C. Lamb-Ruiz, Trans.) New York: Other Press.

Haraway, D. (2004). *The Haraway Reader*. New York: Routledge.

Harper, D. (2018). *Critical (adj.)*. Retrieved from Online Etymology Dictionary: https://www.etymonline.com/word/critical.

Harrington, A. (2019). *Mind Fixers: Psychiatry's Troubled Search for the Biology of Mental Illness*. New York: W. W. Norton and Company.

Harris, J. (2007). *Enhancing Evolution: The Ethical Case for Making Better People*. Princeton, NJ: Princeton University Press.

Harrison, M. (1998). *The Economics of World War II: Six Great Powers in International Comparision*. Cambridge, UK: Cambridge University Press.

Harvey, D. (1990). *The Condition of Postmodernity*. Malden, MA: Blackwell Publishing.

Harvey, D. (2005). *A Brief History of Neoliberalism*. Oxford, UK: Oxford University Press.

Harvey, D. (2018). *Marx, Capital, and the Madness of Economic Reason*. New York: Oxford University Press.

Hassan, I. ([1971] 1982). *The Dismemberment of Orpheus: Toward Postmodern Literature.* Madison, WI: University of Wisconsin Press.

Hegel, G. W. ([1807] 1977). *Phenomenology of Spirit.* (A. V. Miller, Trans.) Oxford, UK: Oxford University Press.

Heidegger, M. ([1927] 2010). *Being and Time.* (J. Stambaugh, Trans.) Albany, NY: State University of New York Press.

Heidegger, M. ([1929/1967] 1998). What is Metaphysics? In M. Heidegger, & W. McNeill (Ed.), *Pathmarks* (D. F. Krell, Trans., pp. 82–96). Cambridge, UK: Cambridge University Press.

Heidegger, M. ([1977] 2013). *The Question Concerning Technology and Other Essays.* (W. Lovitt, Trans.) New York: Harper Perennial.

Heidegger, M. (1998). Heidegger to Marcuse, January 20, 1948. In H. Marcuse, & D. Kellner (Ed.), *Technology, War and Fascism: Collected Papers of Herbert Marcuse, Volume 1* (pp. 265–266). London: Routledge.

Heinlein, R. A. ([1947] 2004). *Rocket Ship Galileo.* New York: Penguin Group.

Herranz-Loncán, A. (2011). *The Contribution of Railways to Economic Growth in Latin America before 1914: A Growth Accounting Approach.* Retrieved from Universitat de Barcelona: http://www.ub.edu/histeco/pdf/herranz-DT01.pdf.

Hewlett, R. G., & Anderson Jr., O. E. (1962). *The New World, 1939/1946: A History of the United States Atomic Energy Commission.* University Park, PA: The Pennsylvania State University Press.

Hillhouse, T. M., & Porter, J. H. (2015). A brief history of the development of antidepressant drugs: From monoamines to glutamate. *Experimental and Clinical Psychopharmachology, 23*(1), 1–21.

Hills, R. L. (2006). Richard Roberts' Contributions to Production Engineering. In I. Inkster (Ed.), *History of Technology, Vol. 26* (pp. 41–62). New York: Continuum.

Hilton, P. (2000). Reminiscences and Reflections of a Codebreaker. In D. Joyner (Ed.), *Coding Theory and Cryptography: From Enigma and Geheimschreiber to Quantum Theory* (pp. 1–8). Berlin: Springer.

Hitler, A. (1941). *Mein Kampf.* New York: Reynal & Hitchcock.

Ho, P.-t. (1959). *Studies on the Population of China, 1368–1953.* Cambridge, MA: Harvard University Press.

Hobsbawm, E. ([1962] 1996). *The Age of Revolution: 1789–1848.* New York: Vintage Books.

Hobsbawm, E. J. (1975). *The Age of Capital: 1848–1875.* London: Abacus.

Hodges, A. (2004). Alan Turing: an Introductory Biography. In C. Teuscher (Ed.), *Alan Turing: Life and Legacy of a Great Thinker* (pp. 3–8). Berlin: Springer.

Hoffman, S. (1963). Paradoxes of the French political community. In S. Hoffman, C. P. Kindleberger, L. W. Wylie, J. R. Pitts, J.-B. Duroselle, & F. Goguel, *In Search of France* (pp. 1–117). Cambridge, MA: Harvard University Press.

Hofmann, R. (2015). *The Fascist Effect: Japan and Italy, 1915–1952*. Ithaca, NY: Cornell University Press.

Honneth, A. (2003). 'Anxiety and Politics': The Strengths and Weaknesses of Franz Neumann's Diagnosis of a Social Pathology. *Constellations, 10*(2), 247–255.

Hook, W. F. (1875). *Lives of the Archbishops of Canterbury* (Vol. x). London: Richard Bentley & Son.

Horkheimer, M. ([1939] 1989). The Jews and Europe. In S. E. Bronner, & D. Kellner (Eds.), *Critical Theory and Society: A Reader* (M. Ritter, Trans., pp. 77–94). New York: Routledge.

Horkheimer, M. ([1947] 2013). *Eclipse of Reason*. Mansfield Centre, CT: Martino Publishing.

Horkheimer, M. ([1968] 1972). *Critical Theory: Selected Essays*. (M. J. O'Connell, Trans.) New York: The Seabury Press.

Horkheimer, M. ([1974] 2012). *Critique of Instrumental Reason*. (M. J. O'Connell, Trans.) London: Verso.

Horkheimer, M. (1978). *Dawn & Decline: Notes 1926–1931 & 1950–1969*. (M. Shaw, Trans.) New York: The Seabury Press.

Horkheimer, M. (1993). *Between Philosophy and Social Science: Selected Early Writings*. (G. F. Hunter, M. S. Kramer, & J. Torpey, Trans.) Cambridge, MA: The MIT Press.

Horkheimer, M., & Adorno, T. W. ([1947] 2002). *Dialectic of Enlightenment: Philosophical Fragments*. (G. S. Noerr, Ed., & E. Jephcott, Trans.) Stanford, CA: Stanford University Press.

Horn, M. (2000). A Private Bank at War: J.P. Morgan & Co. and France, 1914–1918. *The Business History Review, 74*(1), 85–112.

Howard, T. A. (2006). *Protestant Theology and the Making of the Modern German University*. Oxford, UK: Oxford University Press.

Horwitz, A. V. (2003) *Creating Mental Illness*. Chicago, IL: The University of Chicago Press.

Horwitz, A. V. (2013) *Anxiety: A Short History*. Baltimore, MD: The Johns Hopkins University Press.

Horwitz, A. V. (2020). *Between Sanity and Madness: Mental Illness from Ancient Greece to the Neuroscientific Era*. Oxford, UK: Oxford University Press.

Horwitz, A. V. and J. C. Wakefield. (2012a). *The Loss of Sadness: How Psychiatry Transformed Normal Sorrow into Depressive Disorder*. Oxford, UK: Oxford University Press. Oxford, UK: Oxford University Press.

Horwitz, A. V. and J. C. Wakefield. (2012b). *All We Have to Fear: Psychiatry's Transformation of Natural Anxieties into Mental Disorders*. Oxford, UK: Oxford University Press. Oxford, UK: Oxford University Press.

Hughes, E. ([1993] 2001). A Cypherpunk's Manifesto. In P. Ludlow (Ed.), *Crypto Anarchy, Cyberstates, and Pirate Utopias* (pp. 81–83). Cambridge, MA: The MIT Press.

Hughes, T. P. (1979). The Electrification of America: The System Builders. *Technology and Culture, 20*(1), 124–161.

Hughes, T. P. (1983). *Networks of Power: Electrification in Western Society, 1880–1930.* Baltimore, MD: The Johns Hopkins University Press.

Hugill, P. J., & Bachmann, V. (2005). The Route to the Techno-Industrial World Economy and the Transfer of German Organic Chemistry to America Before, During, and Immediately After World War I. *Comparative Technology Transfer and Society, 3*(2), 159–86.

Hume, D. ([1738] 1965). *A Treatise of Human Nature: An Attempt to Introduce the Experimental Method of Reasoning into Moral Subjects.* (L. A. Selby-Bigge, Ed.) Oxford, UK: Clarendon Press.

Hume, D. ([1748] 2007). *An Enquiry Concerning Human Understanding.* (P. Millican, Ed.) Oxford, UK: Oxford University Press.

Hume, D. ([1757] 1889). *The Natural History of Religion.* London: A. and H. Bradlaugh Bonner.

Huxley, A. ([1932] 2005). *Brave New World and Brave New World Revisited.* New York: Harper Perennial.

Huyssen, A. (1984). Mapping the Postmodern. *New German Critique*, 5–52.

Hydrographer of the Navy. (1973). *Ocean Passages for the World.* Taunton, UK: The Hydrographer of the Navy.

International Transport Forum. (2017). *Managing the Transition to Driverless Road Freight Transport.* Paris, France: OECD.

J. P. Morgan Chase and Co. (2018). *The History of JPMorgan Chase & Co.: 200 Years of Leadership in Banking.* Retrieved from About J.P. Morgan Chase: https://www.jpmorganchase.com/corporate/About-JPMC/document/shorthistory.pdf.

Jacoby, R. (1975). *Social Amnesia: A Critique of Contemporary Psychology from Adler to Laing.* Boston, MA: Beacon Press.

Jameson, F. (1991). *Postmodernism or, the Cultural Logic of Late Capitalism.* Durham, NC: Duke University Press.

Jay, M. (1973). *The Dialectical Imagination: A History of the Frankfurt School and the Institute for Social Research, 1923–1950.* Boston, MA: Little, Brown and Company.

Jefferson, M. (1917). Our Trade in the Great War. *Geographical Review, 3*(6), 474–480.

Jensen, K., & Nichols, C. M. (2017). The War to End War One Hundered Years Later: A First World War Roundtable. *Oregon Historical Quarterly, 118*(2), 234–251.

Joas, H. (1999). The Modernity of War: Modernization Theory and the Problem of Violence. *International Sociology*, 457–472.

Johnson, H. A. (2001). *Wingless Eagle: U.S. Army Aviation through World War I.* Chapel Hill, NC: University of North Carolina Press.

Jones, E., & Wessely, S. (2014). Battle for the mind: World War 1 and the birth of military psychology. *Lancet, 384*, 1708–1714.

Jones, G. S. (2016). *Karl Marx: Greatness and Illusion*. Cambridge, MA: Belknap Press.

Kaczynski, T. J. (2010). *Technological Slavery: The Collected Writings of Theodor J. Kaczynski, a.k.a. "The Unabomber"*. Port Townsend, WA: Feral House.

Kalogridis, L. (2018). Altered Carbon. Netflix.

Kant, I. ([1781/1787] 1998). *Critique of Pure Reason*. (P. Guyer, A. W. Wood, Eds., P. Guyer, & A. W. Wood, Trans.) Cambridge, UK: Cambridge University Press.

Kant, I. (1783 [2001]). *Prolegomena To Any Future Metaphysics That Will Be Able to Come Forward as Science*. (J. W. Ellington, Trans.) Indianapolis, IN: Hackett Publishing Company, Inc.

Kant, I. (2007). Was ist Aufklärung? In M. Foucault, *The Politics of Truth* (pp. 29–37). Los Angeles, CA: Semiotext(e).

Kaplan, J. (2015). *Humans Need Not Apply: A Guide to Wealth and Work in the Age of Artificial Intelligence*. New Haven, CT: Yale University Press.

Kaplan, J. (2016). *Artificial Intelligence: What Everyone Needs to Know*. Oxford, UK: Oxford University Press.

Karatani, K. (2014). *The Structure of World History: From Modes of Production to Modes of Exchange*. (M. K. Bourdaghs, Trans.) Durham, NC: Duke University Press.

Karp, D. A. (2002). *The Burden of Sympathy: How Families Cope with Mental Illness*. Oxford, UK: Oxford University Press.

Karp, D. A. (2007). *Is It Me or My Meds?: Living with Antidepressants*. Cambridge, MA: Harvard University Press.

Karp, D. A. (2016). *Speaking of Sadness: Depression, Disconnection, and the Meanings of Illness*, Updated and Expanded Edition. Oxford, UK: Oxford University Press.

Karp, D. A. and G. E. Sisson. (2009). *Voices from the Inside: Readings on the Experiences of Mental Illness*. Oxford, UK: Oxford University Press.

Käsler, D. ([1979] 1988). *Max Weber: An Introduction to His Life and Work*. (P. Hurd, Trans.) Chicago, IL: The University of Chicago Press.

Kātz, B. M. (1982). *Herbert Marcuse and the Art of Liberation: An Intellectual Biography*. London: Verso.

Kātz, B. M. (1987). The Criticism of Arms: The Frankfurt School Goes to War. *The Journal of Modern History, 59*(3), 439–478.

Katz, C. J. (1993). Karl Marx on the transition from feudalism to capitalism. *Theory and Society, 22*, 363–389.

Kautzer, C. (2017). Marx's Influence on the Early Frankfurt School. In M. J. Thompson (Ed.), *The Palgrave Handbook of Critical Theory* (pp. 43–65). New York: Palgrave Macmillan.

Kellner, D. (1984). *Herbert Marcuse and the Crisis of Marxism*. Berkeley, CA: University of California Press.

Kennan, G. (1946). *Telegram, George Kennan to George Marshall ["Long Telegram"], February 22, 1946. Harry S. Truman Administration File, Elsey Papers*. Retrieved from

Truman Library: https://www.trumanlibrary.org/whistlestop/study_collections/coldwar/documents/pdf/6-6.pdf.

Kevels, D. J. (1979). *The Physicists: The History of a Scientific Community in Modern America.* New York: Vintage Books.

Keynes, J. M. (1920). *The Economic Consequences of the Peace.* New York: Harcourt, Brace and Howe.

Kierkegaard, S. ([1844] 2014). *The Concept of Anxiety: A Simple Psychologically Oriented Deliberation in View of the Dogmatic Problem of Hereditary Sin.* (A. Hannay, Trans.) New York: W.W. Norton & Company.

Kim, T. (2017, June 20). *McDonald's Hits All-Time High as Wall Street Cheers Replacement of Cashiers with Kiosks.* Retrieved from CNBC: https://www.cnbc.com/2017/06/20/mcdonalds-hits-all-time-high-as-wall-street-cheers-replacement-of-cashiers-with-kiosks.html.

Kirkbright, S. (2004). *Karl Jaspers: A biography. Navigations in truth.* New Haven, CT: Yale University Press.

Kline, R. R. (2015). *The Cybernetics Moment: Or Why We Call Our Age the Information Age.* Baltimore, MD: Johns Hopkins University Press.

Kloes, A. (2016). Dissembling Orthodoxy in the Age of Enlightenment: Fredrick the Great and his Confession of Faith. *Harvard Theological Review, 109*(1), 102–128.

Kojève, A. ([1947] 1980). *Introduction to the Reading of Hegel: Lectures on the Phenomenology of Spirit.* (A. Bloom, Ed., & J. H. Nichols Jr., Trans.) Ithaca, NY: Cornell University Press.

König, W. (2004). Adolf Hitler vs. Henry Ford: The Volkswagen, the Role of America as a Model, and the Failure of a Nazi Consumer Society. *German Studies Review, 27*(2), 249–268.

Korsch, K. ([1923] 2012). *Marxism and Philosophy.* (F. Halliday, Trans.) London: Verso.

Koselleck, R. ([1959] 1988). *Critique and Crisis: Enlightenment and the Pathogenesis of Modern Society.* Cambridge, MA: The MIT Press.

Kramer, B. M., Kalick, S. M., & Milburn, M. A. (1983). Attitudes Toward Nuclear Weapons and Nuclear War: 1945–1982. *Journal of Social Issues, 39*(1), 7–24.

Kroker, A. (2014). *Exits to the Posthuman Future.* Cambridge, UK: Polity Press.

Kroker, A., & Cook, D. (1988). *The Postmodern Scene: Excremental Culture and Hyper-Aesthetics.* London: Macmillan Education.

Kryzhanovsky, L. N. (1989). Mapping the History of Electricity. *Scientometrics, 17*(1–2), 165–170.

Kuhn, T. S. (1957). *The Copernican Revolution: Planetary Astronomy in the Development of Western Thought.* Cambridge, MA: Harvard University Press.

Kuhn, T. S. ([1962] 2012). *The Structure of Scientific Revolutions.* Chicago, IL: University of Chicago Press.

Kus, B. (2015). Sociology of Debt: States, Credit Markets, and Indebted Citizens. *Sociology Compass, 9*(3), 212–223.

Lacan, J. ([1966] 2006). *Écrits*. (B. Fink, Trans.) New York: W. W. Norton & Company.

Lacan, J. ([1972] 1978). On Psychoanalytic Discourse. In *Lacan in Italia* (J. W. Stone, Trans., pp. 32–55). Milan, Italy: La Salamandra.

Lacan, J. ([1973] 1998). *The Seminar of Jacques Lacan, Book XI: The Four Fundamental Concepts of Psychoanalysis*. (J.-A. Miller, Ed., & A. Sheridan, Trans.) New York: W. W. Norton & Company.

Lacan, J. (1975). *The Seminar of Jacques Lacan: Book XXII: RSI (1974–75)*. (C. Gallagher, Trans.) Unpublished Manuscript. Retrieved from http://www.lacaninireland.com/web/wp-content/uploads/2010/06/RSI-Complete-With-Diagrams.pdf.

Lacan, J. ([1975] 1999). *The Seminar of Jacques Lacan, Book XX: Encore, On Feminine Sexuality, The Limits of Love and Knowledge*. (J.-A. Miller, Ed., & B. Fink, Trans.) New York: W. W. Norton & Company.

Lacan, J. ([1978] 1991). *The Seminar of Jacques Lacan: Book II: The Ego in Freud's Theory and in the Technique of Psychoanalysis, 1954–1955*. (J.-A. Miller, Ed., & S. Tomaselli, Trans.) New York: W. W. Norton & Company.

Lacan, J. ([1991] 2007). *The Seminar of Jacques Lacan, Book XVII: The Other Side of Psychoanalysis*. (J.-A. Miller, Ed., & R. Grigg, Trans.) New York: W. W. Norton & Company.

Lacan, J. ([2004] 2014). *The Seminar of Jacques Lacan, Book X: Anxiety*. (J.-A. Miller, Ed., & A. R. Price, Trans.) Cambridge, UK: Polity Press.

Lash, S. (2002). *Critique of Information*. London: SAGE.

Latour, B. ([1988] 1993). *The Pasteurization of France*. (A. Sheridan, & J. Law, Trans.) Cambridge, MA: Harvard University Press.

Latour, B. ([1991] 1993). *We Have Never Been Modern*. (C. Porter, Trans.) Cambridge, MA: Harvard University Press.

Laurent, J. (1998). Science, Technology and the Economy. In M. Bridgstock, D. Burch, J. Forge, J. Laurent, & I. Lowe, *Technology and Society* (pp. 132–158). Cambridge, UK: Cambridge University Press.

Lederer, E. ([1915] 2006). On the Sociology of World War. *European Journal of Sociology*, 241–268.

Leithäuser, T. (2013). Psychoanalysis, Socialization and Society—The Psychoanalytic Thorugh and Interpretation of Alfred Lorenzer. *Historical Social Research / Historische Sozialforschung, 38*(2), 56–70.

Lemert, C. ([1995] 2004). *Sociology After the Crisis* (Second ed.). Boulder, CO: Paradigm Publishers.

Lemert, C. (2017). *Social Theory: The Multicultural, Global, and Classic Readings*. (C. Lemert, Ed.) Boulder, CO: Westview Press.

Lenin, V. I. (1969). *What Is To Be Done? Burning Questions of Our Movement.* New York: International Publishers.

Lenin, V. I. (1974). A "Scientific" System of Sweating. In V. I. Lenin, *Lenin Collected Works, Volume 18* (S. Apresyan, Trans., pp. 594–595). Moscow: Progress Publishers.

Lepsius, M. R., & Vale, M. (1983). The Development of Sociology in Germany after World War II, (1945–1968). *International Journal of Sociology, 13*(3), 1–88.

Lestition, S. (1993). Kant and the End of the Enlightenment in Prussia. *The Journal of Modern History, 65*(1), 57–112.

Lethem, J. (1995). *Amnesia Moon.* San Diego, CA: Harcourt Brace & Co.

Levesque, H., & Lafont, O. (2000). Aspirin throughout the ages: an historical review. *La Revue de Medecive Interne, 32*(Suppl 1), 8–17.

Lévi-Strauss, C. (1963). *Structural Anthropology.* (C. Jacobson, & B. G. Schoepf, Trans.) New York: Basic Books.

Levitt, C. (2009). Sigmund Freud's Intensive Reading of Ludwig Feuerbach. *Canadian Journal of Psychoanalysis, 17*(1), 14–35.

Li, S. (2017, February 28). *Wendy's Adds Automation to the Fast-Food Menu.* Retrieved from Los Angeles Times: http://www.latimes.com/business/la-fi-wendys-kiosk-20170227-story.html.

Licklider, J. C. (1960). Man-Computer Synthesis. *IRE Transactions on Human Factors in Electronics, HFE-1,* 4–11.

Liedman, S.-E. (2018). *A World to Win: the Life and Works of Karl Marx.* London: Verso.

Liew, A. (2007). Understanding Data, Information, Knowledge and Their Inter-Relationships. *Journal of Knowledge Management Practice, 8*(2).

Lindholm, R. W. (1947). German Finance in World War II. *The American Econmic Review, 37*(1), 121–134.

Loader, C., & Tilman, R. (1995). Thorstein Veblen's Analysis of German Intellectualism: Institutionalism as a Forecasting Method. *American Journal of Economics and Sociology,* 339–355.

Loewenstein, K. (1966). *Max Weber's Political Ideas in the Perspective of Our Time.* (R. Winston, & C. Winston, Trans.) Amherst, MA: University of Massachusetts Press.

Lorde, A. ([1984] 2007). *Sister Outsider: Essays and Speeches.* Berkeley, CA: Crossing Press.

Lorenzer, A. (2016). Language, Life Praxis and Scenic Understanding in Psychoanalytic Therapy. *The International Journal of Psychoanalysis, 97,* 1399–1414.

Loughran, T. (2012). Shell Shock, Trauma, and the First World War: The Making of a Diagnosis and Its Histories. *Journal of the History of Medicine and Allied Sciences, 67*(1), 94–119.

Lozada, C. (2005). *The Economics of World War I.* Retrieved from The National Bureau of Economic Research: http://www.nber.org/digest/jan05/w10580.html.

Luce, H. ([1941] 1994). The American Century. *Society, 31*(5), 4–11.

Luhmann, N. (1993). Deconstruction as Second-Order Observing. *New Literary History,* 24(4), 763–782.

`Luhmann, N. (1994). "What is the Case?" and "What Lies behind It?" The Two Sociologies and the Theory of Society. *Sociological Theory,* 12(2), 126–139.

Lukács, G. ([1923] 1971). *History and Class Consciousness: Studies in Marxist Dialectic.* (R. Livingstone, Trans.) Cambridge, MA: The MIT Press.

Lukes, S. (1971). The Meanings of "Individualism". *Journal of the History of Ideas,* 32(1), 45–66.

Lukes, S. ([1973] 1985). *Emile Durkheim: A Historical and Critical Study.* Stanford, CA: Stanford University Press.

Lupton, D. (1995). The Embodied Computer/User. In M. Featherstone, & R. Burrows (Eds.), *Cyberspace/Cyberbodies/Cyberpunk: Cultures of Technological Embodiment* (pp. 97–112). London: SAGE.

Lyman, S. (1997). *Postmodernism and a Sociology of the Absurd: And Other Essays on the "Nouvelle Vague" in American Social Science.* Fayetteville, AR: The University of Arkansas Press.

Lyotard, J.-F. ([1979] 1984). *The Postmodern Condition: A Report on Knowledge.* (G. Bennington, & B. Massumi, Trans.) Minneapolis, MN: University of Minnesota Press.

Lyotard, J.-F. ([1983] 1988). *The Differend: Phrases in Dispute.* (G. Van Den Abbeele, Trans.) Minneapolis, MN: University of Minnesota.

Lyotard, J.-F. ([1988] 1993). *The Postmodern Explained.* (J. Pefanis, M. Thomas, Eds., D. Barry, B. Maher, J. Pefanis, V. Spate, & M. Thomas, Trans.) Minneapolis, MN: University of Minnesota Press.

Magrini, J. M. (2006). "Anxiety" in Heidegger's Being and Time: The Harbringer of Authenticity. *Dialogue/Philosophy Scholarship, Paper 15,* 77–86.

Malik, S. (2005). Information and Knowledge. *Theory, Culture & Society,* 22(1), 29–49.

Marcuse, Harold. (1997, April 16). *Biographical Notes on Herbert Marcuse.* Retrieved from History Department at University of California Santa Barbara: http://www.history.ucsb.edu/faculty/marcuse/herbert.htm.

Marcuse, H. ([1964] 1991). *One-Dimensional Man.* Boston, MA: Beacon Press.

Marcuse, H. ([1968] 2009). *Negations: Essays in Critical Theory.* (J. J. Shapiro, Trans.) London: Mayfly Books.

Marcuse, H. (1998). *Collected Papers of Herbert Marcuse: Technology, War and Fascism* (Vol. One). (D. Kellner, Ed.) London: Routledge.

Marcuse, H. (2001). *Collected Papers of Herbert Marcuse: Towards a Critical Theory of Society* (Vol. Two). (D. Kellner, Ed.) London: Routledge.

Marinetti, F. T. (2006). *Critical Writings.* (G. Berghaus, Ed., & D. Thompson, Trans.) New York: Farrar, Straus and Giroux.

Markoff, J. (2004). *The Abolition of Feudalism: Peasants, Lords, and Legislators in the French Revolution.* University Park, PA: Penn State University Press.

Marks III, F. W. (1985). Six between Roosevelt and Hitler: America's Role in the Appeasment of Nazi Germany. *The Historical Journal, 28*(4), 969–982.

Martindale, D. (1976). American Sociology Before World War II. *Annual Review of Sociology, 2*, 121–143.

Marx, K. ([1841] 1978). To Make the World Philosophical. In R. C. Tucker (Ed.), *The Marx and Engles Reader,* 2nd Ed. (pp. 9–11). New York: W. W. Norton & Company.

Marx, K. ([1843] 1978). For a Ruthless Criticism of Everything Existing. In R. C. Tucker (Ed.), *The Marx and Engels Reader* (pp. 12–15). New York: W. W. Norton & Company.

Marx, K. ([1844a] 1992). A Contribution to Hegel's Philosophy of Right. In K. Marx, *Early Writings* (R. Livingstone, & G. Benton, Trans., pp. 243–257). New York: Penguin.

Marx, K. ([1844b] 1992). Economic and Philosophic Manuscripts. In K. Marx, *Early Writings* (pp. 279–400). New York: Penguin.

Marx, K. ([1845] 1992). Concerning Feuerbach. In K. Marx, *Early Writings* (R. Livingstone, & G. Benton, Trans., pp. 421–423). New York: Penguin.

Marx, K. ([1848] 1988). *The Communist Manifesto.* (F. L. Bender, Ed.) New York: W. W. Norton & Company.

Marx, K. ([1852] 2017). The Eighteenth Brumaire of Louis Bonaparte. In C. Lemert, *Social Theory* (pp. 37–40). Boulder, CO: Westview Press.

Marx, K. ([1867] 1990). *Capital, Volume 1.* (B. Fowkes, Trans.) New York: Penguin.

Marx, K. ([1939] 1973). *Grundrisse.* (M. Nicolaus, Trans.) London: Penguin.

Marx, K., & Engels, F. (1978). Preface to the German Edition of 1883 (Manifesto of the Communist Party). In R. C. Tucker (Ed.), *The Marx-Engels Reader* (Second ed., p. 472). New York: W. W. Norton & Company.

May, R. ([1950] 2015). *The Meaning of Anxiety.* New York: W. W. Norton & Company.

Mayer, T. F. (2010). The Roman Inquisition's precept to Galileo (1616). *The British Journal for the History of Science, 43*(3), 327–351.

Mayer-Schönberger, V., & Cukier, K. (2013). *Big Data: A Revolution that will Transform How We Live, Work, and Think.* Boston, MA: Houghton Mifflin Harcourt.

McAfee, A., & Brynjolfsson, E. (2017). *Machine, Platform, Crowd: Harnessing Our Digital Future.* New York: W. W. Norton & Company.

McCarthy, G. (1985). Development of the Concept and Method of Critique in Kant, Hegel, and Marx. *Studies in Soviet Thought, 30*, 15–38.

McCarthy, J., Minsky, M. L., Rochester, N., & Shannon, C. (1955, August 31). *A Proposal for the Dartmouth Summer Research Project on Artificial Intelligence.* Retrieved from Stanford University: http://www-formal.stanford.edu/jmc/history/dartmouth/dartmouth.html.

McCorduck, P. (2004). *Machines Who Think: A Personal Inquiry into the History and Prospects of Artificial Intelligence.* Natick, MA: A K Peters, Ltd.

McDougall, W. A. ([1985] 1997). *... the Heavens and the Earth: A Political History of the Space Age.* Baltimore, MD: The Johns Hopkins University Press.

Mckinney, J. B. (2006a). The long prelude (1873–1922): Phase I of the invention of radar. *IEEE Aerospace and Electronic Systems Magazine, 21*(8), 17–25.

Mckinney, J. B. (2006b). The rise of radio (1922–1930): Phase II of the invention of radar. *IEEE Aerospace and Electronic Systems Magazine, 21*(8), 27–39.

Mckinney, J. B. (2006c). The arrival of radar (1930–1935): Phase III of the invention of radar. *IEEE Aerospace and Electronic Systems Magazine, 28*(8), 41–54.

Mckinney, J. B. (2006d). The race with destiny (1935–1939): Phase IV of the invention of radar. *IEEE Aerospace and Electronic Systems Magazine, 21*(8), 55–73.

Mckinney, J. B. (2006e). Radar becomes operational (1939–1941): Phase V of the invention of radar. *IEEE Aerospace and Electronic Systems Magazine, 21*(8), 75–88.

McLellan, D. (2006). *Karl Marx: A Biography* (4th ed.). London: Palgrave Macmillan.

McManners, J. (1969). *The French Revolution and the Church*. Westport, CT: Greenword Press.

McVeigh, S. (2009). A mirror reflecting the image of our own unquiet desperation: America's engagment with the Spanish Civil War. *Journal of War & Culture Studies, 2*(3), 259–274.

Medco Health Solutions, Inc. (2011). *America's State of Mind Report*. Franklin Lakes, NJ: Medco Health Solutions, Inc.

Megargee, G. P. (2009). *Encyclopedia of Camps and Ghettos, 1933–1945*. Bloomington, IN: Indiana University Press.

Meitner, L., & Frisch, O. R. (1939). Disintegration of Uranium by Neutrons: A New Type of Nuclear Reaction. *Nature, 143*, 239–240.

Mellon, A. W. (1924). *Taxation: The People's Business*. New York: The Macmillan Company.

Meyer, P. B. (2013). The Airplane as an Open-Source Invention. *Revue économique, 64*(1), 115–132.

Michaels, M. (2018, May 2). *Jeff Bezos, the richest person in the world, thinks it's possible to blow through his entire $131 billion fortune—and he has one big purchase he plans to spend it on.* Retrieved from Business Insider: http://www.businessinsider.com/amazon-ceo-jeff-bezos-plans-to-spend-fortune-space-travel-2018-5.

Miel, J. (1966). Jacques Lacan and the Structure of the Unconsious. *Yale French Studies, 36/37*, 104–111.

Migone, G. G. ([1980] 2015). *The United States and Fascist Italy: The Rise of American Finance in Europe*. (M. Tambor, Trans.) Cambridge, UK: Cambridge University Press.

Millar, J. R., & Linz, S. J. (1978). The Cost of World War II to the Soviet People: A Research Note. *The Journal of Economic History, 38*(4), 959–962.

Miller, J.-A. (1988). *"A and a in Clinical Structures": Acts of the Paris-New York Psychoanalytic Workshop*. Retrieved from Lacan dot com: http://www.lacan.com/symptom6_articles/miller.html.

Mills, C. W. ([1959] 2000). *The Sociological Imagination*. Oxford: Oxford University Press.

Mitchell, A. (2000). *The Great Train Race: Railways and the Franco-German Rivalry, 1815–1914.* New York: Berghahn Books.

Mitchell, R. (2018, April 17). *Musk has Second Thoughts on Agressive Automation for Telsa Model 3.* Retrieved from Los Angeles Times: http://www.latimes.com/business/autos/la-fi-hy-tesla-model-3-20180417-story.html.

Mitchell, T. (2010). The Resources of Economics. *Journal of Cultural Economy, 3*(2), 189–204.

Mitter, R. (2013). *Forgotten Ally: China's World War II, 1937–1945.* Boston, MA: Houghton Mifflin Harcourt.

Moffett, S. E. (1907). *The Americanization of Canada (Ph.D. Dissertation).* New York: Columbia University.

Monro, A. (1879). *The United States and the Dominion of Canada: Their Future.* St. John, NB: Barnes and Company.

More, M., & Vita-More, N. (2013). *The Transhumanist Reader: Classical and Contemporary Essays on the Science, Technology, and Philosophy of the Human Future.* (M. More, & N. Vita-More, Eds.) Malden, MA: Wiley-Blackwell.

Morgan, R. (2002). *Altered Carbon.* New York: Ballantine Books.

Morgan, R. B. (1991). History of Electricity: The Individuals, Inventions and Companies that Made a Difference. *EC&M Electrical Construction and Maintenance.*

Morgan, S. L., & Lee, J. (2018). Trump Voters and the White Working Class. *Sociological Science, 5,* 234–245.

Müller-Doohm, S. (2005). *Adorno: A Biography.* (R. Livingstone, Trans.) Cambridge, UK: Polity Press.

Mullett, M. (2003). Martin Luter's ninety-five Theses: Michael Mullett defines the role of the 95 Theses in the Lutheran Reformation. *History Review*(September 2003), 46–51.

Murphy, V. M. (1960). Anxiety: Common Ground for Psychology and Sociology. *The American Catholic Sociological Review, 21*(3), 213–220.

Musk, E. (2017a). Making Humans a Multi-Planetary Species. *New Space, 5*(2), 46–61.

Musk, E. (2017b, December 7). *Elon Musk's Twitter Feed.* Retrieved from Twitter: https://twitter.com/elonmusk/status/938816780444745728.

Mutz, D. C. (2018). Status Threat, Not Economic Hardship, Explains the 2016 Presidential Vote. *PNAS,* 1–10.

Nader, L. (1972). Up the Anthropologist: Perspectives Gained from Studying Up. In D. Hymes (Ed.), *Reinventing Anthropology* (pp. 284–311). New York: Pantheon Books.

National Science Foundation, National Center for Science and Engineering Statistics. (2017). *R&D at Colleges and Universities.* Retrieved from American Association for the Advancement of Science: https://www.aaas.org/page/rd-colleges-and-universities.

Nedelkoska, L., & Quintini, G. (2018). Automation, Skills Use and Training. *OECD Social, Employment and Migration Working Papers, 202.*

Neumann, F. L. ([1957] 2017). Anxiety and Politics. *tripleC, 15*(2), 612–636.

Neumann, F., Marcuse, H., & Kirchheimer, O. (2013). *Secret Reports on Nazi Germany: The Frankfurt School Contribution to the War Effort.* (R. Laudani, Ed.) Princeton, NJ: Princeton University Press.

Newsom, C. A. (2010). The Economics of Sin: A Not So Dismal Science. *Harvard Theological Review, 103*(3), 365–371.

Nicholls, A., & Liebscher, M. (2010). *Thinking the Unconscious: Nineteenth Century German Thought.* (A. Nicholls, & M. Liebscher, Eds.) Cambridge, UK: Cambridge University Press.

Nilsson, N. J. (2009). *The Quest for Artificial Intelligence: A History of Ideas and Achievements.* Cambridge, UK: Cambridge University Press.

North, J. D. (1954). *The Rational Behavior of Mechanically Extended Man.* Wolverhapton, UK: Boulton Paul Aircraft Ltd.

Nossal, K. R. (2005). *Anti-Americanism in Canada.* Budapest, Hungry: Center for Policy Studies, Central European University.

Nuvolari, A. (2004). *The Making of Steam Power Technology: A Study of Techincal Change during the British Industrial Revolution.* Eindhoven, The Netherlands: Eindhoven University Press.

Office of the Director of National Intelligence. (2018). *Statistical Transparency Report.* Washington, DC: Office of Civil Liberties, Privacy, and Transparency. Retrieved from https://www.dni.gov/files/documents/icotr/2018-ASTR----CY2017----FINAL-for-Release-5.4.18.pdf.

Olesen, H. S., & Weber, K. (2013). Socialization, Language, and Scenic Understanding. Alfred Lorenzer's Contribution to a Psycho-Societal Methodology. *Historical Social Research / Historische Sozialforschung, 38*(2), 26–55.

One Hundred Tenth Congress of the United States of America. (2007). *Energy Independence and Security Act of 2007.* Washington, D.C.: U.S. Government Publishing Office. Retrieved from https://www.gpo.gov/fdsys/pkg/BILLS-110hr6enr/pdf/BILLS-110hr6enr.pdf.

One Hundred Year Study on Artificial Intelligence. (2016). *Artificial Intelligence and Life in 2030.* Stanford, CA: Stanford University.

O'Regan, G. (2012). *A Brief History of Computing.* London: Springer.

Orwell, G. (1950). *1984.* New York: Penguin.

Overholser, J. C. (2010). Psychotherapy That Strives to Encourage Social Interest: A Simulated Interview With Alfred Adler. *Journal of Psychotherapy Integration, 20*(4), 347–363.

Pacey, A. (1991). *Technology in World Civilization.* Cambridge, MA: The MIT Press.

Pahl, R. (1995). *After Success: Fin-de-Siecle Anxiety and Identity.* Cambridge, UK: Polity Press.

Painter, G. S. (1922). The Idea of Progress. *American Journal of Sociology, 28*(3), 257–282.

Panageotou, S. A. (2017). *The Three Dimensions of Political Action in United States Democracy: Corporations as Political Actors and "Franchise Governments".* Knoxville, TN: University of Tennessee.

Pareto, V. (1922, November 25). An Italian View. *The Living Age (1897–1941),* pp. 447–450.

Pareto, V. (2014). *Manual of Political Economy.* (A. Montesano, A. Zanni, L. Bruni, J. S. Chipman, & M. McLure, Eds.) Oxford, UK: Oxford University Press.

Paris Peace Conference XIII. (1921). *Treaty of Peace with Germany (Treaty of Versailles).* Washington, DC: Library of Congress. Retrieved from https://www.loc.gov/law/help/us-treaties/bevans/m-ust000002-0043.pdf.

Paris, J., & Phillips, J. (2013). *Making the DSM-5: Concepts and Controversies.* New York: Springer.

Parsons, T. (1950). Psychoanalysis and the Social Structure. *The Psychoanalytic Quarterly, 19*(3), 371–384.

Parsons, T. ([1951] 1991). *The Social System.* London: Routledge.

Parsons, T. (1964). *Social Structure and Personality.* New York: The Free Press.

Parsons, T. (1971). *The System of Modern Societies.* Englewood Cliffs, NJ: Prentice-Hall.

Pattillo, D. M. (1998). *Pushing the Envelope: The American Aircraft Industry.* Ann Arbor, MI: University of Michigan Press.

Payne, S. G. (2008). *Franco and Hitler: Spain, Germany, and World War II.* New Haven, CT: Yale University Press.

Pellicani, L. (2012). Fascism, capitalism, modernity. *European Journal of Political Theory,* 394–409.

Pettegree, A. (2015). *Brand Luther: 1517, Printing, and the Making of Reformation.* New York: Penguin Press.

Pew Research Center. (2018a, February 5). *Internet/Broadband Fact Sheet.* Retrieved from Pew Research Center Internet & Technology: http://www.pewinternet.org/fact-sheet/internet-broadband/.

Pew Research Center. (2018b, February 5). *Mobile Fact Sheet.* Retrieved from Pew Research Center Internet & Technology: http://www.pewinternet.org/fact-sheet/mobile/.

Pickering, A. (2010). *The Cybernetic Brain.* Chicago, IL: The University of Chicago Press.

Pius II. (1571). *Opera Omnia.* Basile AE. Ex Officina.

Polanyi, K. ([1944] 2001). *The Great Transformation: The Political Origins of Our Time.* Boston, MA: Beacon Press.

Polish American Journal. (1993, February 1). When The Earth and Sun Were Moved 450 Years Ago Nicholas Copernicus Changed The World: The Father of Modern Astronomy. *Polish American Journal.*

Postone, M. (1993). *Time, Labor, and Social Domination: A Reinterpretation of Marx's Critical Theory.* Cambridge, UK: Cambridge University Press.

Postone, M. (2009). The Subject and Social Theory: Marx and Lukács on Hegel. In A. Chitty, &M. McIvor, *Karl Marx and Contemporary Philosophy* (pp. 205–220). London: Palgrave Macmillan.

Postone, M. (2015). The Task of Critical Theory: Rethinking the Critique of Capitalism and its Futures. (H. F. Dahms, Ed.) *Current Perspectives in Social Theory, 33*, 3–28.

Powaski, R. E. (2006). *Lightning War: Blitzkrieg in the West, 1940.* Edison, NJ: Castle Books.

Prins, S. J., Bates, L. M., Keyes, K. M., & Muntaner, C. (2015). Anxious? Depressed? You Might Be Suffering from Capitalism: Contradictory Class Locations and the Prevalence of Depression and Anxiety in the USA. *Sociology of Health & Illness, 37*(8), 1352–1372.

Pruijt, H. D. (1997). *Job Design and Technology: Taylorism vs. Anti-Taylorism.* Abingdon, UK: Routledge.

Purdy, M., & Daugherty, P. (2016). *Why Artificial Intelligence is the Future of Growth.* Accenture. Retrieved from https://www.accenture.com/t20161031T154852__w__/us-en/_acnmedia/PDF-33/Accenture-Why-ai-is-the-Future-of-Growth.PDF#zoom=50.

Puzder, A. (2014, October 5). *Minimum Wage, Maximum Politics.* Retrieved from *The Wall Street Journal*: https://www.wsj.com/articles/andy-puzder-minimum-wage-maximum-politics-1412543682.

Radkau, J. ([2005] 2009). *Max Weber: A Biography.* (P. Camiller, Trans.) Cambridge, UK: Polity Press.

Raff, D. M. (1988). Wage Determination Theory and the Five-Dollar Day at Ford. *The Journal of Economic History*, 387–399.

Rasmussen, D. C. (2017). *The Infidel and the Professor: David Hume, Adam Smith, and the Friendship That Shaped Modern Thought.* Princeton, NJ: Princeton University Press.

Ratiu, S. (2003). The History of the Internal Combustion Engine. *Annals of the Faculty of Engineering Hunedoara*, 145–148.

Rawls, A. W. (2018). The Wartime Narrative in US Sociology, 1940–1947: Stigmatizing Qualitative Research in the Name of 'Science'. *European Journal of Social Theory, Online First*, 1–21.

Reed, B. C. (2014). The Manhattan Project. *Physica Scripta, 89*, 1–26.

Reimer, J. (2015, December 15). *Total Share: 30 Years of Personal Computer Market Share Figures.* Retrieved from Ars Technica: https://arstechnica.com/features/2005/12/total-share/.

Rhodes, R. (2004). The Atomic Bomb in the Second World War. In *Remembering the Manhattan Project: Perspectives on the Making of the Atomic Bomb and Its Legacy* (pp. 17–29). Hackensack, NJ: World Scientific Publishing.

Richard, M. G. (2012, December 26). *How fast could you travel across the U.S. in the 1800s?* Retrieved May 9, 2018, from Mother Nature Netwrok: https://www.mnn.com/green-tech/transportation/stories/how-fast-could-you-travel-across-the-us-in-the-1800s.

Riesman, D., Glazer, N., & Denny, R. ([1950] 1989). *The Lonely Crowd*. New Haven, CT: Yale University Press.

Ringer, F. (2004). *Max Weber: An Intellectual Biography*. Chicago, IL: University of Chicago Press.

Ritzer, G. (2007). *The Globalization of Nothing 2*. Thousand Oaks, CA: Pine Forge Press.

Rizzuto, A.-M. (1998). *Why Did Freud Reject God?: A Psychodynamic Interpretation*. New Haven, CT: Yale University Press.

Roberts, L. (1986). The ARPANET & Computer Networks. *HPW '86 Proceedings of the ACM Conference on the History of Personal Workstations* (pp. 51–58). Palo Alto, CA: Association for Computing Machinery.

Rodden, J. (2015). Warfare, form Cold to Cyber. *Society, 52*, 504–409.

Rose, N. (2007). *The Politics of Life Itself: Biomedicine, Power, and Subjectivity in the Twenty-First Century*. Princeton, NJ: Princeton University Press.

Rosen, J. (1987). The Printed Photograph and the Logic of Progress in Nineteenth-Century France. *Art Journal, 46*(4), 305–311.

Ross, A. (2017). *The Industries of the Future*. New York: Simon and Schuster.

Roudometof, V. (2001). *Nationalism, Globalization, and Orthodoxy: The Social Origins of Ethnic Conflict in the Balkins*. Westport, CT: Greenwood Press.

Rowland, I. (2013). A Catholic Reader of Giordano Bruno in Counter-Reformation Rome: Athanasius Kircher, SJ and Panspermia Rerum. In A. Eusterschulte, & H. S. Hufnagel, *Turning Traditions Upside Down: Rethinking Giordano Bruno's Enlightenment* (pp. 221–236). Budapest, Hungary: Central European University Press.

Sahara, T. (2016). The Making of "Black Hand" Reconsidered. *Istorija 20, 34*(1), 9–29.

Saint-Simon, H. (1825). *Nouveau Christianisme: dialogues entre un conservateur et un novateur*. Paris: Bossange Père.

Sarbin, T. R. (1964). Anxiety: Reification Of a Metaphor. *Archives of General Psychiatry*, 630–8.

Sarkar, T. K., Mailloux, R. J., Oliner, A. A., Salazar-Palma, M., & Sengupta, D. L. (2006). *History of Wireless*. Hoboken, NJ: John Wiley & Sons.

Schelling, F. W. ([1800] 1978). *System of Transcendental Idealism.* (P. Heath, Trans.) Charlottesville, VA: University Press of Virginia.

Schnee, C. (2016). Images of weakness and the fall of Rome—an analysis of reputation management's impact on political history. *Management & Organizational History, 11*(1), 1–18.

Schorshe, C. E. ([1961] 1981). *Fin-De-Siecle Vienna: Politics and Culture*. New York: Vintage Books.

Schroyer, T. (1973). *The Critique of Domination: The Origins and Development of Critical Theory*. Boston, MA: The Beacon Press.

Schulze, M.-S. (1996). *Engineering and Economic Growth: The Development of Austria-Hungary's Machine Building Industry in the late Nineteenth Century.* Frankfurt, Germany: Peter Lang GmbH.

Schumpeter, J. A. ([1943] 2003). *Capitalism, Socialism & Democracy.* London: Routledge.

Schwab, K. (2017). *The Fourth Industrial Revolution.* New York: Crown Publishing.

Scientific American (1845–1908). (1861, August 3). Sir Peter Fairbairns' Patent. p. 71.

Scott-Heron, G. (1971). The Revolution Will Not Ne Televised. On *Pieces of Man*. New York: RCA Studios.

Scull, A. (2010). The Mental Health Sector and the Social Sciences in post-World War II USA. Part I: Total War and its Aftermath. *History of Psychiatry, 22*(1), 3–19.

Scull, A. (2011). The Mental Health Sector and the Social Sciences in post-World War II USA. Part 2: The Impact of Federal Research Funding and the Drugs Revolution. *History of Psychiatry, 22*(3), 268–284.

Scull, A. (2016). *Madness in Civilization: A Cultural History of Insanity, from the Bible to Freud, from the Madhouse to Modern Medicine.* Princeton, NJ: Princeton University Press.

Scull, A. (2019). *Psychiatry and Its Discontents.* Oakland, CA: University of California Press.

Semuels, A. (2018). *Robots Will Transform Fast Food.* Retrieved from The Atlantic: https://www.theatlantic.com/magazine/archive/2018/01/iron-chefs/546581/.

Serban, S. (2016). Cinematography?, Lumiere's Cinematographe—The Starting Point of the History of. *Annals of Spiru Haret University, Journalism Studies, 17*(2), 49–52.

Serber, R. (1943). *The Los Alamos Primer.* Retrieved from http://extremal-mechanics.org/wp-content/uploads/2012/09/Primer.pdf.

Seymour, T., & Shaheen, A. (2011). History of Wireless Communication. *Review of Business Information Systems, 15*(2), 37–42.

Shannon, C. E. (1940). *A Symbolic Analysis of Relay and Switching Circuits* (MS Thesis). Boston, MA: Massachusetts Institute of Technology.

Shapira, M. (2017). Interpersonal Rivalries, Gender and the Intellectual and Scientific Making of Psychoanalysis in 1940s Britain. *History of Psychology, 20*(2), 172–194.

Shillony, B.-A. ([1981] 2001). *Politics and Culture in Wartime Japan.* Oxford, UK: Oxford University Press.

Siboni, J. (2014a, March 9). *The Four Discourses & the Ethics of Psychoanalysis: Lecture delivered at the Wright Institute.* Retrieved from YouTube: https://www.youtube.com/watch?v=7bxYe14BF-Q.

Siboni, J. (2014b, March 9). *The Four Discourses & the Ethics of Psychoanalysis: Lecture delivered at GIFRIC.* Retrieved from YouTube: https://www.youtube.com/watch?v=EIIyWiYa8YY.

Sica, A. ([2004] 2017). *Max Weber and the New Century.* London: Routledge.

Silver, D., Schrittwieser, J., Simonyan, K., Antonoglou, I., Huang, A., Guez, A., Hubert, T., Baker, L., Lai, M., Bolton, A., Chen, Y., Lillicrap, T., Hui, F., Sifre, L., van den Driessche,

G., Graepel, T., and Hassabis, D. (2017). Mastering the Game of Go Without Human Knowledge. *Nature, 550*, 354–359.

Singman, J. L. (1999). *Daily Life in Medieval Europe.* Westport, CT: Greenwood Press.

Slavin, P. (2014). Market Failure during The Great Famine in England and Wales (1315–1317). *Past and Present, 222*(1), 9–50.

Slichter, S. (1919). *The Turnover of Factory Labor.* New York: D. Appleton and Company.

Sloterdijk, P. ([1983] 1987). *Critique of Cynical Reason.* (M. Eldred, Trans.) Minneapolis, MN: University of Minnesota Press.

Sloterdijk, P. ([2001] 2009). Rules for the Human Zoo: A Response to the Letter on Humanism. *Environment and Planning, 27*, 12–28.

Smelser, N., & Wallerstein, R. S. (1998). Psychoanalysis and Sociology: Articulations and Applications. In N. Smelser, *The Social Edges of Psychoanalysis* (pp. 3–35). Berkeley, CA: University of California Press.

Smirnov, S. (2015). Economic Fluctuations in Russia (form the late 1920s to 2015). *Russian Journal of Economics, 1*(2), 130–153.

Smith, A. ([1776] 1981). *An Inquiry into the Nature and Causes of the Wealth of Nations.* (R. H. Campbell, A. S. Skinner, & W. B. Todd, Eds.) Indianapolis, IN: LibertyClassics.

Smith, D. M. (1965). *The Great Departure: The United States and World War I, 1914–1920.* New York: John Wiley and Sons.

Smith, G. (1891). *Canada and the Canadian Question.* London: Macmillan and Co.

Smith, P. (1913). *Luther's Correspondence and Other Contemporary Letters: Volume 1, 1507–1521.* (P. Smith, Ed., & P. Smith, Trans.) Philadelphia, PA: The Lutheran Publication Society.

Solon, O. (2018, March 11). *Elon Musk: We Must Colonise Mars to Preserve Our Species in a Third World War.* Retrieved from The Guardian: https://www.theguardian.com/technology/2018/mar/11/elon-musk-colonise-mars-third-world-war.

Sousanis, N. (2015). *Unflattening.* Cambridge, MA: Harvard University Press.

Soviet Union Information Bureau. (1929). *The Soviet Union: Facts, Descriptions, Statistics.* Washington, DC: Soviet Union Information Bureau.

Spaight, J. M. (1914). *Aircraft in War.* London: Macmillan and Co.

Sperber, J. (2014). *Karl Marx: A Nineteenth-Century Life.* New York: Liveright.

Srnicek, N., & Wiliams, A. (2015). *Inventing the Future: Postcapitalism and a World Without Work.* London: Verso.

Stanford, J. (2017). The Resurgance of Gig Work: Historical and Theoretical Perspectives. *The Economic and Labor Relations Review*, 382–401.

Stein, J. (2018, June 12). *Seattle council votes to repeal tax to help homeless amid opposition from Amazon, other businesses.* Retrieved from The Washington Post: https://www.washingtonpost.com/news/wonk/wp/2018/06/12/seattle-backs-off-tax-to-help-homeless-after-amazon-business-groups-mount-fierce-opposition/?utm_term=.60ea4aea453e.

Steinberg, J. (2011). *Bismark: A Life*. Oxford, UK: Oxford University Press.

Stenmark, D. (2001). The Relationship Between Information and Knowledge. *IRIS 24*. Ulvik, Norway. Retrieved from http://citeseerx.ist.psu.edu/viewdoc/download?doi=10.1.1.21.965&rep=rep1&type=pdf.

Stephens-Davidowitz, S. (2016, August 6). *Fifty States of Anxiety*. Retrieved from The New York Times: https://www.nytimes.com/2016/08/07/opinion/sunday/fifty-states-of-anxiety.html.

Stevenson, R. D., & Wassersug, R. J. (1993, July 15). Horsepower from a horse. *Nature*, *364*(6434), 195.

Susskind, R., & Susskind, D. (2015). *The Future of the Professions: How Technology Will Transform the Work of Human Experts*. Oxford, UK: Oxford University Press.

Sussman, G. (2016). Nineteenth-Century Telegraphy: Wiring the Emerging Urban Corporate Economy. *Media History*, 40–66.

Tanin, O., & Iogan, E. (1934). *Militarism and Fascism in Japan*. New York: International Publishers.

Taylor, C. C. 1942. "Summary Statement," *American Sociological Review* 7(2): 157–159.

Tayler, F. W. (1895). A Piece-Rate System, Being a Step Toward Partial Solution of the Labor Problem. *American Engineer and Railroad Journal (1895–1911)*, *69*(8), 354–356.

Taylor, F. W. ([1911] 1919). *The Principles of Scientific Management*. New York: Harper & Brothers Publishers.

Taylor, J., & Selgin, G. (1999). By our bootstraps: Origins and effects of the high-wage doctrine and the minimum wage. *Journal of Labor Research*, *20*(4), 447–462.

Terrell, E. (2016, September 30). *World War I: The Tech of the Tank*. Retrieved from Library of Congress: https://blogs.loc.gov/inside_adams/2016/09/world-war-i-the-tech-of-the-tank/.

Thackray, A., Brock, D. C., & Jones, R. (2015). *Moore's Law: The Life of Gordon Moore, Silicon Valley's Quiet Revolutionary*. New York: Basic Books.

The Frankfurt Institute of Social Research. ([1956] 1972). *Aspects of Sociology*. (J. Viertel, Trans.) Boston, MA: Beacon Press.

The National Museum of American History. (2014). *Emergence of Electrical Utilities in America*. Retrieved May 9, 2018, from Powering a Generation: http://americanhistory.si.edu/powering/past/h1main.htm.

The National Museum of American History. (2018). *Transportation Infrastructure Videos*. Retrieved from America on the Move: http://amhistory.si.edu/onthemove/themes/story_47_1.html.

The New–York Mirror: a Weekly Sazette of Literature and the Fine Arts (1823–1842). (1838, November 17). Invention of the Telescope. p. 163.

The Wall Street Journal. (2018, April 3). *Auto Sales*. Retrieved from Market Data Center: http://www.wsj.com/mdc/public/page/2_3022-autosales.html.

The World Bank and Institute for Health Metrics and Evaluation. (2016). *The Cost of Air Pollution: Strengthening the Economic Case for Action*. Washington, DC: International Bank for Reconstruction and Development/The World Bank.

Thompson, E. P. (1966). *The Making of the English Working Class*. New York: Vintage.

Thompson, E. P. (1967). Time, Work-Discipline, and Industrial Capitalism. *Past & Present, 38*, 56–97.

Thompson, M. J. (2016). *The Domestication of Critical Theory*. London: Rowman & Littlefield International.

Thompson, M. J. (2017). Introduction: What Is Critical Theory? In M. J. Thompson (Ed.), *The Palgrave Handbook of Critical Theory* (pp. 1–14). New York: Palgrave Macmillan.

Thurston, H. (1911). Ecclesiastical Property. *The Catholic Encyclopedia*. New York: Robert Appleton Company. Retrieved October 7, 2017, from New Advent: http://www.newadvent.org/cathen/12466a.htm.

Tomšič, S. (2015). *The Capitalist Unconscious*. London: Verso.

Tooze, A. (2016, August 18). *When We Loved Mussolini*. Retrieved from The New York Review of Books: http://www.nybooks.com/articles/2016/08/18/when-we-loved-mussolini/.

Touraine, A. ([1969] 1971). *The Post-Industrial Society: Tomorrow's Social History: Classes, Conflicts and Culture in the Programmed Society*. (L. F. Mayhew, Trans.) New York: Random House.

Treese, J. B. (2013). A Historic Shift; 10 Ways the 1973 Oil Embargo Changed the Industry. *Automotive News*, 3.

Turkle, S. (2011). *Alone Together: Why We Expect More from Technology and Less from Each Other*. New York: Basic Books.

Turkle, S. (2015). *Reclaiming Conversation: The Power of Talk in a Digital Age*. New York: Penguin.

Tutte, W. T. (2000). FISH and I. In D. Joyner (Ed.), *Coding Theory and Cryptography: From Enigma and Geheimschreiber to Quantum Theory* (pp. 9–17). Berlin: Springer.

Tutton, R. (2017). Wicked futures: Meaning, matter and the sociology of the future. *The Sociological Review, 65*(3), 478–492.

Twarog, S. (1997). Heights and Living Standards in Germany, 1850–1939: The Case of Wurttemberg. In R. H. Steckel, & R. Floud (Eds.), *Health and Welfare during Industrialization* (pp. 285–330). Chicago, IL: University of Chicago Press.

Tyson, N. d. (2016, February 5). *Neil deGrasse Tyson's Twitter Feed*. Retrieved from Twitter: https://twitter.com/neiltyson/status/695759776752496640?lang=en.

United Nations. (2018). *For Deserts and the Fight Against Desertification*. Retrieved from United Nations Decade: http://www.un.org/en/events/desertification_decade/whynow.shtml.

United States Census Bureau. (2002). *Population by Region and Country: 1950–2050*. Retrieved from United States Census Bureau: https://www2.census.gov/programs-surveys/international-programs/tables/time-series/glob-pop-app-a/tab-04.pdf.

United States Census Bureau. (2018, June 15). *U.S. and World Population Clock*. Retrieved from United States Census Bureau: https://www.census.gov/popclock/world.

Valéry, P. (1919). *Paul Valéry's Crisis of the Mind*. Retrieved from The History Guide: Lectures on Twentieth Century Europe: http://www.historyguide.org/europe/valery.html.

Van Helden, A., Dupré, S., van Gent, R., & Zuidervaart, H. (2010). *The Origins of the Telescope*. Amsterdam, Netherlands: KNAW Press.

Veblen, T. ([1915] 2003). *Imperial Germany and the Industrial Revolution*. Kitchener, Canada: Batoche Books.

Veblen, T. (1917). *The Nature of Peace and the Terms of its Perpetuation*. New York: The Macmillan Company.

Veblen, T. (1919). *The Vested Interests and the Common Man ("The Modern Point of View and the New Order")*. New York: B. W. Huebsch, Inc.

Verne, J. ([1865/1869] 2011). *From the Earth to the Moon & Around the Moon*. Hertfordshire, UK: Wordsworth Editions.

Verrone, P. (Writer), Hughart, R., & Vanzo, G. (Directors). (1999). *Futurama: A Fishful of Dollars (S1/E6)* [Motion Picture].

Viola, F., & Barna, G. (2002). *Pagan Christianity?: Exploring the Roots of Our Church Practices*. Carol Stream, IL: Tyndale House Publishers.

Virilio, P. ([1977] 2006). *Speed and Politics*. (M. Polizzotti, Trans.) Los Angeles: Semiotext(e).

Virilio, P. ([1990] 2000). *Polar Inertia*. (P. Camiller, Trans.) London: Sage.

Virilio, P. ([2005] 2007). *The Original Accident*. (J. Rose, Trans.) London: Polity.

Virilio, P. (2009). *Grey Ecology*. (D. Burk, Trans.) New York: Atropos Press.

Virilio, P. ([2009] 2010). *The Futurism of the Instant: Stop-Eject*. (J. Rose, Trans.) Cambridge, UK: Polity Press.

Virilio, P. ([2010] 2012). *The Great Accelerator*. (J. Rose, Trans.) Cambridge, UK: Polity Press.

Virilio, P., & Lotringer, S. (2002). *Crepuscular Dawn*. Los Angeles: Semiotext(e).

Voltaire, F. M. (1771). *Questions sur L'Encyclopedie, distribuées en forme de dictionnaire* (Second ed., Vol. 7). London: M.DCC.LXXI. Retrieved from <http://find.galegroup.com/ecco/infomark.do?&source=gale&prodId=ECCO&userGroupName=knox-61277&tabID=T001&docId=CW125914135&type=multipage&contentSet=ECCOArticles&version=1.0&docLevel=FASCIMILE>.

Von Hartmann, E. ([1869] 1884). *Philosophy of the Unconscious*. (W. C. Coupland, Trans.) London: Trubner & Co. Ludgate Hill.

Vouros, D. (2014). Hegel, "Totality," and "Abstract Universality" in the Philosophy of Theodor Adorno. *Parrhesia, 21,* 174–186.

Wakefield, J. C. (2013). DSM-5: An Overview of Changes and Controversies. *Clinical Social Work Journal, 41*(2), 139–54.

Wallerstein, R. S. (2002). The Growth and Transformation of American Ego Psychology. *Journal of the American Psychoanalytic Association, 50*(1), 135–168.

Walter, G., & Zeller, A. (1957). International Comparison of Unemployment Rates. In U.-N. Bureau, *The Measurement and Behavior of Unemployment* (pp. 439–584). Cambridge, MA: National Bureau of Economic Research.

Warlouzet, L. (2017). *Governing Europe in a Globalizing World: Neoliberalism and Its Alternatives Following the 1973 Oil Crisis.* London: Routledge.

Watson Health. (2018, June 1). *Watson Health: Get the Facts.* Retrieved from IBM: https://www.ibm.com/blogs/watson-health/watson-health-get-facts/.

Watson, I. (2012). *The Universal Machine: From the Dawn of Computing to Digital Consciousness.* Berlin, Germany: Springer.

Watson, R., McCarthy, J. J., & Hisas, L. (2017). *The Economic Case for Climate Action in the United States.* Alexandria, Virginia: Universal Ecological Fund (FEU-US).

Wawro, G. (1995). The Habsburg Flucht nach vorne in 1866: Domestic Political Origins of the Austro-Prussian War. *The International History Review, 17*(2), 221–248.

Weber, M. ([1904–05] 2011). *The Protestant Ethic and the Spirit of Capitalism: The Revised 1920 Edition.* (S. Kalberg, Trans.) New York: Oxford University Press.

Weber, M. ([1920] 2011). Prefatory Remarks to Collected Essays in the Sociology of Religion. In M. Weber, *The Protestant Ethic and the Spirit of Capitalism* (S. Kalberg, Trans., pp. 233–250). Oxford, UK: Oxford University Press.

Weber, M. (1949). *The Methodology of the Social Sciences.* (E. A. Shils, H. A. Finch, Eds., E. A. Shils, & H. A. Finch, Trans.) Glencoe, IL: The Free Press.

Weber, M. (1978). *Economy and Society.* (G. Roth, & C. Wittich, Eds.) Berkeley, CA: University of California Press.

Weber, M. ([1985] 2011). "Churches" and "Sects" in North America: An Ecclesiastical Sociopolitical Sketch. In M. Weber, *The Protestant Ethic and the Spirit of Capital* (C. Loader, & S. Kalberg, Trans., pp. 227–232). Oxford, UK: Oxford University Press.

Weber, M. (2004). *The Vocation Lectures.* (D. Owen, T. B. Strong, Eds., & R. Livingstone, Trans.) Indianapolis, IN: Hackett Publishing Group.

Wehler, H. U. ([1973] 1985). *The German Empire, 1871–1918.* Oxford, UK: Berg.

Weinberg, A. M. (2014). *Plaque at X-10 Nuclear Reactor.* Oak Ridge National Laboratory, Oak Ridge, TN.

Weiner, C., & Hart, E. (1972). *Exploring the History of Nuclear Physics.* New York: American Institute of Physics.

Wells, H. G. ([1901] 2017). *The First Men in the Moon.* Oxford, UK: Oxford University Press.

Welsh, S., Klassen, C., Borisova, O., & Clothier, H. (2013). The DSM-5 Controversies: How Should Psychologists Respond? *Canadian Psychology, 54*(3), 166–75.

Wheatcroft, S. G. (1999). Victims of Stalinism and the Soviet Secret Police: The Comparability and Reliability of the Archival Data—Not the Last Word. *Europe-Asia Studies, 51*(2), 315–345.

Whitebook, J. (1995). *Perversion and Utopia: A Study in Psychoanalysis and Critical Theory.* Cambridge, MA: The MIT Press.

Whitebook, J. (2017). *Freud: An Intellectual Biography.* Cambridge, UK: Cambridge University Press.

Whitehead, A. N. ([1925] 1948). *Science and the Modern World.* New York: The New American Library.

WHO and UNICEF. (2017). *Progress on drinking water, sanitation and hygiene: 2017 update and SDG baselines.* Geneva, Switzerland: World Health Organization (WHO) and the United Nations Children's Fund (UNICEF).

Whooley, O. (2017). "Defining Mental Disorders: Sociological Investigations into the Classification of Mental Disorders," in *A Handbook for the Study of Mental Health: Social Contexts, Theories, and Systems,* 3rd Ed., edited by T. L. Scheid and E. R. Wright, pp. 45–65. Cambridge, UK: Cambridge University Press.

Wiener, N. ([1948] 1965). *Cybernetics: Or Control and Communication in the Animal and the Machine.* Cambridge, MA: The MIT Press.

Wiener, N. (1954). *The Human Use of Human Beings: Cybernetics and Soceity.* Cambridge, MA: Da Capo Press.

Wiggershaus, R. ([1986] 1994). *The Frankfurt School: It's History, Theories, and Political Significance.* (M. Robertson, Trans.) Cambridge, MA: The MIT Press.

Wilken, R. L. (2012). *The First Thousand Years: A Global History of Christianity.* New Haven, CT: Yale University Press.

Wilkinson, I. (1999). Where is the Novelty in our Current 'Age of Anxiety'? *European Journal of Social Theory, 2*(4), 445–467.

Wilson, W. (1916, July 10). *Address to the Salesmanship Congress in Detroit, Michigan.* Retrieved from The American Presidency Project: http://www.presidency.ucsb.edu/ws/index.php?pid=117701.

Wistrich, R. S. (2006). *The Jews of Vienna in the Age of Franz Joseph.* Lexington, MA: Plunkett Lake Press.

Wittgenstein, L. (2009). *Major Works.* New York: Harper Perennial.

Wolfe, A. (1991). Mind, Self, Society, and Computer: Artificial Intelligence and the Sociology of Mind. *American Journal of Sociology, 96*(5), 1073–1096.

Wolmar, C. (2010). *Blood, Iron, and Gold: How Railroads Transformed the World.* New York: Public Affairs.

Wong, R. (2018, April 14). *Elon Musk blames Tesla Model 3 'production hell' on excessive automation*. Retrieved from Mashable: https://mashable.com/2018/04/14/elon-musk-fixes-model-3-production-problems-more-humans/#bFV5gCltcPqY.

Wood, I. (1987). The Fall of the Western Empire and the End of Roman Britiain. *Britannia, 18*, 251–262.

Wood, R. E. (1986). *From Marshall Plan to Debt Crisis: Foreign Aid and Development Choices in the World Economy*. Berkeley, CA: University of California Press.

World Bank. (2018, April 11). *Water*. Retrieved from World Bank: http://www.worldbank.org/en/topic/water/overview.

Woytinsky, W. S. (1945). Post War Economic Perspectives: 1. Experience After World War I. *Social Security Bulletin*, 18–29.

Wuthnow, R. (2018). *The Left Behind*. Princeton, NJ: Princeton University Press.

WWF. (2016). *Living Planet Report 2016: Risk and Resilience in a New Era*. Gland, Switzerland: WWF International.

Wyman, D. S. (1984). *The Abandonment of the Jews: America and the Holocaust 1941–1945*. New York: Pantheon Books.

Yamamura, E. (2013). Atomic bombs and the long-run effect on trust: Experiences in Hiroshima and Nagasaki. *The Journal of Socio-Economics, 46*, 17–24.

Young America. (1845, May 31). Downfall of Feudalism. *Young America, 2*(10), 4.

Zagorin, P. (2003). On humanism past & present. *Daedalus, 132*(4), 87–92.

Zhao, G. (2017, February 28). *ai in News Reporting: Machines are Now Writing Dialogues, Q&As, and News Articles*. Retrieved from Medium: https://medium.com/syncedreview/ai-in-news-reporting-machines-are-now-writing-dialogues-q-as-and-news-articles-6e9e0da30a61.

Zinn, H. (1980). *A People's History of the United States*. London: Longman.

Žižek, S. (2006). *How to Read Lacan*. New York: W. W. Norton & Company.

Index

absurdist sociology   11–12, 257–258, 261
accident   128–129, 306–307
Accursed Share   161–162
Adorno, Theodor W.   203–206, 210–211, 224–225, 227–233, 237–239, 253, 300–302
agriculture   25–26, 184
aircraft   see aviation
alienation   vii, 55–65, 78, 86–87, 91–92, 158–159, 197–198, 233, 234, 270–272, 273–274, 296
America   12–13, 41–42, 84–85, 164, 167–168, 171–174, 176–178, 181–182, 187–188, 192–195, 207–208, 231, 276–278,
    See also United States
American Century   192–193
American exceptionalism   156, 192–193
American Journal of Sociology   190–191
American Sociological Review   191–192
analysand   x, 20, 64–65, 105–107, 121–122, 234–235, 319–321,
    resistance of   324–325
analysis   106–107, 121–122
analytic   44
anomie   76–78, 91–92, 128
anti-Semitism   69, 98–99, 180–182
anxiety   vii–viii, 1–2, 3–7, 10–11, 19–20, 29, 32, 35–36, 39–41, 45–47, 56, 63, 64–65, 69, 80, 89–90, 91–92, 95, 109–110, 111–117, 121–122, 127, 128, 146–147, 158–159, 169–170, 184–186, 193–194, 195–196, 221–222, 238–245, 249, 254, 265–266, 284–285, 300, 301–302, 324, 359–360,
    and critical theorists   218–219
    and critique   43
    and employment   157–158, 175–176, 305–306, 309–310
    and fascism   179
    and fear   3–4, 21–22, 114–116, 239–240
    definition of   11–12, 20–21, 242–243, 244–245
    modern   22–23
    of the longue durée   21–22
    premodern   21–22
    religious   22–23, 33–34, 38
    sources of   21–22, 114
Anzadula, Gloria   12–13

appearance   33, 45–46
ARPANET   293–294
artificial intelligence   284–285, 309–315, 317
asceticism   86–87, 89, 90–91
assembly line   154–156
atemporal   282–285, 339–340
Austria-Hungary   97–99, 162–163
authoritarianism   133–134, 153–154
automation   141–144, 306–310
automobile   150, 153–155, 275–276
aviation   164–165, 185–186, 276

basic research   82–83, 147–148, 217–218, 276–278
Bataille, Georges   161–162, 194–195
Baudrillard, Jean   287–293, 297–299, 300–302
being   219–220, 239–242
Bell, Daniel   129–130, 272–274
Bezos, Jeff   315–317
Bible   31, 87, 88–89, 96–97
binary Code   186–187
Bion, W. R.   ix–x
Bismark, Otto von   79, 81, 97, 98–99
Black Death   28
Black Hand   162–163
borderlands   12–13
Bostrom, Nick   156–157
bourgeois   40–41, 67, 102–103, 107–108, 109–111, 119, 135–136, 141–143, 159–160, 179, 197–198, 200–201
bourgeoisie   60, 129, 134–136, 154–155, 197–199, 246–248
Braverman, Harry   132–133

calling   88–89, 91–92
Calvin, John   89–90
Canada   160–161
capital   59, 60–61, 91, 110–111, 143–144, 161–163, 174–175, 197, 218, 221, 246–248, 284–285, 315–317, 338–339,
    and fascism   181
    and information   280
    spirit of   84–86, 91–92, 102–103
capitalism   173–175, 179–180, 226–227

Catholic Church   22–23, 26–28, 33–34, 35–36, 86–88,
    and land   27–28, 29–30, 41–42
    and the printing press   31
    and the state   27–28, 29–30, 40–41
    compared to Protestantism   39–40, 90
    in France   67–68
change   25–26, 54–55, 58, 129–130, 287–288
Christianity   26–28, 57–58
Churchill, Winston   193
citations   347–348
civilization   108–110, 297–298
class   60, 97, 174–175, 182–183, 197–198, 303–307
code   234–235, 247, 251–252, 265–266
commodity   60–61, 342
communication   41, 216–218, 290–291, 293–295, 345–347
communism   182–184, 193–194, 205, 225
comparative-historical research   12, 14–15
computer   278–279, 291–294, 297–298, 307–308
Comte, Auguste   69–71
conscious   102–105
Constantine   27
consumerism   140, 141, 157–159
consumption   60, 66, 90–91, 153–155, 174–175, 247, 251–252, 279–280
conversation   302, 319–321
Copernicus   32–35
Council of Nicaea   27
critical   204
critical method   15, 23–24, 42–43, 52–53, 55, 56–57, 58–59, 63–64, 69, 92–94, 95–96, 117–118, 121–122, 129–131, 133–134, 136–137, 205–206, 208–214, 224–225, 231–232, 336–337, 357
critique   9–10, 14–15, 23, 43, 46–47, 55, 58, 85, 91–92, 100–101, 318, 353
    one-dimensional   48–49
    see also immanent critique
cryptoanalysis   186–187
culture   26–28, 31, 84–85, 92, 96–98, 100–101, 109–111, 127, 130, 135–136, 140, 157–158, 199, 208–209, 220–221, 237–238, 244–245, 250, 256–257, 282–284
cyborg   285–287, 291–293, 296–302, 349–351

Dahms, Harry F.   vii, 12, 93–94, 207–208, 217–218
data   30–32, 233–234,
    big-   307–308
death   115–116, 246–248
Declaration of Independence   39–42
Defense Advanced Research Project Agency (DARPA)   293–294
democracy   67–68, 133–134, 172, 179–180, 313–315
denotation   283–284
depression   20–21
depth hermeneutics   2–3, 13–15,
    see also hermeneutics
Descartes, René   35–37, 50
desire   323–325, 331
Diagnostic and Statistical Manual of Mental Disorders (DSM)   3–4, 169
dialectic   50, 51, 52–55, 144, 212, 227–229, 243–244, 253, 300–301, 304–305,
    see also negative dialectic
disaster   128–129
discourse   283–284, 296, 320–321
division of labor   60–61, 65–66, 70–71, 72–74, 76–77, 89, 246–247
doctor   100–101, 106–107
domination   19–20, 22–23, 35, 38–39, 58–59, 94–95, 221, 265–266, 360–361
double-movement   177
Douglas, Fredrick   300
dreams   114
drives   101–102
drugs   6, 157–159,
    see also medication
Du Bois, W. E. B.   165–166
Durkheim, Emile   65–66, 68–71, 74–78, 165–166, 200–201

Edict of Milan   27
Edison, Thomas   148–149
education   26, 31, 41, 82
ego   101–104, 115–116
Eichmann, Adolph   229
electricity   147–150, 158–159
Ellul, Jacques   25, 85
empiricism   35, 39, 43–45
Energy Independence and Security Act of 2007   77–78

England 28–30, 40–41, 59, 66
engine
   steam 51–52, 143, 144–148
   combustion 150
Enlightenment 40–42, 46–47, 69–70, 255–256,
   see also Haskalah
entertainment 250
environment 287–288
epistemology 37, 43, 46
estrangement see alienation
Europe 25–26, 27–29, 160, 163–164, 177–178, 192–193

factory 59–60
fascism 177–182, 241–242, 243–244
fear 21–22, 115–116, 239–240
feminist 299
Fermi, Enrico 187–189
feudalism 59–60
Feuerbach, Ludwig 55, 57–58
finance 280
Fink, Bruce 106–107, 325–326
flexible accumulation 275–276, 279–280
Ford, Henry 153–157, 181
Fordism 156–157
Ford Motor Company 153–154
France 28–29, 41–42, 66–70, 164–165, 194
Franco, Francisco 182
Frankfurt School 136, 199–200, 214–216, 239–240, 243–244, 245–246, 247, 255–258,
   see also Institut für Sozialforschung
Franklin, Benjamin 84–85
Franz Joseph I 97, 98–99
Fredrick II 42
Fredrick William II 48
Fredrick William III 48
freedom 41–42, 46–47, 59–60, 68–69, 108–110, 133, 152–153, 197–198, 219–221, 231, 240–242, 251–252
Freud, Sigmund 95–98, 99–103, 105, 107–108, 112–118, 120–121, 129, 321–323, 345
Friedan, Betty 158–159
Fromm, Erich 89–90, 239, 241
future 19, 20–21, 22–23, 63–64, 134–138, 198–199, 222–223, 284–285, 360
Futurist movement 177–178

Galileo 34–35
General Electric 149–150
Germany 79, 81–83, 97, 98–99, 164, 165–167, 173–174, 180–182, 187–188, 207–208,
   see also Prussia
Gerth, Hans 190–191
God 36, 42–43, 45–46, 57–58, 86–87, 88–90
Gramsci, Antonio 156–157
Great Depression 177
Great European Famine 28
Gutenberg, Johannes 30–31

Habermas, Jürgen 264n1
   communicative action 290–291
happiness 55
Haraway, Donna 299–300
Harvey, David 161, 275–276, 279–280
Haskalah 96–97,
   see also Enlightenment
Hegel 48, 50–56, 57–58, 135–136, 210–211, 327–328
Heidegger, Martin 202–203, 210–211, 239–242, 243–244
Henry VIII 29–30
hermeneutic 13–15,
   see also depth hermeneutics
history 25, 26, 37, 46, 48–51, 54–56, 58–59, 92–93, 134–135, 212
Hitler, Adolph 180–182, 185–186
Horkheimer, Max 200–202, 204–205, 208–209, 214–216, 218
horse-power 145–146, 150
Hughes, Thomas P. 148
human 10–11, 33–34, 35–36, 48–49, 55–56, 57, 70–71, 108–109, 131–133, 139–140, 143, 146–147, 152, 198–199, 254, 265–266, 286–288, 291, 296, 338–339
humanism 31–32
Hume, David 35, 37–39, 42–43, 50
Hundred Years' War 28
Huxley, Aldus 157

IBM 278–279, 294–295
id 101–102, 103–104
idea 54
ideal 25, 49–50, 53–55, 57–58, 224, 229–230
identity 63–64, 76, 229–230,
   structure 7, 14, 19, 20–21
   thinking 230–231

INDEX

illusion   55–56, 57–59, 67–68
image   298–299
imaginary   265–266
immanent critique   14–15, 135–136,
   see also critique
individual   65–66, 70, 72, 74–75, 76–77,
   102–103, 108–109, 111–112, 119, 139, 171,
   200–201, 207–208, 220–221, 228–229,
   237–238, 245–246, 250–252
individualism   39–41, 66, 197–198
individuality   61–62, 64–65, 69, 74–75
indulgences   86–87
industrialization   54–55, 66, 81, 98–99, 127–
   128, 139, 143, 184, 272–273
inflation   274–276
information   30–31, 215–216, 265–267, 280–
   285, 296–297, 345–347
informationalization   8–9, 251–252, 266–267
Institut für Sozialforschung   130–131, 136,
   199, 205–209, 214
   see also Frankfurt School
institution   27, 29–30
internet   293–295
inter-War years   169–178
Italy   179–180

Japan   185, 188–189, 275–276
Jewish   180–181, 189, 199–200, 202–203
jouissance   324
Judaism   27–28, 96–99, 200–201

Kant, Immanuel   42–47
Karatani, Kojin   27–28, 71–72
Kennan, George F.   193–194
Keynes, John Maynard   173–175, 177
Kierkegaard, Soren   90, 111–113
knowledge   43–46, 48–50, 100–101, 102–103,
   105, 144–145, 147–148, 214–215, 233–235,
   265–267,
   a posteriori   43–45
   a priori   43–45
   legitimacy crisis of   283–284
Koselleck, Reinhart   40–41

labor   25, 59–62, 77–78, 91–92, 127–128,
   139–140, 143, 146–147, 151, 152, 154–155,
   158–159, 273–276, 305–310,
   see also automation
Lacan, Jacques   265–266, 321

lack   231, 323–325, 331, 332, 339–344, 348–
   352, 361–362
language   203–204, 265–266, 321
language games   94–95, 283–284
Latour, Bruno   264n1
Lederer, Emil   168–170
Lenin, V. I.   153, 182–184, 225
literacy   31, 41
Little Hans   113–114
Lonely Crowd, The   270–272
Lorde, Audre   299
Lorenzer, Alfred   2–3, 13–14
Luddite   77
Luhmann, Niklas   32, 216–217
Luther, Martin   85–89
Lyotard, Jean-François   282–285

Magna Carta   27–28, 29–30
Manhattan Project   187–189
Marcuse, Herbert   202–203, 219–223
Marinetti, Filippo Tommaso   177–178
Mars   315–317
Marxism   205
Marx, Karl   51–52, 53–55, 56, 58–59, 60–65,
   119, 127–128, 136, 146–147, 224–225, 228–
   229, 337–339
masculinist   177–178
mass, massify, massification   139, 168, 250–
   252, 288–291,
   conceptualization of   288–289
   -society   10–11, 196, 250–251, 270–272,
   297–299, 300–301, 303–305, 317, 321,
   332–333, 359–362
material, materialist, materiality   53–55, 224,
   225, 227–228, 261–263, 265–266, 284–
   285, 298–300
matheme   323–324, 326
   barred Other   334, 341
   barred subject   323–324
   capital   337–340
   information   340–341
   knowledge   327
   master signifier   327
   object a   324–325, 341–342
May, Rollo   242–245
meaning   266–267, 282–285, 290–291,
   345–346
mechanthropomorphic   62, 64, 110–111, 261
mechanthropomorphism   249, 266–267, 296

medication 4, 6
    see also drugs
Mellon, Andrew 176–177
Mexico 160
Mills, C. Wright 8–9
mind 35, 36, 39, 69, 85, 92–93, 101–107, 121–122, 198–199, 200–201, 223, 227–228, 298–299
modern 19, 39–43, 47, 70–71, 83–85, 96–97, 100, 102–103, 119–120, 141–144, 150, 157–159, 169–171, 261, 263–266, 269–270, 284–285
modern society vii–viii, 2–3, 5–7, 8–10, 12, 25–26, 63, 65–66, 68, 70–71, 73, 92–93, 94–95, 109–110, 129–131, 133–136, 169–171, 177–178, 190–191, 192, 195–196, 198, 206–209, 214, 216–218, 221–222, 232–233, 238–239, 244–245, 254–256
Morgan, J. P. 149–150, 163–164
Murphy, Vincent M. 5–6
Musk, Elon 308–309, 315–317
Mussolini, Benito 179–180
myth 255–256, 297–299

Napoleon III 67–68
narrative 283–285,
    cultural 26, 27–28
    historical 85–86, 134–136
    religious 22–23, 29, 36
    traditional 31–32
National Aeronautics and Space Administration (NASA) 276–278
national socialism 179, 241–242
nature 60–61, 62, 145–147, 287–288
navies 164
negative dialectic 229–232,
    see also dialectic
neoliberalism 280
Neumann, Franz 239
nuclear bomb 187–189, 276–278

oil 275–276
ontology 37
Oppenheimer, J. Robert 187–189

Painter, George 170–171
Pareto, Vilfredo 172, 174n9
Parsons, Talcott 236–237
patent 148–149

patient 100–101, 105–107
perception 32–33
personality 72, 79, 236–237, 241–244
perspective 32, 39
pessimism 231–232
petit bourgeoisie 142n1
philosophy 42–43, 45–46, 48–49, 50–51, 53–55, 58–59, 69–70, 207–208, 209–210, 211–213, 219–221
photograph 128–129
piece-rate system 151
political economy 51–52, 55, 60–61, 149–150, 182–183,
    of speed 306–307
Pollock, Friedrich 200–201
pope 29
posthuman vii–viii, 64, 261, 265–266, 297–298
post-industrial society 272–274, 278–279
postmodern 263–266, 272–273, 273n4, 278–279, 280–285
post-War years 192–196, 226–227
preconscious 102–104
predestination 89–90
premodern 25
printing press 30–31, 41, 87–88
private property 27–28, 60, 62
production 60–61, 66, 153, 246–247, 279–280, 307–308
productivity 60–61
progress 46–47, 48–50, 170–171
proletarian 60, 182–184, 199, 226, 246–247
proletariat 182–183, 224–225
Protestantism 83–92,
    Protestant ethic 84–85, 90–91, 174–175
    Protestant sects 48, 90–91
    Protestant Reformation 39–40, 85–89
Prussia 42, 48, 97, 98
    see also Germany
psychoanalysis vii, 1–2, 13, 95–96, 102–103, 105–106, 120–121, 136–137, 236–238, 321–322,
    critiques of mainstream 333–334
    discourses of 321
        the analyst 332–333
        the capitalist 335, 337
        the hysteric 330–332
        the master 327–328
        the university 328–330

INDEX

psychoanalyst   105–107
psychology   1, 3–6, 100–101, 102–103, 105–106, 235–237

racism   172, 315
radar   186
radio   128–129, 186
railroad   51–52, 67, 81, 97, 107–108, 127–128, 160
rationalism   35–39, 42–43, 44–45, 85–86, 87–88, 90–91, 153
rationality   92, 139–140, 198, 211, 237–238, 249, 251–252, 254, 256–257, 273–274
real   265–267, 298–299
reason   43–44, 46–47, 210–211, 219–221
recognition   55–56, 57–58
recombinant innovation   30–32, 95–96
reflection   93–94, 251–252, 283, 296–297, 319–320
religion   26, 41–42, 52, 55, 57–59, 72–73, 84–85, 91–92, 96–98, 108–110, 200–201, 263
Renaissance   31–32
repression   102–104, 113–117
revolution   41–42, 67, 153–154, 182–184, 224–226, 252
roads   153–154, 155
rockets   185–186, 276–278, 293n11
Roosevelt, Franklin Delano   177, 181–182, 187–188
Russia   *see* Soviet Union

salvation   22–23, 86–87, 88–90
satellites   276–278
scenic landscape   2–3, 13–14, 15, 23–24
science   43, 52–53, 136–137, 211, 218–219, 221–223, 269–270, 283–284, 321–322
scientific management   151–153
scientific revolution   32
self   viii–x, 5, 11–12, 20–21, 32, 33–34, 35–36, 37, 38, 40–41, 43, 55–56, 111–112, 245, 320–321, 324–325, 361–362,
   -awareness   243–244
   -consciousness   55–56
Serbian   162–163
sex   111–113, 292–293
Shannon, Claude   186–187
simulation   251–252, 297–298
sin   90

Sloterdijk, Peter   57–58, 226
smart phone   294–295, 346–347
Smith, Adam   60–61
social   55–56, 62–63, 65–66, 69–70, 74–76, 109, 129–131, 139, 167, 179–180, 209–210, 246–247, 249–250, 252, 280, 288–289, 321–322,
   character   270–272
   concepts   20
   facts   74–76
   forecasting   302–304
   power   29–30
   structure   7, 19, 20–21, 76, 228–229, 272–273
society   70, 74–75, 135–136, 244–245, 320–321
   see also modern society; post-industrial society
socioanalysis   vii, x, 1–2, 9–10, 14, 20, 58–59, 63–65, 69, 75–76, 78, 92–95, 110–111, 117–118, 121–122, 234–235, 237–239, 245, 266–267, 301–302, 317, 336–337, 358, 361–363, 365–366,
   and critique   353–354
   and traditional theory   352–353
   Australian Institute of Socio-Analysis   ix–x
   discourses of   335
      Capital   337, 342–344, 348–349
      the archive   345–349
      the cyborg   349–352
      the socioanalyst   354–356
   setting   319–321
   training   318–319
sociology   vii, 1, 5–9, 13, 19, 20–21, 55, 69–70, 74–76, 93–95, 129–130, 136–137, 190–151, 207–208, 209–213, 214–218, 233–239, 289, 304–305, 321–322
sociological imagination   6–7, 8, 121–122, 233–234
solidarity   70–71,
   mechanical   71–73
   organic   73, 89
Soviet Union   182–184, 193–195, 207–208, 225–227, 276–278
space   45–46, 48–49, 217–218
space race   276–278
space travel   315–317
Spain   182

sports   157–158
stalemate society   67–68
Stalin, Joseph   183–184, 225
state   27–28, 40–41, 52, 280
   dynastic   166–167
statistics   151, 214–215, 297–298
studying up   304–305
subconscious   104
superego   101–102, 103–104, 115–116
surveillance   314–315
symbolic   265–267
symptom   1, 2–3, 6, 56, 100–101, 105, 116, 131, 221–222, 226–227, 253, 309–310, 318–319, 363–364
synthetic   44
systems   148

talk therapy   1–2
taxes   176–177
Taylor, Fredrick Winslow   151–153
technique   82, 127–128, 133–134, 139–140, 150–157
technological adoption   140–141
technology   10–11, 66–67, 71–72, 77, 82–83, 87–88, 107–108, 127–129, 132–133, 137, 139–144, 146–148, 149–151, 155–156, 167, 177–178, 221–222, 249, 250–251, 253–255, 263–265, 272–274, 276–280, 291–296, 302–303, 364–365,
   consumerist   141, 157–159
   military   164–165, 168–169, 185–189
Teilhard de Chardin, Pierre   33
telegraph   51–52
telephone   97–98, 128
   see also smart phone
telescope   34–35
television   250–252
Tesla, Nikola   128–129, 148–149
time   45–46, 48–49, 50–51, 61–62, 141, 217–218,
   travel-   147
   women's   158–159
   see also atemporality
theory   31–32, 55, 209–210, 214–215, 216–217, 232–233, 357–358,
   critical   12, 214–224, 232–233, 257–258
   of geocentrism   32
   of heliocentrism   32, 34–35

social   51, 54–55, 58–59, 75–76
sociological   75–76
systems   236–237
traditional   214–216
thinking   26–27, 37, 38–39, 54, 94–95, 137, 216–217, 229–231, 233, 234–235, 243–244
thought   32, 33, 48–51, 55–56, 80, 94–95, 106–107, 133, 137–138, 203–204, 210–211, 212–213, 218, 227–228, 231–232, 261–265, 269
   enlightenment   45–47
   negative   50
   rational   35–36, 37–38, 39, 43
   traditional   48–49
totality   49–54
totalizing logic   22–23, 46–47, 84–85, 95, 116, 136–137, 211–212, 221, 234, 246–248, 253, 261–266, 303–304, 364–365,
   definition   26–27
   of capital   51–52, 54–55, 59, 60–63, 67–68, 85, 91–92, 108–111, 129–130, 133, 134–135, 151, 155–157, 167, 169–171, 197–198, 221–222, 245–246, 250–251
   see also matheme
   of information   143–144, 151, 186–187, 198–199, 233, 248–253, 263–265, 280–285, 314–315, 340
   see also matheme
   of religion   26–27, 29, 31–32, 35–36, 38, 50, 55, 57–59, 69–70, 83–84, 86–87, 92, 108–110, 197–198
traditional life   25, 30, 71–73, 134–135
trench warfare   164
Turing, Alan   186–187

unconscious   102–105, 237–238, 291, 321–323, 324–325
United States   160, 163–164, 165, 175–176, 226–227,
   economy   274–281, 287–288, 303–304
   see also America
utopian fantasy   218

Valéry, Paul   169–171
vanguard   182–183
Veblen, Thorstein   165–168, 261–265
Versailles, Treaty of   173–175, 180–181

Vienna   97–99
Virilio, Paul   306–307, 313–314
voting   222n6

war   28–29, 41–42, 159, 161–162, 170–171, 177–178, 193–194, 228–229, 247–248,
   Cold War   256–257
   definition of world war   162
   sociology of   147–192
World War I   115–116, 162–169,
   costs   169
   deaths   169
   mental illness   169
World War II   184–190,
   costs   189–190
   deaths   189
   Northfield experiment   ix–x
   see also inter-War years; post-War years

Watt, James   51–52, 145–146
wealth   91, 161–162, 339
Weber, Max   78–80, 83–85, 91–93, 128–129
Weil, Felix   205
Weinberg, Alvin M.   269
Western Europe   28
Western Roman Empire   25–26, 27–28
Westinghouse Electric & Manufacturing Company   148–149
Wiener, Norbert   349–350, 358
Wilson, Woodrow   156, 167–168, 173–175
wireless   294–295
Wissenschaft   82–83
women   4, 72, 111–113, 157–159
work   *see* labor
writing   299–300, 345

Xerox   278–279

Printed in the United States
By Bookmasters